THE AUTOBIOGRAPHY OF

CHARLES DARWIN

AND SELECTED LETTERS

EDITED BY FRANCIS DARWIN

DOVER PUBLICATIONS INC., NEW YORK

This new Dover edition first published in 1958 is an unabridged and unaltered republictaion of the work first published in 1892 in the United States by D. Appleton and Company under the title, *CHARLES DARWIN, His Life Told in an Autobiographical Chapter and in a Selected Series of his Published Letters.*

Standard Book Number: 486-20479-0
Library of Congress Catalog Card Number: 58-13934

Manufactured in the United States of America

Dover Publications, Inc.
180 Varick Street
New York, N.Y. 10014

TO DR. HOLLAND, ST. MORITZ.

13th July, 1892.

DEAR HOLLAND,

This book is associated in my mind with St. Moritz (where I worked at it), and therefore with you.

I inscribe your name on it, not only in token of my remembrance of your many acts of friendship, but also as a sign of my respect for one who lives a difficult life well.

Yours gratefully,

FRANCIS DARWIN.

PREFACE.

In preparing this volume, which is practically an abbreviation of the *Life and Letters* (1887), my aim has been to retain as far as possible the personal parts of those volumes. To render this feasible, large numbers of the more purely scientific letters are omitted, or represented by the citation of a few sentences.* In certain periods of my father's life the scientific and the personal elements run a parallel course, rising and falling together in their degree of interest. Thus the writing of the *Origin of Species*, and its publication, appeal equally to the reader who follows my father's career from interest in the man, and to the naturalist who desires to know something of this turning point in the history of Biology. This part of the story has therefore been told with nearly the full amount of available detail.

In arranging my material I have followed a roughly chronological sequence, but the character and variety of my father's researches make a strictly chronological order an impossibility. It was his habit to work more or less simultaneously at several subjects. Experimental work was often carried on as a refreshment or variety, while books entailing reasoning and the marshalling of large bodies of facts were being written. Moreover many of his researches were dropped only to be resumed after years had elapsed. Thus

---

* I have not thought it necessary to indicate all the omissions in the abbreviated letters.

a chronological record of his work would be a patchwork, from which it would be difficult to disentangle the history of any given subject. The Table of Contents will show how I have tried to avoid this result. It will be seen, for instance, that after Chapter VIII. a break occurs; the story turns back from 1854 to 1831 in order that the Evolutionary chapters which follow may tell a continuous story. In the same way the Botanical Work which occupied so much of my father's time during the latter part of his life is treated separately in Chapters XVI. and XVII.

With regard to Chapter IV., in which I have attempted to give an account of my father's manner of working, I may be allowed to say that I acted as his assistant during the last eight years of his life, and had therefore an opportunity of knowing something of his habits and methods.

It is pleasure to me to acknowledge the kindness of Mr. Cameron who has allowed me to reproduce the late Mrs. Cameron's fine photograph of my father as a frontispiece. My acknowledgments, too, are gladly made to the publishers of the *Century Magazine*, who have courteously given me the use of one of their illustrations for the heading of Chapter IV.

FRANCIS DARWIN.

WYCHFIELD, CAMBRIDGE,
August, 1892.

TABLE OF CONTENTS.

"For myself I found that I was fitted for nothing so well as for the study of Truth; . . . as being gifted by nature with desire to seek, patience to doubt, fondness to meditate, slowness to assert, readiness to reconsider, carefulness to dispose and set in order; and as being a man that neither affects what is new nor admires what is old, and that hates every kind of imposture. So I thought my nature had a kind of familiarity and relationship with Truth."—BACON (Proem to the *Interpretatio Naturæ*).

Facsimile of a page from a notebook of 1837 (see transcript opposite)

FROM A NOTE–BOOK OF 1837.

TRANSCRIPT FROM FACSIMILE OF HANDWRITING ON OPPOSITE PAGE.

———

led to comprehend true affinities. My theory would give zest to recent & Fossil Comparative Anatomy: it would lead to study of instincts, heredity, & mind heredity, whole metaphysics, it would lead to closest examination of hybridity & generation, causes of change in order to know what we have come from & to what we tend, to what circumstances favour crossing & what prevents it, this and direct examination of direct passages of structure in species, might lead to laws of change, which would then be main object of study, to guide our speculations.

THE AUTOBIOGRAPHY OF

CHARLES
DARWIN

AND SELECTED LETTERS

CHARLES DARWIN.

CHAPTER I.

THE DARWINS.

CHARLES ROBERT DARWIN was the second son of Dr. Robert Waring Darwin, of Shrewsbury, where he was born on February 12, 1809. Dr. Darwin was a son of Erasmus Darwin, sometimes described as a poet, but more deservedly known as physician and naturalist. Charles Darwin's mother was Susannah, daughter of Josiah Wedgwood, the well-known potter of Etruria, in Staffordshire.

If such speculations are permissible, we may hazard the guess that Charles Darwin inherited his sweetness of disposition from the Wedgwood side, while the character of his genius came rather from the Darwin grandfather.*

Robert Waring Darwin was a man of well-marked character. He had no pretensions to being a man of science, no tendency to generalise his knowledge, and though a successful physician he was guided more by intuition and everyday observation than by a deep knowledge of his subject. His chief mental characteristics were his keen powers of observation, and his knowledge of men, qualities which led him to " read the characters and even the thoughts of those whom he saw even for a short time." It is not therefore surprising that his help should have been sought, not merely in illness, but in cases of family trouble and sorrow. This was largely the case, and his wise sympathy, no less than his medical skill, obtained for him a strong influence over the lives of a large number of people. He

---

* See Charles Darwin's biographical sketch of his grandfather, prefixed to Ernst Krause's *Erasmus Darwin*. (Translated from the German by W. S. Dallas, 1878.) Also Miss Meteyard's *Life of Josiah Wedgwood*.

was a man of a quick, vivid temperament, with a lively interest in even the smaller details in the lives of those with whom he came in contact. He was fond of society, and entertained a good deal, and with his large practice and many friends, the life at Shrewsbury must have been a stirring and varied one—very different in this respect to the later home of his son at Down.*

We have a miniature of his wife, Susannah, with a remarkably sweet and happy face, bearing some resemblance to the portrait by Sir Joshua Reynolds of her father; a countenance expressive of the gentle and sympathetic nature which Miss Meteyard ascribes to her.† She died July 15, 1817, thirty-two years before her husband, whose death occurred on November 13, 1848. Dr. Darwin lived before his marriage for two or three years on St. John's Hill, afterwards at the Crescent, where his eldest daughter Marianne was born, lastly at the " Mount," in the part of Shrewsbury known as Frankwell, where the other children were born. This house was built by Dr. Darwin about 1800, it is now in the possession of Mr. Spencer Phillips, and has undergone but little alteration. It is a large, plain, square, red-brick house, of which the most attractive feature is the pretty green-house, opening out of the morning-room.

The house is charmingly placed, on the top of a steep bank leading down to the Severn. The terraced bank is traversed by a long walk, leading from end to end, still called " the Doctor's Walk." At one point in this walk grows a Spanish chestnut, the branches of which bend back parallel to themselves in a curious manner, and this was Charles Darwin's favourite tree as a boy, where he and his sister Catharine had each their special seat.

The Doctor took great pleasure in his garden, planting it with ornamental trees and shrubs, and being especially successful with fruit trees; and this love of plants was, I think, the only taste kindred to natural history which he possessed.

Charles Darwin had the strongest feeling of love and respect for his father's memory. His recollection of everything that was connected with him was peculiarly distinct, and he spoke of him frequently, generally prefacing an an-

---

* The above passage is, by permission of Messrs. Smith & Elder, taken from my article *Charles Darwin*, in the *Dictionary of National Biography*.
† *A Group of Englishmen*, by Miss Meteyard, 1871.

ecdote with some such phrase as, " My father, who was the wisest man I ever knew," &c. It was astonishing how clearly he remembered his father's opinions, so that he was able to quote some maxim or hint of his in many cases of illness. As a rule he put small faith in doctors, and thus his unlimited belief in Dr. Darwin's medical instinct and methods of treatment was all the more striking.

His reverence for him was boundless, and most touching. He would have wished to judge everything else in the world dispassionately, but anything his father had said was received with almost implicit faith. His daughter, Mrs. Litchfield, remembers him saying that he hoped none of his sons would ever believe anything because he said it, unless they were themselves convinced of its truth—a feeling in striking contrast with his own manner of faith.

A visit which Charles Darwin made to Shrewsbury in 1869 left on the mind of the daughter who accompanied him a strong impression of his love for his old home. The tenant of the Mount at the time, showed them over the house, and with mistaken hospitality remained with the party during the whole visit. As they were leaving, Charles Darwin said, with a pathetic look of regret, " If I could have been left alone in that green-house for five minutes, I know I should have been able to see my father in his wheel-chair as vividly as if he had been there before me."

Perhaps this incident shows what I think is the truth, that the memory of his father he loved the best, was that of him as an old man. Mrs. Litchfield has noted down a few words which illustrate well his feeling towards his father. She describes him as saying with the most tender respect, " I think my father was a little unjust to me when I was young; but afterwards, I am thankful to think I became a prime favourite with him." She has a vivid recollection of the expression of happy reverie that accompanied these words, as if he were reviewing the whole relation, and the remembrance left a deep sense of peace and gratitude.

Dr. Darwin had six children, of whom none are now living : Marianne, married Dr. Henry Parker ; Caroline, married Josiah Wedgwood ; Erasmus Alvey ; Susan, died unmarried ; Charles Robert ; Catharine, married Rev. Charles Langton.

The elder son, Erasmus, was born in 1804, and died unmarried at the age of seventy-seven.

His name, not known to the general public, may be re-

membered from a few words of description occurring in
Carlyle's *Reminiscences* (vol. ii. p. 208). A truer and more
sympathetic sketch of his character, by his cousin, Miss
Julia Wedgwood, was published in the *Spectator*, September 3, 1881.

There was something pathetic in Charles Darwin's affection for his brother Erasmus, as if he always recollected his
solitary life, and the touching patience and sweetness of his
nature. He often spoke of him as " Poor old Ras," or
" Poor dear old Philos." I imagine Philos (Philosopher)
was a relic of the days when they worked at chemistry in
the tool-house at Shrewsbury—a time of which he always
preserved a pleasant memory. Erasmus was rather more
than four years older than Charles Darwin, so that they
were not long together at Cambridge, but previously at
Edinburgh they shared the same lodgings, and after the
Voyage they lived for a time together in Erasmus' house in
Great Marlborough Street. In later years Erasmus Darwin
came to Down occasionally, or joined his brother's family in
a summer holiday. But gradually it came about that he
could not, through ill health, make up his mind to leave
London, and thus they only saw each other when Charles
Darwin went for a week at a time to his brother's house in
Queen Anne Street.

This brief sketch of the family to which Charles Darwin
belonged may perhaps suffice to introduce the reader to the
autobiographical chapter which follows.

CHAPTER II.

AUTOBIOGRAPHY.

[My father's autobiographical recollections, given in the present chapter, were written for his children,—and written without any thought that they would ever be published. To many this may seem an impossibility; but those who knew my father will understand how it was not only possible, but natural. The autobiography bears the heading, *Recollections of the Development of my Mind and Character*, and ends with the following note:—" Aug. 3, 1876. This sketch of my life was begun about May 28th at Hopedene,* and since then I have written for nearly an hour on most afternoons." It will easily be understood that, in a narrative of a personal and intimate kind written for his wife and children, passages should occur which must here be omitted; and I have not thought it necessary to indicate where such omissions are made. It has been found necessary to make a few corrections of obvious verbal slips, but the number of such alterations has been kept down to the minimum.—F. D.]

A GERMAN Editor having written to me for an account of the development of my mind and character with some sketch of my autobiography, I have thought that the attempt would amuse me, and might possibly interest my children or their children. I know that it would have interested me greatly to have read even so short and dull a sketch of the mind of my grandfather, written by himself, and what he thought and did, and how he worked. I have attempted to write the following account of myself, as if I were a dead man in another world looking back at my own life. Nor have I found this difficult, for life is nearly over with me. I have taken no pains about my style of writing.

I was born at Shrewsbury on February 12th, 1809, and my earliest recollection goes back only to when I was a few months over four years old, when we went to near Abergele for sea-bathing, and I recollect some events and places there with some little distinctness.

My mother died in July 1817, when I was a little over eight years old, and it is odd that I can remember hardly

---

* The late Mr. Hensleigh Wedgwood's house in Surrey.

anything about her except her deathbed, her black velvet
gown, and her curiously constructed work-table. In the
spring of this same year I was sent to a day-school in
Shrewsbury, where I stayed a year. I have been told that I
was much slower in learning than my younger sister Catherine, and I believe that I was in many ways a naughty boy.

By the time I went to this day-school* my taste for
natural history, and more especially for collecting, was well
developed. I tried to make out the names of plants, and
collected all sorts of things, shells, seals, franks, coins, and
minerals. The passion for collecting which leads a man to
be a systematic naturalist, a virtuoso, or a miser, was very
strong in me, and was clearly innate, as none of my sisters
or brother ever had this taste.

One little event during this year has fixed itself very
firmly in my mind, and I hope that it has done so from my
conscience having been afterwards sorely troubled by it;
it is curious as showing that apparently I was interested at
this early age in the variability of plants! I told another
little boy (I believe it was Leighton,† who afterwards became a well-known lichenologist and botanist), that I could
produce variously coloured polyanthuses and primroses by
watering them with certain coloured fluids, which was of
course a monstrous fable, and had never been tried by me.
I may here also confess that as a little boy I was much given
to inventing deliberate falsehoods, and this was always done
for the sake of causing excitement. For instance, I once
gathered much valuable fruit from my father's trees and
hid it in the shrubbery, and then ran in breathless haste to
spread the news that I had discovered a hoard of stolen
fruit.‡

* Kept by Rev. G. Case, minister of the Unitarian Chapel in the High
Street. Mrs. Darwin was a Unitarian and attended Mr. Case's chapel, and
my father as a little boy went there with his elder sisters. But both he and
his brother were christened and intended to belong to the Church of England; and after his early boyhood he seems usually to have gone to church
and not to Mr. Case's. It appears (*St. James's Gazette*, December 15, 1883)
that a mural tablet has been erected to his memory in the chapel, which is
now known as the " Free Christian Church."—F. D.

† Rev. W. A. Leighton remembers his bringing a flower to school and saying that his mother had taught him how by looking at the inside of the blos-
som the name of the plant could be discovered. Mr. Leighton goes on, " This
greatly roused my attention and curiosity, and I inquired of him repeatedly
how this could be done ? "—but his lesson was naturally enough not trans-
missible.—F. D.

‡ His father wisely treated this tendency not by making crimes of the
fibs, but by making light of the discoveries.—F. D.

I must have been a very simple little fellow when I first went to the school. A boy of the name of Garnett took me into a cake shop one day, and bought some cakes for which he did not pay, as the shopman trusted him. When we came out I asked him why he did not pay for them, and he instantly answered, " Why, do you not know that my uncle left a great sum of money to the town on condition that every tradesman should give whatever was wanted without payment to any one who wore his old hat and moved [it] in a particular manner? " and he then showed me how it was moved. He then went into another shop where he was trusted, and asked for some small article, moving his hat in the proper manner, and of course obtained it without payment. When we came out he said, " Now if you like to go by yourself into that cake shop (how well I remember its exact position), I will lend you my hat, and you can get whatever you like if you move the hat on your head properly." I gladly accepted the generous offer, and went in and asked for some cakes, moved the old hat, and was walking out of the shop, when the shopman made a rush at me, so I dropped the cakes and ran for dear life, and was astonished by being greeted with shouts of laughter by my false friend Garnett.

I can say in my own favour that I was as a boy humane, but I owed this entirely to the instruction and example of my sisters. I doubt indeed whether humanity is a natural or innate quality. I was very fond of collecting eggs, but I never took more than a single egg out of a bird's nest, except on one single occasion, when I took all, not for their value, but from a sort of bravado.

I had a strong taste for angling, and would sit for any number of hours on the bank of a river or pond watching the float; when at Maer * I was told that I could kill the worms with salt and water, and from that day I never spitted a living worm, though at the expense probably of some loss of success.

Once as a very little boy whilst at the day school, or before that time, I acted cruelly, for I beat a puppy, I believe, simply from enjoying the sense of power; but the beating could not have been severe, for the puppy did not howl, of which I feel sure as the spot was near the house. This act lay heavily on my conscience, as is shown by my remember-

---

* The house of his uncle, Josiah Wedgwood, the younger.

2

ing the exact spot where the crime was committed. It probably lay all the heavier from my love of dogs being then, and for a long time afterwards, a passion. Dogs seemed to know this, for I was an adept in robbing their love from their masters.

I remember clearly only one other incident during this year whilst at Mr. Case's daily school,—namely, the burial of a dragoon soldier; and it is surprising how clearly I can still see the horse with the man's empty boots and carbine suspended to the saddle, and the firing over the grave. This scene deeply stirred whatever poetic fancy there was in me.*

In the summer of 1818 I went to Dr. Butler's great school in Shrewsbury, and remained there for seven years till Midsummer 1825, when I was sixteen years old. I boarded at this school, so that I had the great advantage of living the life of a true schoolboy; but as the distance was hardly more than a mile to my home, I very often ran there in the longer intervals between the callings over and before locking up at night. This, I think, was in many ways advantageous to me by keeping up home affections and interests. I remember in the early part of my school life that I often had to run very quickly to be in time, and from being a fleet runner was generally successful; but when in doubt I prayed earnestly to God to help me, and I well remember that I attributed my success to the prayers and not to my quick running, and marvelled how generally I was aided.

I have heard my father and elder sister say that I had, as a very young boy, a strong taste for long solitary walks; but what I thought about I know not. I often became quite absorbed, and once, whilst returning to school on the summit of the old fortifications round Shrewsbury, which had been converted into a public foot-path with no parapet on one side, I walked off and fell to the ground, but the height was only seven or eight feet. Nevertheless, the number of thoughts which passed through my mind during this very short, but sudden and wholly unexpected fall, was as-

---

* It is curious that another Shrewsbury boy should have been impressed by this military funeral; Mr. Gretton, in his *Memory's Harkback*, says that the scene is so strongly impressed on his mind that he could " walk straight to the spot in St. Chad's churchyard where the poor fellow was buried." The soldier was an Inniskilling Dragoon, and the officer in command had been recently wounded at Waterloo, where his corps did good service against the French Cuirassiers.

tonishing, and seem hardly compatible with what physiologists have, I believe, proved about each thought requiring quite an appreciable amount of time.

Nothing could have been worse for the development of my mind than Dr. Butler's school, as it was strictly classical, nothing else being taught, except a little ancient geography and history. The school as a means of education to me was simply a blank. During my whole life I have been singularly incapable of mastering any language. Especial attention was paid to verse-making, and this I could never do well. I had many friends, and got together a good collection of old verses, which by patching together, sometimes aided by other boys, I could work into any subject. Much attention was paid to learning by heart the lessons of the previous day; this I could effect with great facility, learning forty or fifty lines of Virgil or Homer, whilst I was in morning chapel; but this exercise was utterly useless, for every verse was forgotten in forty-eight hours. I was not idle, and with the exception of versification, generally worked conscientiously at my classics, not using cribs. The sole pleasure I ever received from such studies, was from some of the odes of Horace, which I admired greatly.

When I left the school I was for my age neither high nor low in it; and I believe that I was considered by all my masters and by my father as a very ordinary boy, rather below the common standard in intellect. To my deep mortification my father once said to me, " You care for nothing but shooting, dogs, and rat-catching, and you will be a disgrace to yourself and all your family." But my father, who was the kindest man I ever knew, and whose memory I love with all my heart, must have been angry and somewhat unjust when he used such words.

Looking back as well as I can at my character during my school life, the only qualities which at this period promised well for the future, were, that I had strong and diversified tastes, much zeal for whatever interested me, and a keen pleasure in understanding any complex subject or thing. I was taught Euclid by a private tutor, and I distinctly remember the intense satisfaction which the clear geometrical proofs gave me. I remember with equal distinctness the delight which my uncle gave me (the father of Francis Galton) by explaining the principle of the vernier of a barometer. With respect to diversified tastes, independently of science, I was fond of reading various books, and I used to sit for hours reading the historical plays of Shakespeare, generally in an old window in the thick walls of the school. I read also other poetry, such as Thomson's *Seasons*, and the recently published poems of Byron and Scott. I mention this because later in life I wholly lost, to my great regret, all pleasure from poetry of any kind, including Shakespeare. In connection with pleasure from poetry, I may add that in 1822 a vivid delight in scenery was first awakened in my mind, during a riding tour on the borders of Wales, and this has lasted longer than any other æsthetic pleasure.

Early in my school-days, a boy had a copy of the *Wonders of the World*, which I often read, and disputed with other boys about the veracity of some of the statements; and I believe that this book first gave me a wish to travel in remote countries, which was ultimately fulfilled by the voyage of the *Beagle*. In the latter part of my school life I became passionately fond of shooting; I do not believe that any one could have shown more zeal for the most holy cause than I did for shooting birds. How well I remember killing my first snipe, and my excitement was so great that I had much difficulty in reloading my gun from the trembling of my hands. This taste long continued, and I became a very good shot. When at Cambridge I used to practice throwing up my gun to my shoulder before a looking glass to see that I threw it up straight. Another and better plan was to get a friend to wave about a lighted candle, and then to fire at it with a cap on the nipple, and if the aim was accurate the little puff of air would blow out the candle. The explosion of the cap caused a sharp crack, and I was told that the tutor of the college remarked, " What an extraordinary thing it is, Mr. Darwin seems to spend hours in cracking a horse-whip in his room, for I often hear the crack when I pass under his windows."

I had many friends amongst the schoolboys, whom I loved dearly, and I think that my disposition was then very affectionate.

With respect to science, I continued collecting minerals with much zeal, but quite unscientifically—all that I cared about was a new-named mineral, and I hardly attempted to classify them. I must have observed insects with some little care, for when ten years old (1819) I went for three weeks to Plas Edwards on the sea-coast in Wales, I was very much interested and surprised at seing a large black and scarlet Hemipterous insect, many moths (Zygœna), and a Cicindela, which are not found in Shropshire. I almost made up my mind to begin collecting all the insects which I could find dead, for on consulting my sister, I concluded that it was not right to kill insects for the sake of making a collection. From reading White's *Selborne*, I took much pleasure in watching the habits of birds, and even made notes on the subject. In my simplicity, I remember wondering why every gentleman did not become an ornithologist.

Towards the close of my school life, my brother worked hard at chemistry, and made a fair laboratory with proper apparatus in the tool-house in the garden, and I was allowed to aid him as a servant in most of his experiments. He made all the gases and many compounds, and I read with care several books on chemistry, such as Henry and Parkes' *Chemical Catechism*. The subject interested me greatly, and we often used to go on working till rather late at night. This was the best part of my education at school, for it showed me practically the meaning of experimental science. The fact that we worked at chemistry somehow got known at school, and as it was an unprecedented fact, I was nicknamed " Gas." I was also once publicly rebuked by the head-master, Dr. Butler, for thus wasting my time on such useless subjects; and he called me very unjustly a " poco curante," and as I did not understand what he meant, it seemed to me a fearful reproach.

As I was doing no good at school, my father wisely took me away at a rather earlier age than usual, and sent me (October 1825) to Edinburgh * University with my brother, where I stayed for two years or sessions. My brother was completing his medical studies, though I do not believe he ever really intended to practise, and I was sent there to commence them. But soon after this period I became convinced from various small circumstances that my father would leave me property enough to subsist on with some comfort, though I never imagined that I should be so rich a man as I am; but my belief was sufficient to check any strenuous effort to learn medicine.

The instruction at Edinburgh was altogether by lectures,

---

* He lodged at Mrs. Mackay's, 11, Lothian Street. What little the records of Edinburgh University can reveal has been published in the *Edinburgh Weekly Dispatch*, May 22, 1888; and in the *St. James's Gazette*, February 16, 1888. From the latter journal it appears that he and his brother Erasmus made more use of the library than was usual among the students of their time.

and these were intolerably dull, with the exception of those on chemistry by Hope ; but to my mind there are no advantages and many disadvantages in lectures compared with reading. Dr. Duncan's lectures on Materia Medica at 8 o'clock on a winter's morning are something fearful to remember. Dr. Munro made his lectures on human anatomy as dull as he was himself, and the subject disgusted me. It has proved one of the greatest evils in my life that I was not urged to practise dissection, for I should soon have got over my disgust, and the practice would have been invaluable for all my future work. This has been an irremediable evil, as well as my incapacity to draw. I also attended regularly the clinical wards in the hospital. Some of the cases distressed me a good deal, and I still have vivid pictures before me of some of them ; but I was not so foolish as to allow this to lessen my attendance. I cannot understand why this part of my medical course did not interest me in a greater degree ; for during the summer before coming to Edinburgh, I began attending some of the poor people, chiefly children and women in Shrewsbury : I wrote down as full an account as I could of the case with all the symptoms, and read them aloud to my father, who suggested further inquiries and advised me what medicines to give, which I made up myself. At one time I had at least a dozen patients, and I felt a keen interest in the work.* My father, who was by far the best judge of character whom I ever knew, declared that I should make a successful physician,— meaning by this, one who would get many patients. He maintained that the chief element of success was exciting confidence ; but what he saw in me which convinced him that I should create confidence I know not. I also attended on two occasions the operating theatre in the hospital at Edinburgh, and saw two very bad operations, one on a child, but I rushed away before they were completed. Nor did I ever attend again, for hardly any inducement would have been strong enough to make me do so ; this being long before the blessed days of chloroform. The two cases fairly haunted me for many a long year.

My brother stayed only one year at the University, so that during the second year I was left to my own resources ; and this was an advantage, for I became well acquainted

---

* I have heard him call to mind the pride he felt at the results of the successful treatment of a whole family with tartar emetic.—F. D.

with several young men fond of natural science. One of these was Ainsworth, who afterwards published his travels in Assyria; he was a Wernerian geologist, and knew a little about many subjects. Dr. Coldstream * was a very different young man, prim, formal, highly religious, and most kind-hearted; he afterwards published some good zoological articles. A third young man was Hardie, who would, I think, have made a good botanist, but died early in India. Lastly, Dr. Grant, my senior by several years, but how I became acquainted with him I cannot remember; he published some first-rate zoological papers, but after coming to London as Professor in University College, he did nothing more in science, a fact which has always been inexplicable to me. I knew him well; he was dry and formal in manner, with much enthusiasm beneath this outer crust. He one day, when we were walking together, burst forth in high admiration of Lamarck and his views on evolution. I listened in silent astonishment, and as far as I can judge, without any effect on my mind. I had previously read the *Zoonomia* of my grandfather, in which similar views are maintained, but without producing any effect on me. Nevertheless it is probable that the hearing rather early in life such views maintained and praised may have favoured my upholding them under a different form in my *Origin of Species*. At this time I admired greatly the *Zoonomia;* but on reading it a second time after an interval of ten or fifteen years, I was much disappointed; the proportion of speculation being so large to the facts given.

Drs. Grant and Coldstream attended much to marine Zoology, and I often accompanied the former to collect animals in the tidal pools, which I dissected as well as I could. I also became friends with some of the Newhaven fishermen, and sometimes accompanied them when they trawled for oysters, and thus got many specimens. But from not having had any regular practice in dissection, and from possessing only a wretched microscope, my attempts were very poor. Nevertheless I made one interesting little discovery, and read, about the beginning of the year 1826, a short paper on the subject before the Plinian Society. This was that the so-called ova of Flustra had the power of independent movement by means of cilia, and were in fact

---

* Dr. Coldstream died September 17, 1863 ; see Crown 16mo. Book Tract, No. 19 of the Religious Tract Society (no date).

larvæ. In another short paper, I showed that the little globular bodies, which had been supposed to be the young state of *Fucus loreus*, were the egg-cases of the worm-like *Pontobdella muricata*.

The Plinian Society * was encouraged and, I believe, founded by Professor Jameson : it consisted of students, and met in an underground room in the University for the sake of reading papers on natural science and discussing them. I used regularly to attend, and the meetings had a good effect on me in stimulating my zeal and giving me new congenial acquaintances. One evening a poor young man got up, and after stammering for a prodigious length of time, blushing crimson, he at last slowly got out the words, " Mr. President, I have forgotten what I was going to say." The poor fellow looked quite overwhelmed, and all the members were so surprised that no one could think of a word to say to cover his confusion. The papers which were read to our little society were not printed, so that I had not the satisfaction of seeing my paper in print ; but I believe Dr. Grant noticed my small discovery in his excellent memoir on Flustra.

I was also a member of the Royal Medical Society, and attended pretty regularly ; but as the subjects were exclusively medical, I did not much care about them. Much rubbish was talked there, but there were some good speakers, of whom the best was the [late] Sir J. Kay-Shuttleworth. Dr. Grant took me occasionally to the meetings of the Wernerian Society, where various papers on natural history were read, discussed, and afterwards published in the Transactions. I heard Audubon deliver there some interesting discourses on the habits of N. American birds, sneering somewhat unjustly at Waterton. By the way, a negro lived in Edinburgh, who had travelled with Waterton, and gained his livelihood by stuffing birds, which he did excellently : he gave me lessons for payment, and I used often to sit with him, for he was a very pleasant and intelligent man.

Mr. Leonard Horner also took me once to a meeting of the Royal Society of Edinburgh, where I saw Sir Walter Scott in the chair as President, and he apologised to the meeting as not feeling fitted for such a position. I looked

---

* The society was founded in 1823, and expired about 1848 (*Edinburgh Weekly Dispatch*, May 22, 1888).

at him and at the whole scene with some awe and reverence, and I think it was owing to this visit during my
youth, and to my having attended the Royal Medical Society, that I felt the honour of being elected a few years ago
an honorary member of both these Societies, more than any
other similar honour. If I had been told at that time that
I should one day have been thus honoured, I declare that I
should have thought it as ridiculous and improbable, as if I
had been told that I should be elected King of England.

During my second year at Edinburgh I attended Jameson's lectures on Geology and Zoology, but they were incredibly dull. The sole effect they produced on me was the
determination never as long as I lived to read a book on
Geology, or in any way to study the science. Yet I feel
sure that I was prepared for a philosophical treatment of
the subject; for an old Mr. Cotton, in Shropshire, who knew
a good deal about rocks, had pointed out to me two or three
years previously a well-known large erratic boulder in the
town of Shrewsbury, called the "bell-stone;" he told me
that there was no rock of the same kind nearer than Cum-
berland or Scotland, and he solemnly assured me that the
world would come to an end before any one would be able
to explain how this stone came where it now lay. This produced a deep impression on me, and I meditated over this
wonderful stone. So that I felt the keenest delight when I
first read of the action of icebergs in transporting boulders,
and I gloried in the progress of Geology. Equally striking
is the fact that I, though now only sixty-seven years old,
heard the Professor, in a field lecture at Salisbury Craigs,
discoursing on a trap-dyke, with amygdaloidal margins and
the strata indurated on each side, with volcanic rocks all
around us, say that it was a fissure filled with sediment
from above, adding with a sneer that there were men who
maintained that it had been injected from beneath in a
molten condition. When I think of this lecture, I do not
wonder that I determined never to attend to Geology.

From attending Jameson's lectures, I became acquainted
with the curator of the museum, Mr. MacGillivray, who afterwards published a large and excellent book on the birds
of Scotland. I had much interesting natural-history talk
with him, and he was very kind to me. He gave me some
rare shells, for I at that time collected marine mollusca, but
with no great zeal.

My summer vacations during these two years were wholly

given up to amusements, though I always had some book in hand, which I read with interest. During the summer of 1826, I took a long walking tour with two friends with knapsacks on our backs through North Wales. We walked thirty miles most days, including one day the ascent of Snowdon. I also w nt with my sister a riding tour in North Wales, a servant with saddle-bags carrying our clothes. The autumns were devoted to shooting, chiefly at Mr. Owen's, at Woodhouse, and at my Uncle Jos's,* at Maer. My zeal was so great that I used to place my shooting-boots open by my bed-side when I went to bed, so as not to lose half a minute in putting them on in the morning; and on one occasion I reached a distant part of the Maer estate, on the 20th of August for black-game shooting, before I could see: I then toiled on with the gamekeeper the whole day through thick heath and young Scotch firs.

I kept an exact record of every bird which I shot throughout the whole season. One day when shooting at Woodhouse with Captain Owen, the eldest son, and Major Hill, his cousin, afterwards Lord Berwick, both of whom I liked very much, I thought myself shamefully used, for every time after I had fired and thought that I had killed a bird, one of the two acted as if loading his gun, and cried out, " You must not count that bird, for I fired at the same time," and the gamekeeper, perceiving the joke, backed them up. After some hours they told me the joke, but it was no joke to me, for I had shot a large number of birds, but did not know how many, and could not add them to my list, which I used to do by making a knot in a piece of string tied to a button-hole. This my wicked friends had perceived.

How I did enjoy shooting! but I think that I must have been half-consciously ashamed of my zeal, for I tried to persuade myself that shooting was almost an intellectual employment; it required so much skill to judge where to find most game and to hunt the dogs well.

One of my autumnal visits to Maer in 1827 was memorable from meeting there Sir J. Mackintosh, who was the best converser I ever listened to. I heard afterwards with a glow of pride that he had said, " There is something in that young man that interests me." This must have been chiefly due to his perceiving that I listened with much interest to

---

* Josiah Wedgwood, the son of the founder of the Etruria Works.

everything which he said, for I was as ignorant as a pig about his subjects of history, politics, and moral philosophy. To hear of praise from an eminent person, though no doubt apt or certain to excite vanity, is, I think, good for a young man, as it helps to keep him in the right course.

My visits to Maer during these two or three succeeding years were quite delightful, independently of the autumnal shooting. Life there was perfectly free; the country was very pleasant for walking or riding; and in the evening there was much very agreeable conversation, not so personal as it generally is in large family parties, together with music. In the summer the whole family used often to sit on the steps of the old portico with the flower-garden in front, and with the steep wooded bank opposite the house reflected in the lake, with here and there a fish rising or a water-bird paddling about. Nothing has left a more vivid picture on my mind than these evenings at Maer. I was also attached to and greatly revered my Uncle Jos; he was silent and reserved, so as to be a rather awful man; but he sometimes talked openly with me. He was the very type of an upright man, with the clearest judgment. I do not believe that any power on earth could have made him swerve an inch from what he considered the right course. I used to apply to him in my mind the well-known ode of Horace, now forgotten by me, in which the words " nec vultus tyranni, &c.," * come in.

Cambridge, 1828–1831.—After having spent two sessions in Edinburgh, my father perceived, or he heard from my sisters, that I did not like the thought of being a physician, so he proposed that I should become a clergyman. He was very properly vehement against my turning into an idle sporting man, which then seemed my probable destination. I asked for some time to consider, as from what little I had heard or thought on the subject I had scruples about declaring my belief in all the dogmas of the Church of England; though otherwise I liked the thought of being a country clergyman. Accordingly I read with great care *Pearson on the Creed*, and a few other books on divinity; and as I did not then in the least doubt the strict and literal

---

* Justum et tenacem propositi virum
　Non civium ardor prava jubentium,
　Non vultus instantis tyranni
　Mente quatit solidâ.

truth of every word in the Bible, I soon persuaded myself
that our Creed must be fully accepted.

Considering how fiercely I have been attacked by the
orthodox, it seems ludicrous that I once intended to be a
clergyman. Nor was this intention and my father's wish
ever formally given up, but died a natural death when, on
leaving Cambridge, I joined the *Beagle* as naturalist. If
the phrenologists are to be trusted, I was well fitted in one
respect to be a clergyman. A few years ago the secretaries
of a German psychological society asked me earnestly by
letter for a photograph of myself; and some time afterwards I received the proceedings of one of the meetings,
in which it seemed that the shape of my head had been
the subject of a public discussion, and one of the speakers
declared that I had the bump of reverence developed enough
for ten priests.

As it was decided that I should be a clergyman, it was
necessary that I should go to one of the English universities and take a degree; but as I had never opened a
classical book since leaving school, I found to my dismay,
that in the two intervening years, I had actually forgotten, incredible as it may appear, almost everything which
I had learnt, even to some few of the Greek letters. I did
not therefore proceed to Cambridge at the usual time in
October, but worked with a private tutor in Shrewsbury,
and went to Cambridge after the Christmas vacation, early
in 1828. I soon recovered my school standard of knowledge, and could translate easy Greek books, such as Homer
and the Greek Testament, with moderate facility.

During the three years which I spent at Cambridge my
time was wasted, as far as the academical studies were concerned, as completely as at Edinburgh and at school. I
attempted mathematics, and even went during the summer
of 1828 with a private tutor to Barmouth, but I got on
very slowly. The work was repugnant to me, chiefly from
my not being able to see any meaning in the early steps in
algebra. This impatience was very foolish, and in after
years I have deeply regretted that I did not proceed far
enough at least to understand something of the great leading principles of mathematics, for men thus endowed seem
to have an extra sense. But I do not believe that I should
ever have succeeded beyond a very low grade. With respect to Classics I did nothing except attend a few compul-
sory college lectures, and the attendance was almost nomi-

nal. In my second year I had to work for a month or two
to pass the Little-Go, which I did easily. Again, in my last
year I worked with some earnestness for my final degree of
B.A., and brushed up my Classics, together with a little
Algebra and Euclid, which latter gave me much pleasure,
as it did at school. In order to pass the B.A. examination,
it was also necessary to get up Paley's *Evidences of Christianity*, and his *Moral Philosophy*. This was done in a
thorough manner, and I am convinced that I could have
written out the whole of the *Evidences* with perfect cor-
rectness, but not of course in the clear language of Paley.
The logic of this book and, as I may add, of his *Natural
Theology*, gave me as much delight as did Euclid. The
careful study of these works, without attempting to learn
any part by rote, was the only part of the academical course
which, as I then felt, and as I still believe, was of the least
use to me in the education of my mind. I did not at that
time trouble myself about Paley's premises; and taking
these on trust, I was charmed and convinced by the long
line of argumentation. By answering well the examination
questions in Paley, by doing Euclid well, and by not failing miserably in Classics, I gained a good place among the
οἱ πολλοὶ or crowd of men who do not go in for honours.
Oddly enough, I cannot remember how high I stood, and
my memory fluctuates between the fifth, tenth, or twelfth,
name on the list.*

Public lectures on several branches were given in the
University, attendance being quite voluntary; but I was so
sickened with lectures at Edinburgh that I did not even attend Sedgwick's eloquent and interesting lectures. Had I
done so I should probably have become a geologist earlier
than I did. I attended, however, Henslow's lectures on Botany, and liked them much for their extreme clearness, and
the admirable illustrations; but I did not study botany.
Henslow used to take his pupils, including several of the
older members of the University, field excursions, on foot
or in coaches, to distant places, or in a barge down the river,
and lectured on the rarer plants and animals which were observed. These excursions were delightful.

Although, as we shall presently see, there were some redeeming features in my life at Cambridge, my time was
sadly wasted there, and worse than wasted. From my pas-

---

* Tenth in the list of January 1831.

sion for shooting and for hunting, and, when this failed, for riding across country, I got into a sporting set, including some dissipated low-minded young men. We used often to dine together in the evening, though these dinners often included men of a higher stamp, and we sometimes drank too much, with jolly singing and playing at cards afterwards. I know that I ought to feel ashamed of days and evenings thus spent, but as some of my friends were very pleasant, and we were all in the highest spirits, I cannot help looking back to these times with much pleasure.*

But I am glad to think that I had many other friends of a widely different nature. I was very intimate with Whitley,† who was afterwards Senior Wrangler, and we used continually to take long walks together. He inoculated me with a taste for pictures and good engravings, of which I bought some. I frequently went to the Fitzwilliam Gallery, and my taste must have been fairly good, for I certainly admired the best pictures, which I discussed with the old curator. I read also with much interest Sir Joshua Reynolds' book. This taste, though not natural to me, lasted for several years, and many of the pictures in the National Gallery in London gave me much pleasure; that of Sebastian del Piombo exciting in me a sense of sublimity.

I also got into a musical set, I believe by means of my warm-hearted friend, Herbert,‡ who took a high wrangler's degree. From associating with these men, and hearing them play, I acquired a strong taste for music, and used very often to time my walks so as to hear on week days the anthem in King's College Chapel. This gave me intense pleasure, so that my backbone would sometimes shiver. I am sure that there was no affectation or mere imitation in this taste, for I used generally to go by myself to King's College, and I sometimes hired the chorister boys to sing in my rooms. Nevertheless I am so utterly destitute of an ear, that I cannot perceive a discord, or keep time and hum a tune correctly; and it is a mystery how I could possibly have derived pleasure from music.

My musical friends soon perceived my state, and some-

---

* I gather from some of my father's contemporaries that he has exaggerated the Bacchanalian nature of these parties.—F. D.
† Rev. C. Whitley, Hon. Canon of Durham, formerly Reader in Natural Philosophy in Durham University.
‡ The late John Maurice Herbert, County Court Judge of Cardiff and the Monmouth Circuit.

times amused themselves by making me pass an examination, which consisted in ascertaining how many tunes I could recognise, when they were played rather more quickly or slowly than usual. 'God save the King,' when thus played, was a sore puzzle. There was another man with almost as bad an ear as I had, and strange to say he played a little on the flute. Once I had the triumph of beating him in one of our musical examinations.

But no pursuit at Cambridge was followed with nearly so much eagerness or gave me so much pleasure as collecting beetles. It was the mere passion for collecting, for I did not dissect them, and rarely compared their external characters with published descriptions, but got them named anyhow. I will give a proof of my zeal : one day, on tearing off some old bark, I saw two rare beetles, and seized one in each hand ; then I saw a third and new kind, which I could not bear to lose, so that I popped the one which I held in my right hand into my mouth. Alas ! it ejected some intensely acrid fluid, which burnt my tongue so that I was forced to spit the beetle out, which was lost, as was the third one.

I was very successful in collecting, and invented two new methods ; I employed a labourer to scrape, during the winter, moss off old trees and place it in a large bag, and likewise to collect the rubbish at the bottom of the barges in which reeds are brought from the fens, and thus I got some very rare species. No poet ever felt more delighted at seeing his first poem published than I did at seeing, in Stephens' *Illustrations of British Insects*, the magic words, " captured by C. Darwin, Esq." I was introduced to entomology by my second cousin, W. Darwin Fox, a clever and most pleasant man, who was then at Christ's College, and with whom I became extremely intimate. Afterwards I became well acquainted, and went out collecting, with Albert Way of Trinity, who in after years became a well-known archæologist ; also with H. Thompson,* of the same College, afterwards a leading agriculturist, chairman of a great railway, and Member of Parliament. It seems, therefore, that a taste for collecting beetles is some indication of future success in life !

I am surprised what an indelible impression many of the beetles which I caught at Cambridge have left on my mind.

---

* Afterwards Sir H. Thompson, first baronet.

I can remember the exact appearance of certain posts, old trees and banks where I made a good capture. The pretty *Panagœus crux-major* was a treasure in those days, and here at Down I saw a beetle running across a walk, and on picking it up instantly perceived that it differed slightly from *P. crux-major*, and it turned out to be *P. quadripunctatus*, which is only a variety or closely allied species, differing from it very slightly in outline. I had never seen in those old days Licinus alive, which to an uneducated eye hardly differs from many of the black Carabidous beetles; but my sons found here a specimen, and I instantly recognised that it was new to me; yet I had not looked at a British beetle for the last twenty years.

I have not yet mentioned a circumstance which influenced my whole career more than any other. This was my friendship with Professor Henslow. Before coming up to Cambridge, I had heard of him from my brother as a man who knew every branch of science, and I was accordingly prepared to reverence him. He kept open house once every week * when all undergraduates and some older members of the University, who were attached to science, used to meet in the evening. I soon got, through Fox, an invitation, and went there regularly. Before long I became well acquainted with Henslow, and during the latter half of my time at Cambridge took long walks with him on most days; so that I was called by some of the dons " the man who walks with Henslow;" and in the evening I was very often asked to join his family dinner. His knowledge was great in botany, entomology, chemistry, mineralogy, and geology. His strongest taste was to draw conclusions from long-continued minute observations. His judgment was excellent, and his whole mind well-balanced; but I do not suppose that any one would say that he possessed much original genius.

He was deeply religious, and so orthodox, that he told me one day he should be grieved if a single word of the Thirty-nine Articles were altered. His moral qualities were in every way admirable. He was free from every tinge of vanity or other petty feeling; and I never saw a man who thought so little about himself or his own concerns. His temper was imperturbably good, with the most winning and

---

* The *Cambridge Ray Club*, which in 1887 attained its fiftieth anniversary, is the direct descendant of these meetings, having been founded to fill the blank caused by the discontinuance, in 1836, of Henslow's Friday evenings. See Professor Babington's pamphlet, *The Cambridge Ray Club*, 1887.

courteous manners; yet, as I have seen, he could be roused by any bad action to the warmest indignation and prompt action.

I once saw in his company in the streets of Cambridge almost as horrid a scene as could have been witnessed during the French Revolution. Two body-snatchers had been arrested, and whilst being taken to prison had been torn from the constable by a crowd of the roughest men, who dragged them by their legs along the muddy and stony road. They were covered from head to foot with mud, and their faces were bleeding either from having been kicked or from the stones; they looked like corpses, but the crowd was so dense that I got only a few momentary glimpses of the wretched creatures. Never in my life have I seen such wrath painted on a man's face as was shown by Henslow at this horrid scene. He tried repeatedly to penetrate the mob; but it was simply impossible. He then rushed away to the mayor, telling me not to follow him, but to get more policemen. I forget the issue, except that the two men were got into the prison without being killed.

Henslow's benevolence was unbounded, as he proved by his many excellent schemes for his poor parishioners, when in after years he held the living of Hitcham. My intimacy with such a man ought to have been, and I hope was, an inestimable benefit. I cannot resist mentioning a trifling incident, which showed his kind consideration. Whilst examining some pollen-grains on a damp surface, I saw the tubes exserted, and instantly rushed off to communicate my surprising discovery to him. Now I do not suppose any other professor of botany could have helped laughing at my coming in such a hurry to make such a communication. But he agreed how interesting the phenomenon was, and explained its meaning, but made me clearly understand how well it was known; so I left him not in the least mortified, but well pleased at having discovered for myself so remarkable a fact, but determined not to be in such a hurry again to communicate my discoveries.

Dr. Whewell was one of the older and distinguished men who sometimes visited Henslow, and on several occasions I walked home with him at night. Next to Sir J. Mackintosh he was the best converser on grave subjects to whom I ever listened. Leonard Jenyns,* who afterwards published some

---

* Mr. Jenyns (now Blomefield) described the fish for the *Zoology of the Voyage of H. M. S. Beagle;* and is author of a long series of papers, chiefly

good essays in Natural History, often stayed with Henslow, who was his brother-in-law. I visited him at his parsonage on the borders of the Fens [Swaffham Bulbeck], and had many a good walk and talk with him about Natural History. I became also acquainted with several other men older than me, who did not care much about science, but were friends of Henslow. One was a Scotchman, brother of Sir Alexander Ramsay, and tutor of Jesus College ; he was a delightful man, but did not live for many years. Another was Mr. Dawes, afterwards Dean of Hereford, and famous for his success in the education of the poor. These men and others of the same standing, together with Henslow, used sometimes to take distant excursions into the country, which I was allowed to join, and they were most agreeable.

Looking back, I infer that there must have been something in me a little superior to the common run of youths, otherwise the above-mentioned men, so much older than me and higher in academical position, would never have allowed me to associate with them. Certainly I was not aware of any such superiority, and I remember one of my sporting friends, Turner, who saw me at work with my beetles, saying that I should some day be a Fellow of the Royal Society, and the notion seemed to me preposterous.

During my last year at Cambridge, I read with care and profound interest Humboldt's *Personal Narrative.* This work, and Sir J. Herschel's *Introduction to the Study of Natural Philosophy*, stirred up in me a burning zeal to add even the most humble contribution to the noble structure of Natural Science. No one or a dozen other books influenced me nearly so much as these two. I copied out from Humboldt long passages about Teneriffe, and read them aloud on one of the above-mentioned excursions, to (I think) Henslow, Ramsay, and Dawes, for on a previous occasion I had talked about the glories of Teneriffe, and some of the party declared they would endeavour to go there ; but I think they were only half in earnest. I was, however, quite in earnest, and got an introduction to a merchant in London to enquire about ships ; but the scheme was, of course, knocked on the head by the voyage of the *Beagle.*

My summer vacations were given up to collecting bee-

Zoological. In 1887 he printed, for private circulation, an autobiographical sketch, *Chapters in my Life*, and subsequently some (undated) addenda. The well-known Soame Jenyns was cousin to Mr. Jenyns' father.

tles, to some reading, and short tours. In the autumn my whole time was devoted to shooting, chiefly at Woodhouse and Maer, and sometimes with young Eyton of Eyton. Upon the whole the three years which I spent at Cambridge were the most joyful in my happy life; for I was then in excellent health, and almost always in high spirits.

As I had at first come up to Cambridge at Christmas, I was forced to keep two terms after passing my final examination, at the commencement of 1831; and Henslow then persuaded me to begin the study of geology. Therefore on my return to Shropshire I examined sections, and coloured a map of parts round Shrewsbury. Professor Sedgwick intended to visit North Wales in the beginning of August to pursue his famous geological investigations amongst the older rocks, and Henslow asked him to allow me to accompany him.* Accordingly he came and slept at my father's house.

A short conversation with him during this evening produced a strong impression on my mind. Whilst examining an old gravel-pit near Shrewsbury, a labourer told me that he had found in it a large worn tropical Volute shell, such as may be seen on chimney-pieces of cottages; and as he would not sell the shell, I was convinced that he had really found it in the pit. I told Sedgwick of the fact, and he at once said (no doubt truly) that it must have been thrown away by some one into the pit; but then added, if really embedded there it would be the greatest misfortune to geology, as it would overthrow all that we know about the superficial deposits of the Midland Counties. These gravelbeds belong in fact to the glacial period, and in after years I found in them broken arctic shells. But I was then utterly astonished at Sedgwick not being delighted at so wonderful a fact as a tropical shell being found near the surface in the middle of England. Nothing before had ever made me thoroughly realise, though I had read various scientific books, that science consists in grouping facts so that general laws or conclusions may be drawn from them.

Next morning we started for Llangollen, Conway, Ban-

---

* In connection with this tour my father used to tell a story about Sedgwick: they had started from their inn one morning, and had walked a mile or two, when Sedgwick suddenly stopped, and vowed that he would return, being certain "that damned scoundrel" (the waiter) had not given the chambermaid the sixpence intrusted to him for the purpose. He was ultimately persuaded to give up the project, seeing that there was no reason for suspecting the waiter of perfidy.—F. D.

gor, and Capel Curig. This tour was of decided use in teaching me a little how to make out the geology of a country. Sedgwick often sent me on a line parallel to his, telling me to bring back specimens of the rocks and to mark the stratification on a map. I have little doubt that he did this for my good, as I was too ignorant to have aided him. On this tour I had a striking instance how easy it is to overlook phenomena, however conspicuous, before they have been observed by any one. We spent many hours in Cwm Idwal, examining all the rocks with extreme care, as Sedgwick was anxious to find fossils in them; but neither of us saw a trace of the wonderful glacial phenomena all around us; we did not notice the plainly scored rocks, the perched boulders, the lateral and terminal moraines. Yet these phenomena are so conspicuous that, as I declared in a paper published many years afterwards in the *Philosophical Magazine*,* a house burnt down by fire did not tell its story more plainly than did this valley. If it had still been filled by a glacier, the phenomena would have been less distinct than they now are.

At Capel Curig I left Sedgwick and went in a straight line by compass and map across the mountains to Barmouth, never following any track unless it coincided with my course. I thus came on some strange wild places, and enjoyed much this manner of travelling. I visited Barmouth to see some Cambridge friends who were reading there, and thence returned to Shrewsbury and to Maer for shooting; for at that time I should have thought myself mad to give up the first days of partridge-shooting for geology or any other science.

Voyage of the ' Beagle': from December 27, 1831, to October 2, 1836.

On returning home from my short geological tour in North Wales, I found a letter from Henslow, informing me that Captain Fitz-Roy was willing to give up part of his own cabin to any young man who would volunteer to go with him without pay as naturalist to the Voyage of the *Beagle*. I have given, as I believe, in my MS. Journal an account of all the circumstances which then occurred; I will here only say that I was instantly eager to accept the

---

* *Philosophical Magazine*, 1842.

offer, but my father strongly objected, adding the words, fortunate for me, " If you can find any man of common-sense who advises you to go I will give my consent." So I wrote that evening and refused the offer. On the next morning I went to Maer to be ready for September 1st, and whilst out shooting, my uncle * sent for me, offering to drive me over to Shrewsbury and talk with my father, as my uncle thought it would be wise in me to accept the offer. My father always maintained that [my uncle] was one of the most sensible men in the world, and he at once consented in the kindest manner. I had been rather extravagant at Cambridge, and to console my father, said, " that I should be deuced clever to spend more than my allowance whilst on board the *Beagle ;*" but he answered with a smile, " But they tell me you are very clever."

Next day I started for Cambridge to see Henslow, and thence to London to see Fitz-Roy, and all was soon arranged. Afterwards, on becoming very intimate with Fitz-Roy, I heard that I had run a very narrow risk of being rejected on account of the shape of my nose ! He was an ardent disciple of Lavater, and was convinced that he could judge of a man's character by the outline of his features; and he doubted whether any one with my nose could possess sufficient energy and determination for the voyage. But I think he was afterwards well satisfied that my nose had spoken falsely.

Fitz-Roy's character was a singular one, with very many noble features : he was devoted to his duty, generous to a fault, bold, determined, and indomitably energetic, and an ardent friend to all under his sway. He would undertake any sort of trouble to assist those whom he thought deserved assistance. He was a handsome man, strikingly like a gentleman, with highly-courteous manners, which resembled those of his maternal uncle, the famous Lord Castlereagh, as I was told by the Minister at Rio. Nevertheless he must have inherited much in his appearance from Charles II., for Dr. Wallich gave me a collection of photographs which he had made, and I was struck with the resemblance of one to Fitz-Roy ; and on looking at the name, I found it Ch. E. Sobieski Stuart, Count d'Albanie,† a descendant of the same monarch.

---

* Josiah Wedgwood.

† The Count d'Albanie's claim to Royal descent has been shown to be

Fitz-Roy's temper was a most unfortunate one. It was usually worst in the early morning, and with his eagle eye he could generally detect something amiss about the ship, and was then unsparing in his blame. He was very kind to me, but was a man very difficult to live with on the intimate terms which necessarily followed from our messing by ourselves in the same cabin. We had several quarrels; for instance, early in the voyage at Bahia, in Brazil, he defended and praised slavery, which I abominated, and told me that he had just visited a great slave-owner, who had called up many of his slaves and asked them whether they were happy, and whether they wished to be free, and all answered " No." I then asked him, perhaps with a sneer, whether he thought that the answer of slaves in the presence of their master was worth anything? This made him excessively angry, and he said that as I doubted his word we could not live any longer together. I thought that I should have been compelled to leave the ship; but as soon as the news spread, which it did quickly, as the captain sent for the first lieutenant to assuage his anger by abusing me, I was deeply gratified by receiving an invitation from all the gun-room officers to mess with them. But after a few hours Fitz-Roy showed his usual magnanimity by sending an officer to me with an apology and a request that I would continue to live with him.

His character was in several respects one of the most noble which I have ever known.

The voyage of the *Beagle* has been by far the most important event in my life, and has determined my whole career; yet it depended on so small a circumstance as my uncle offering to drive me thirty miles to Shrewsbury, which few uncles would have done, and on such a trifle as the shape of my nose. I have always felt that I owe to the voyage the first real training or education of my mind; I was led to attend closely to several branches of natural history, and thus my powers of observation were improved, though they were always fairly developed.

The investigation of the geology of all the places visited was far more important, as reasoning here comes into play. On first examining a new district, nothing can appear more hopeless than the chaos of rocks; but by recording

---

based on a myth. See the *Quarterly Review*, 1847, vol. lxxxi. p. 83; also Hayward's *Biographical and Critical Essays*, 1873, vol. ii. p. 201.

the stratification and nature of the rocks and fossils at many points, always reasoning and predicting what will be found elsewhere, light soon begins to dawn on the district, and the structure of the whole becomes more or less intelligible. I had brought with me the first volume of Lyell's *Principles of Geology*, which I studied attentively; and the book was of the highest service to me in many ways. The very first place which I examined, namely, St. Jago, in the Cape de Verde islands, showed me clearly the wonderful superiority of Lyell's manner of treating geology, compared with that of any other author whose works I had with me or ever afterwards read.

Another of my occupations was collecting animals of all classes, briefly describing and roughly dissecting many of the marine ones; but from not being able to draw, and from not having sufficient anatomical knowledge, a great pile of MS. which I had made during the voyage has proved almost useless. I thus lost much time, with the exception of that spent in acquiring some knowledge of the Crustaceans, as this was of service when in after years I undertook a monograph of the Cirripedia.

During some part of the day I wrote my Journal, and took much pains in describing carefully and vividly all that I had seen; and this was good practice. My Journal served also, in part, as letters to my home, and portions were sent to England whenever there was an opportunity.

The above various special studies were, however, of no importance compared with the habit of energetic industry and of concentrated attention to whatever I was engaged in, which I then acquired. Everything about which I thought or read was made to bear directly on what I had seen or was likely to see; and this habit of mind was continued during the five years of the voyage. I feel sure that it was this training which has enabled me to do whatever I have done in science.

Looking backwards, I can now perceive how my love for science gradually preponderated over every other taste. During the first two years my old passion for shooting survived in nearly full force, and I shot myself all the birds and animals for my collection; but gradually I gave up my gun more and more, and finally altogether, to my servant, as shooting interfered with my work, more especially with making out the geological structure of a country. I discovered, though unconsciously and insensibly, that the pleasure

of observing and reasoning was a much higher one than
that of skill and sport. That my mind became developed
through my pursuits during the voyage is rendered probable
by a remark made by my father, who was the most acute observer whom I ever saw, of a sceptical disposition, and far
from being a believer in phrenology; for on first seeing
me after the voyage, he turned round to my sisters, and
exclaimed, "Why, the shape of his head is quite altered."

To return to the voyage. On September 11th (1831), I
paid a flying visit with Fitz-Roy to the *Beagle* at Plymouth.
Then to Shrewsbury to wish my father and sisters a long
farewell. On October 24th I took up my residence at
Plymouth, and remained there until December 27th, when
the *Beagle* finally left the shores of England for her circumnavigation of the world. We made two earlier attempts
to sail, but were driven back each time by heavy gales.
These two months at Plymouth were the most miserable
which I ever spent, though I exerted myself in various ways.
I was out of spirits at the thought of leaving all my family
and friends for so long a time, and the weather seemed to
me inexpressibly gloomy. I was also troubled with palpitation and pain about the heart, and like many a young ignorant man, especially one with a smattering of medical knowledge, was convinced that I had heart disease. I did not consult any doctor, as I fully expected to hear the verdict that I
was not fit for the voyage, and I was resolved to go at all
hazards.

I need not here refer to the events of the voyage—where
we went and what we did—as I have given a sufficiently
full account in my published Journal. The glories of the
vegetation of the Tropics rise before my mind at the present
time more vividly than anything else; though the sense of
sublimity, which the great deserts of Patagonia and the for-
est-clad mountains of Tierra del Fuego excited in me, has
left an indelible impression on my mind. The sight of a
naked savage in his native land is an event which can never
be forgotten. Many of my excursions on horseback through
wild countries, or in the boats, some of which lasted several
weeks, were deeply interesting; their discomfort and some
degree of danger were at that time hardly a drawback, and
none at all afterwards. I also reflect with high satisfaction
on some of my scientific work, such as solving the problem
of coral islands, and making out the geological structure of
certain islands, for instance, St. Helena. Nor must I pass

over the discovery of the singular relations of the animals
and plants inhabiting the several islands of the Galapagos
archipelago, and of all of them to the inhabitants of South
America.

As far as I can judge of myself, I worked to the utmost
during the voyage from the mere pleasure of investigation,
and from my strong desire to add a few facts to the great
mass of facts in Natural Science. But I was also ambitious
to take a fair place among scientific men,—whether more
ambitious or less so than most of my fellow-workers, I can
form no opinion.

The geology of St. Jago is very striking, yet simple:
a stream of lava formerly flowed over the bed of the sea,
formed of triturated recent shells and corals, which it has
baked into a hard white rock. Since then the whole island
has been upheaved. But the line of white rock revealed to
me a new and important fact, namely, that there had been
afterwards subsidence round the craters, which had since
been in action, and had poured forth lava. It then first
dawned on me that I might perhaps write a book on the
geology of the various countries visited, and this made me
thrill with delight. That was a memorable hour to me,
and how distinctly I can call to mind the low cliff of lava
beneath which I rested, with the sun glaring hot, a few
strange desert plants growing near, and with living corals
in the tidal pools at my feet. Later in the voyage, Fitz-
Roy asked me to read some of my Journal, and declared it
would be worth publishing; so here was a second book in
prospect!

Towards the close of our voyage I received a letter
whilst at Ascension, in which my sisters told me that Sedgwick had called on my father, and said that I should take a
place among the leading scientific men. I could not at the
time understand how he could have learnt anything of my
proceedings, but I heard (I believe afterwards) that Henslow
had read some of the letters which I wrote to him before
the Philosophical Society of Cambridge,* and had printed
them for private distribution. My collection of fossil bones,
which had been sent to Henslow, also excited considerable
attention amongst palæontologists. After reading this letter,
I clambered over the mountains of Ascension with a bound-

---

* Read at the meeting held November 16, 1835, and printed in a pamphlet
of 31 pp. for distribution among the members of the Society.

ing step, and made the volcanic rocks resound under my geological hammer. All this shows how ambitious I was; but I think that I can say with truth that in after years, though I cared in the highest degree for the approbation of such men as Lyell and Hooker, who were my friends, I did not care much about the general public. I do not mean to say that a favourable review or a large sale of my books did not please me greatly, but the pleasure was a fleeting one, and I am sure that I have never turned one inch out of my course to gain fame.

From my return to England (October 2, 1836) to my marriage (January 29, 1839).

These two years and three months were the most active ones which I ever spent, though I was occasionally unwell, and so lost some time. After going backwards and forwards several times between Shrewsbury, Maer, Cambridge, and London, I settled in lodgings at Cambridge * on December 13th, where all my collections were under the care of Henslow. I stayed here three months, and got my minerals and rocks examined by the aid of Professor Miller.

I began preparing my *Journal of Travels*, which was not hard work, as my MS. Journal had been written with care, and my chief labour was making an abstract of my more interesting scientific results. I sent also, at the request of Lyell, a short account of my observations on the elevation of the coast of Chili to the Geological Society.†

On March 7th, 1837, I took lodgings in Great Marlborough Street in London, and remained there for nearly two years, until I was married. During these two years I finished my Journal, read several papers before the Geological Society, began preparing the MS. for my *Geological Observations*, and arranged for the publication of the *Zoology of the Voyage of the Beagle*. In July I opened my first note-book for facts in relation to the *Origin of Species*, about which I had long reflected, and never ceased working for the next twenty years.

During these two years I also went a little into society, and acted as one of the honorary secretaries of the Geological Society. I saw a great deal of Lyell. One of his chief

---

* In Fitzwilliam Street.
† *Geolog. Soc. Proc.* ii. 1838, pp. 446–449.

characteristics was his sympathy with the work of others, and I was as much astonished as delighted at the interest which he showed when, on my return to England, I explained to him my views on coral reefs. This encouraged me greatly, and his advice and example had much influence on me. During this time I saw also a good deal of Robert Brown; I used often to call and sit with him during his breakfast on Sunday mornings, and he poured forth a rich treasure of curious observations and acute remarks, but they almost always related to minute points, and he never with me discussed large or general questions in science.

During these two years I took several short excursions as a relaxation, and one longer one to the parallel roads of Glen Roy, an account of which was published in the *Philosophical Transactions.** This paper was a great failure, and I am ashamed of it. Having been deeply impressed with what I had seen of the elevation of the land in South America, I attributed the parallel lines to the action of the sea; but I had to give up this view when Agassiz propounded his glacier-lake theory. Because no other explanation was possible under our then state of knowledge, I argued in favour of sea-action; and my error has been a good lesson to me never to trust in science to the principle of exclusion.

As I was not able to work all day at science, I read a good deal during these two years on various subjects, including some metaphysical books; but I was not well fitted for such studies. About this time I took much delight in Wordsworth's and Coleridge's poetry; and can boast that I read the *Excursion* twice through. Formerly Milton's *Paradise Lost* had been my chief favourite, and in my excursions during the voyage of the *Beagle*, when I could take only a single volume, I always chose Milton.

From my marriage, January 29, 1839, and residence in Upper Gower Street, to our leaving London and settling at Down, September 14, 1842.

[After speaking of his happy married life, and of his children, he continues:]

During the three years and eight months whilst we resided in London, I did less scientific work, though I

---

* 1839, pp. 39–82.

worked as hard as I possibly could, than during any other equal length of time in my life. This was owing to frequently recurring unwellness, and to one long and serious illness. The greater part of my time, when I could do anything, was devoted to my work on *Coral Reefs*, which I had begun before my marriage, and of which the last proof-sheet was corrected on May 6th, 1842. This book, though a small one, cost me twenty months of hard work, as I had to read every work on the islands of the Pacific and to consult many charts. It was thought highly of by scientific men, and the theory therein given is, I think, now well established.

No other work of mine was begun in so deductive a spirit as this, for the whole theory was thought out on the west coast of South America, before I had seen a true coral reef. I had therefore only to verify and extend my views by a careful examination of living reefs. But it should be observed that I had during the two previous years been incessantly attending to the effects on the shores of South America of the intermittent elevation of the land, together with denudation and the deposition of sediment. This necessarily led me to reflect much on the effects of subsidence, and it was easy to replace in imagination the continued deposition of sediment by the upward growth of corals. To do this was to form my theory of the formation of barrier-reefs and atolls.

Besides my work on coral-reefs, during my residence in London, I read before the Geological Society papers on the Erratic Boulders of South America,* on Earthquakes,† and on the Formation by the Agency of Earth-worms of Mould.‡ I also continued to superintend the publication of the *Zoology of the Voyage of the Beagle*. Nor did I ever intermit collecting facts bearing on the origin of species; and I could sometimes do this when I could do nothing else from illness.

In the summer of 1842 I was stronger than I had been for some time, and took a little tour by myself in North Wales, for the sake of observing the effects of the old glaciers which formerly filled all the larger valleys. I published a short account of what I saw in the *Philosophical Magazine*.# This excursion interested me greatly, and it was the last time I was ever strong enough to climb mountains or to take long walks such as are necessary for geological work.

---

* *Geolog. Soc. Proc.* iii. 1842.
† *Geolog. Trans.* v. 1840.
‡ *Geolog. Soc. Proc.* ii. 1838.
# *Philosophical Magazine*, 1842.

During the early part of our life in London, I was strong enough to go into general society, and saw a good deal of several scientific men and other more or less distinguished men. I will give my impressions with respect to some of them, though I have little to say worth saying.

I saw more of Lyell than of any other man, both before and after my marriage. His mind was characterised, as it appeared to me, by clearness, caution, sound judgment, and a good deal of originality. When I made any remark to him on Geology, he never rested until he saw the whole case clearly, and often made me see it more clearly than I had done before. He would advance all possible objections to my suggestion, and even after these were exhausted would long remain dubious. A second characteristic was his hearty sympathy with the work of other scientific men. *

On my return from the voyage of the *Beagle*, I explained to him my views on coral-reefs, which differed from his, and I was greatly surprised and encouraged by the vivid interest which he showed. His delight in science was ardent, and he felt the keenest interest in the future progress of mankind. He was very kind-hearted, and thoroughly liberal in his religious beliefs, or rather disbeliefs; but he was a strong theist. His candour was highly remarkable. He exhibited this by becoming a convert to the Descent theory, though he had gained much fame by opposing Lamarck's views, and this after he had grown old. He reminded me that I had many years before said to him, when discussing the opposition of the old school of geologists to his new views, "What a good thing it would be if every scientific man was to die when sixty years old, as afterwards he would be sure to oppose all new doctrines." But he hoped that now he might be allowed to live.

The science of Geology is enormously indebted to Lyell —more so, as I believe, than to any other man who ever lived. When [I was] starting on the voyage of the *Beagle*, the sagacious Henslow, who, like all other geologists, believed at that time in successive cataclysms, advised me to get and study the first volume of the *Principles*, which had then just been published, but on no account to accept the views therein advocated. How differently would any one now

---

* The slight repetition here observable is accounted for by the notes on Lyell, &c., having been added in April, 1881, a few years after the rest of the *Recollections* were written.—F. D.

speak of the *Principles !*　I am proud to remember that the
first place, namely, St. Jago, in the Cape de Verde Archipelago, in which I geologised, convinced me of the infinite
superiority of Lyell's views over those advocated in any
other work known to me.

The powerful effects of Lyell's works could formerly be
plainly seen in the different progress of the science in France
and England.　The present total oblivion of Elie de Beaumont's wild hypotheses, such as his *Craters of Elevation* and
Lines of Elevation (which latter hypothesis I heard Sedgwick at the Geological Society lauding to the skies), may be
largely attributed to Lyell.

I saw a good deal of Robert Brown, "facile Princeps
Botanicorum," as he was called by Humboldt.　He seemed
to me to be chiefly remarkable for the minuteness of his observations and their perfect accuracy.　His knowledge was
extraordinarily great, and much died with him, owing to
his excessive fear of ever making a mistake.　He poured
out his knowledge to me in the most unreserved manner,
yet was strangely jealous on some points.　I called on him
two or three times before the voyage of the *Beagle*, and on
one occasion he asked me to look through a microscope and
describe what I saw.　This I did, and believe now that it
was the marvellous currents of protoplasm in some vegetable cell.　I then asked him what I had seen ; but he answered me, "That is my little secret."

He was capable of the most generous actions.　When
old, much out of health, and quite unfit for any exertion,
he daily visited (as Hooker told me) an old man-servant,
who lived at a distance (and whom he supported), and read
aloud to him.　This is enough to make up for any degree
of scientific penuriousness or jealousy.

I may here mention a few other eminent men whom I
have occasionally seen, but I have little to say about them
worth saying.　I felt a high reverence for Sir J. Herschel,
and was delighted to dine with him at his charming house
at the Cape of Good Hope and afterwards at his London
house.　I saw him, also, on a few other occasions.　He never
talked much, but every word which he uttered was worth
listening to.

I once met at breakfast, at Sir R. Murchison's house, the
illustrious Humboldt, who honoured me by expressing a
wish to see me.　I was a little disappointed with the great
man, but my anticipations probably were too high.　I can

remember nothing distinctly about our interview, except that Humboldt was very cheerful and talked much.

X.* reminds me of Buckle, whom I once met at Hensleigh Wedgwood's. I was very glad to learn from [Buckle] his system of collecting facts. He told me that he bought all the books which he read, and made a full index to each, of the facts which he thought might prove serviceable to him, and that he could always remember in what book he had read anything, for his memory was wonderful. I asked him how at first he could judge what facts would be serviceable, and he answered that he did not know, but that a sort of instinct guided him. From this habit of making indices, he was enabled to give the astonishing number of references on all sorts of subjects which may be found in his *History of Civilisation.* This book I thought most interesting, and read it twice, but I doubt whether his generalisations are worth anything. Buckle was a great talker; and I listened to him, saying hardly a word, nor indeed could I have done so, for he left no gaps. When Mrs. Farrer began to sing, I jumped up and said that I must listen to her. After I had moved away, he turned round to a friend, and said (as was overheard by my brother), " Well, Mr. Darwin's books are much better than his conversation."

Of other great literary men, I once met Sydney Smith at Dean Milman's house. There was something inexplicably amusing in every word which he uttered. Perhaps this was partly due to the expectation of being amused. He was talking about Lady Cork, who was then extremely old. This was the lady who, as he said, was once so much affected by one of his charity sermons, that she *borrowed* a guinea from a friend to put in the plate. He now said, " It is generally believed that my dear old friend Lady Cork has been overlooked "; and he said this in such a manner that no one could for a moment doubt that he meant that his dear old friend had been overlooked by the devil. How he managed to express this I know not.

I likewise once met Macaulay at Lord Stanhope's (the historian's) house, and as there was only one other man at dinner, I had a grand opportunity of hearing him converse, and he was very agreeable. He did not talk at all too much, nor indeed could such a man talk too much, as long

---

* A passage referring to X. is here omitted.—F. D.

as he allowed others to turn the stream of his conversation, and this he did allow.

Lord Stanhope once gave me a curious little proof of the accuracy and fulness of Macaulay's memory. Many historians used often to meet at Lord Stanhope's house; and, in discussing various subjects, they would sometimes differ from Macaulay, and formerly they often referred to some book to see who was right; but latterly, as Lord Stanhope noticed, no historian ever took this trouble, and whatever Macaulay said was final.

On another occasion I met at Lord Stanhope's house one of his parties of historians and other literary men, and amongst them were Motley and Grote. After luncheon I walked about Chevening Park for nearly an hour with Grote, and was much interested by his conversation and pleased by the simplicity and absence of all pretension in his manners.

Long ago I dined occasionally with the old Earl, the father of the historian. He was a strange man, but what little I knew of him I liked much. He was frank, genial, and pleasant. He had strongly-marked features, with a brown complexion, and his clothes, when I saw him, were all brown. He seemed to believe in everything which was to others utterly incredible. He said one day to me, " Why don't you give up your fiddle-faddle of geology and zoology, and turn to the occult sciences?" The historian, then Lord Mahon, seemed shocked at such a speech to me, and his charming wife much amused.

The last man whom I will mention is Carlyle, seen by me several times at my brother's house and two or three times at my own house. His talk was very racy and interesting, just like his writings, but he sometimes went on too long on the same subject. I remember a funny dinner at my brother's, where, amongst a few others, were Babbage and Lyell, both of whom liked to talk. Carlyle, however, silenced every one by haranguing during the whole dinner on the advantages of silence. After dinner, Babbage, in his grimmest manner, thanked Carlyle for his very interesting lecture on silence.

Carlyle sneered at almost every one: One day in my house he called Grote's *History* " a fetid quagmire, with nothing spiritual about it." I always thought, until his *Reminiscences* appeared, that his sneers were partly jokes, but this now seems rather doubtful. His expression was that of a depressed, almost despondent, yet benevolent man, and it is notorious how heartily he laughed. I believe that his benevolence was real, though stained by not a little jealousy. No one can doubt about his extraordinary power of drawing pictures of things and men—far more vivid, as it appears to me, than any drawn by Macaulay. Whether his pictures of men were true ones is another question.

He has been all-powerful in impressing some grand moral truths on the minds of men. On the other hand, his views about slavery were revolting. In his eyes might was right. His mind seemed to me a very narrow one; even if all brances of science, which he despised, are excluded. It is astonishing to me that Kingsley should have spoken of him as a man well fitted to advance science. He laughed to scorn the idea that a mathematician, such as Whewell, could judge, as I maintained he could, of Goethe's views on light. He thought it a most ridiculous thing that any one should care whether a glacier moved a little quicker or a little slower, or moved at all. As far as I could judge, I never met a man with a mind so ill adapted for scientific research.

Whilst living in London, I attended as regularly as I could the meetings of several scientific societies, and acted as secretary to the Geological Society. But such attendance, and ordinary society, suited my health so badly that we resolved to live in the country, which we both preferred and have never repented of.

Residence at Down, from September 14, 1842, to the present time, 1876.

After several fruitless searches in Surrey and elsewhere, we found this house and purchased it. I was pleased with the diversified appearance of the vegetation proper to a chalk district, and so unlike what I had been accustomed to in the Midland counties; and still more pleased with the extreme quietness and rusticity of the place. It is not, however, quite so retired a place as a writer in a German periodical makes it, who says that my house can be approached only by a mule-track! Our fixing ourselves here has answered admirably in one way which we did not anticipate, namely, by being very convenient for frequent visits from our children.

Few persons can have lived a more retired life than we

4

have done. Besides short visits to the houses of relations, and occasionally to the seaside or elsewhere, we have gone nowhere. During the first part of our residence we went a little into society, and received a few friends here; but my health almost always suffered from the excitement, violent shivering and vomiting attacks being thus brought on. I have therefore been compelled for many years to give up all dinner-parties; and this has been somewhat of a deprivation to me, as such parties always put me into high spirits. From the same cause I have been able to invite here very few scientific acquaintances.

My chief enjoyment and sole employment throughout life has been scientific work, and the excitement from such work makes me for the time forget, or drives quite away, my daily discomfort. I have therefore nothing to record during the rest of my life, except the publication of my several books. Perhaps a few details how they arose may be worth giving.

My several Publications.—In the early part of 1844, my observations on the volcanic islands visited during the voyage of the *Beagle* were published. In 1845, I took much pains in correcting a new edition of my *Journal of Researches*, which was originally published in 1839 as part of Fitz-Roy's work. The success of this my first literary child always tickles my vanity more than that of any of my other books. Even to this day it sells steadily in England and the United States, and has been translated for the second time into German, and into French and other languages. This success of a book of travels, especially of a scientific one, so many years after its first publication, is surprising. Ten thousand copies have been sold in England of the second edition. In 1846 my *Geological Observations on South America* were published. I record in a little diary, which I have always kept, that my three geological books (*Coral Reefs* included) consumed four and a half years' steady work; "and now it is ten years since my return to England. How much time have I lost by illness?" I have nothing to say about these three books except that to my surprise new editions have lately been called for.*

In October, 1846, I began to work on 'Cirripedia' (Barnacles). When on the coast of Chile, I found a most curious form, which burrowed into shells of Concholepas, and

---

* *Geological Observations*, 2nd Edit. 1876. *Coral Reefs*, 2nd Edit. 1874.

which differed so much from all other Cirripedes that I had
to form a new suborder for its sole reception. Lately an
allied burrowing genus has been found on the shores of
Portugal. To understand the structure of my new Cirripede I had to examine and dissect many of the common
forms; and this gradually led me on to take up the whole
group. I worked steadily on the subject for the next eight
years, and ultimately published two thick volumes,* describing all the known living species, and two thin quartos on
the extinct species. I do not doubt that Sir E. Lytton Bul-
wer had me in his mind when he introduced in one of his
novels a Professor Long, who had written two huge volumes
on limpets.

Although I was employed during eight years on this
work, yet I record in my diary that about two years out of
this time was lost by illness. On this account I went in
1848 for some months to Malvern for hydropathic treatment, which did me much good, so that on my return home
I was able to resume work. So much was I out of health
that when my dear father died on November 13th, 1848, I
was unable to attend his funeral or to act as one of his ex-
ecutors.

My work on the Cirripedia possesses, I think, considerable value, as besides describing several new and remarkable forms, I made out the homologies of the various parts
—I discovered the cementing apparatus, though I blundered dreadfully about the cement glands—and lastly I
proved the existence in certain genera of minute males
complemental to and parasitic on the hermaphrodites. This
latter discovery has at last been fully confirmed; though
at one time a German writer was pleased to attribute the
whole account to my fertile imagination. The Cirripedes
form a highly varying and difficult group of species to class;
and my work was of considerable use to me, when I had to
discuss in the *Origin of Species* the principles of a natural
classification. Nevertheless, I doubt whether the work was
worth the consumption of so much time.

From September 1854 I devoted my whole time to arranging my huge pile of notes, to observing, and to experimenting in relation to the transmutation of species. During the voyage of the *Beagle* I had been deeply impressed
by discovering in the Pampean formation great fossil ani-

---

* Published by the Ray Society.

mals covered with armour like that on the existing arma-
dillos; secondly, by the manner in which closely allied animals replace one another in proceeding southwards over the
Continent; and thirdly, by the South American character
of most of the productions of the Galapagos archipelago,
and more especially by the manner in which they differ
slightly on each island of the group; none of the islands
appearing to be very ancient in a geological sense.

It was evident that such facts as these, as well as many
others, could only be explained on the supposition that
species gradually become modified; and the subject haunted
me. But it was equally evident that neither the action of
the surrounding conditions, nor the will of the organisms
(especially in the case of plants) could account for the innumerable cases in which organisms of every kind are beautifully adapted to their habits of life—for instance, a woodpecker or a tree-frog to climb trees, or a seed for dispersal
by hooks or plumes. I had always been much struck by
such adaptations, and until these could be explained it
seemed to me almost useless to endeavour to prove by indi-
rect evidence that species have been modified.

After my return to England it appeared to me that by
following the example of Lyell in Geology, and by collecting all facts which bore in any way on the variation of animals and plants under domestication and nature, some light
might perhaps be thrown on the whole subject. My first
note-book was opened in July 1837. I worked on true Ba-
conian principles, and without any theory collected facts on
a wholesale scale, more especially with respect to domesticated productions, by printed enquiries, by conversation
with skilful breeders and gardeners, and by extensive reading. When I see the list of books of all kinds which I read
and abstracted, including whole series of Journals and
Transactions, I am surprised at my industry. I soon perceived that selection was the keystone of man's success in
making useful races of animals and plants. But how selection could be applied to organisms living in a state of
nature remained for some time a mystery to me.

In October 1838, that is, fifteen months after I had begun my systematic enquiry, I happened to read for amusement Malthus on *Population*, and being well prepared to
appreciate the struggle for existence which everywhere goes
on from long-continued observation of the habits of animals
and plants, it at once struck me that under these circum-

stances favourable variations would tend to be preserved and
unfavourable ones to be destroyed. The result of this would
be the formation of new species. Here, then, I had at last
got a theory by which to work; but I was so anxious to
avoid prejudice, that I determined not for some time to
write even the briefest sketch of it. In June 1842 I first
allowed myself the satisfaction of writing a very brief abstract of my theory in pencil in 35 pages; and this was enlarged during the summer of 1844 into one of 230 pages,
which I had fairly copied out and still possess.

But at that time I overlooked one problem of great importance; and it is astonishing to me, except on the principle of Columbus and his egg, how I could have overlooked
it and its solution. This problem is the tendency in organic
beings descended from the same stock to diverge in character as they become modified. That they have diverged
greatly is obvious from the manner in which species of all
kinds can be classed under genera, genera under families,
families under sub-orders, and so forth; and I can remember the very spot in the road, whilst in my carriage, when
to my joy the solution occurred to me; and this was long
after I had come to Down. The solution, as I believe, is
that the modified offspring of all dominant and increasing
forms tend to become adapted to many and highly diversified places in the economy of nature.

Early in 1856 Lyell advised me to write out my views
pretty fully, and I began at once to do so on a scale three or
four times as extensive as that which was afterwards followed
in my *Origin of Species;* yet it was only an abstract of the
materials which I had collected, and I got through about
half the work on this scale. But my plans were overthrown,
for early in the summer of 1858 Mr. Wallace, who was then
in the Malay archipelago, sent me an essay *On the Tendency
of Varieties to depart indefinitely from the Original Type;*
and this essay contained exactly the same theory as mine.
Mr. Wallace expressed the wish that if I thought well of his
essay, I should send it to Lyell for perusal.

The circumstances under which I consented at the request of Lyell and Hooker to allow of an abstract from my
MS., together with a letter to Asa Gray, dated September 5,
1857, to be published at the same time with Wallace's Essay,
are given in the *Journal of the Proceedings of the Linnean
Society,* 1858, p. 45. I was at first very unwilling to consent, as I thought Mr. Wallace might consider my doing so

unjustifiable, for I did not then know how generous and
noble was his disposition. The extract from my MS. and
the letter to Asa Gray had neither been intended for publication, and were badly written. Mr. Wallace's essay, on
the other hand, was admirably expressed and quite clear.
Nevertheless, our joint productions excited very little attention, and the only published notice of them which I can remember was by Professor Haughton of Dublin, whose verdict
was that all that was new in them was false, and what was
true was old. This shows how necessary it is that any new
view should be explained at considerable length in order to
arouse public attention.

In September 1858 I set to work by the strong advice of
Lyell and Hooker to prepare a volume on the transmutation
of species, but was often interrupted by ill-health, and short
visits to Dr. Lane's delightful hydropathic establishment at
Moor Park. I abstracted the MS. begun on a much larger
scale in 1856, and completed the volume on the same reduced
scale. It cost me thirteen months and ten days' hard labour.
It was published under the title of the *Origin of Species*, in
November 1859. Though considerably added to and corrected in the later editions, it has remained substantially the
same book.

It is no doubt the chief work of my life. It was from
the first highly successful. The first small edition of 1250
copies was sold on the day of publication, and a second edition of 3000 copies soon afterwards. Sixteen thousand
copies have now (1876) been sold in England; and considering how stiff a book it is, this is a large sale. It has
been translated into almost every European tongue, even
into such languages as Spanish, Bohemian, Polish, and
Russian. It has also, according to Miss Bird, been translated into Japanese,* and is there much studied. Even an
essay in Hebrew has appeared on it, showing that the theory
is contained in the Old Testament! The reviews were very
numerous; for some time I collected all that appeared on
the *Origin* and on my related books, and these amount (excluding newspaper reviews) to 265; but after a time I gave
up the attempt in despair. Many separate essays and books
on the subject have appeared ; and in Germany a catalogue
or bibliography on "Darwinismus" has appeared every year
or two.

---

* Miss Bird is mistaken, as I learn from Professor Mitsukuri.—F. D.

The success of the *Origin* may, I think, be attributed in large part to my having long before written two condensed sketches, and to my having finally abstracted a much larger manuscript, which was itself an abstract. By this means I was enabled to select the more striking facts and conclusions. I had, also, during many years, followed a golden rule, namely, that whenever a published fact, a new observation or thought came across me, which was opposed to my general results, to make a memorandum of it without fail and at once : for I had found by experience that such facts and thoughts were far more apt to escape from the memory than favourable ones. Owing to this habit, very few objections were raised against my views which I had not at least noticed and attempted to answer.

It has sometimes been said that the success of the *Origin* proved " that the subject was in the air," or " that men's minds were prepared for it." I do not think that this is strictly true, for I occasionally sounded not a few naturalists, and never happened to come across a single one who seemed to doubt about the permanence of species. Even Lyell and Hooker, though they would listen with interest to me, never seemed to agree. I tried once or twice to explain to able men what I meant by Natural selection, but signally failed. What I believe was strictly true is that innumerable well-observed facts were stored in the minds of naturalists ready to take their proper places as soon as any theory which would receive them was sufficiently explained. Another element in the success of the book was its moderate size; and this I owe to the appearance of Mr. Wallace's essay; had I published on the scale in which I began to write in 1856, the book would have been four or five times as large as the *Origin*, and very few would have had the patience to read it.

I gained much by my delay in publishing from about 1839, when the theory was clearly conceived, to 1859; and I lost nothing by it, for I cared very little whether men attributed most originality to me or Wallace; and his essay no doubt aided in the reception of the theory. I was forestalled in only one important point, which my vanity has always made me regret, namely, the explanation by means of the Glacial period of the presence of the same species of plants and of some few animals on distant mountain summits and in the arctic regions. This view pleased me so much that I wrote it out *in extenso*, and I believe that it was read by Hooker some years before E. Forbes published his celebrated memoir * on the subject. In the very few points in which we differed, I still think that I was in the right. I have never, of course, alluded in print to my having independently worked out this view.

Hardly any point gave me so much satisfaction when I was at work on the *Origin*, as the explanation of the wide difference in many classes between the embryo and the adult animal, and of the close resemblance of the embryos within the same class. No notice of this point was taken, as far as I remember, in the early reviews of the *Origin*, and I recollect expressing my surprise on this head in a letter to Asa Gray. Within late years several reviewers have given the whole credit to Fritz Müller and Häckel, who undoubtedly have worked it out much more fully, and in some respects more correctly than I did. I had materials for a whole chapter on the subject, and I ought to have made the discussion longer; for it is clear that I failed to impress my readers; and he who succeeds in doing so deserves, in my opinion, all the credit.

This leads me to remark that I have almost always been treated honestly by my reviewers, passing over those without scientific knowledge as not worthy of notice. My views have often been grossly misrepresented, bitterly opposed and ridiculed, but this has been generally done, as I believe, in good faith. On the whole I do not doubt that my works have been over and over again greatly overpraised. I rejoice that I have avoided controversies, and this I owe to Lyell, who many years ago, in reference to my geological works, strongly advised me never to get entangled in a controversy, as it rarely did any good and caused a miserable loss of time and temper.

Whenever I have found out that I have blundered, or that my work has been imperfect, and when I have been contemptuously criticised, and even when I have been overpraised, so that I have felt mortified, it has been my greatest comfort to say hundreds of times to myself that "I have worked as hard and as well as I could, and no man can do more than this." I remember when in Good Success Bay, in Tierra del Fuego, thinking (and I believe that I wrote home to the effect) that I could not employ my life better than in adding a little to Natural Science. This I have

---

* *Geolog. Survey Mem.*, 1846.

done to the best of my abilities, and critics may say what they like, but they cannot destroy this conviction.

During the two last months of 1859 I was fully occupied in preparing a second edition of the *Origin*, and by an enormous correspondence. On January 1st, 1860, I began arranging my notes for my work on the *Variation of Animals and Plants under Domestication ;* but it was not published until the beginning of 1868; the delay having been caused partly by frequent illnesses, one of which lasted seven months, and partly by being tempted to publish on other subjects which at the time interested me more.

On May 15th, 1862, my little book on the *Fertilisation of Orchids*, which cost me ten months' work, was published : most of the facts had been slowly accumulated during several previous years. During the summer of 1839, and, I believe, during the previous summer, I was led to attend to the cross-fertilisation of flowers by the aid of insects, from having come to the conclusion in my speculations on the origin of species, that crossing played an important part in keeping specific forms constant. I attended to the subject more or less during every subsequent summer; and my interest in it was greatly enhanced by having procured and read in November 1841, through the advice of Robert Brown, a copy of C. K. Sprengel's wonderful book, *Das entdeckte Geheimniss der Natur*. For some years before 1862 I had specially attended to the fertilisation of our British orchids; and it seemed to me the best plan to prepare as complete a treatise on this group of plants as well as I could, rather than to utilise the great mass of matter which I had slowly collected with respect to other plants.

My resolve proved a wise one; for since the appearance of my book, a surprising number of papers and separate works on the fertilisation of all kinds of flowers have appeared; and these are far better done than I could possibly have effected. The merits of poor old Sprengel, so long overlooked, are now fully recognised many years after his death.

During the same year I published in the *Journal of the Linnean Society*, a paper *On the Two Forms, or Dimorphic Condition of Primula*, and during the next five years, five other papers on dimorphic and trimorphic plants. I do not think anything in my scientific life has given me so much satisfaction as making out the meaning of the structure of these plants. I had noticed in 1838 or 1839 the dimorphism of *Linum flavum*, and had at first thought that it was merely a case of unmeaning variability. But on examining the common species of Primula, I found that the two forms were much too regular and constant to be thus viewed. I therefore became almost convinced that the common cowslip and primrose were on the high-road to become diœcious;— that the short pistil in the one form, and the short stamens in the other form were tending towards abortion. The plants were therefore subjected under this point of view to trial; but as soon as the flowers with short pistils fertilised with pollen from the short stamens, were found to yield more seeds than any other of the four possible unions, the abortion-theory was knocked on the head. After some additional experiment, it became evident that the two forms, though both were perfect hermaphrodites, bore almost the same relation to one another as do the two sexes of an ordinary animal. With Lythrum we have the still more wonderful case of three forms standing in a similar relation to one another. I afterwards found that the offspring from the union of two plants belonging to the same forms presented a close and curious analogy with hybrids from the union of two distinct species.

In the autumn of 1864 I finished a long paper on *Climbing Plants*, and sent it to the Linnean Society. The writing of this paper cost me four months; but I was so unwell when I received the proof-sheets, that I was forced to leave them very badly and often obscurely expressed. The paper was little noticed, but when in 1875 it was corrected and published as a separate book it sold well. I was led to take up this subject by reading a short paper by Asa Gray, published in 1858. He sent me seeds, and on raising some plants I was so much fascinated and perplexed by the revolving movements of the tendrils and stems, which movements are really very simple, though appearing at first sight very complex, that I procured various other kinds of climbing plants, and studied the whole subject. I was all the more attracted to it, from not being at all satisfied with the explanation which Henslow gave us in his lectures, about twining plants, namely, that they had a natural tendency to grow up in a spire. This explanation proved quite erroneous. Some of the adaptations displayed by climbing plants are as beautiful as those of Orchids for ensuring cross-fertilisation.

My *Variation of Animals and Plants under Domestica-*

tion was begun, as already stated, in the beginning of 1860, but was not published until the beginning of 1868. It was a big book, and cost me four years and two months' hard labour. It gives all my observations and an immense number of facts collected from various sources, about our domestic productions. In the second volume the causes and laws of variation, inheritance, &c., are discussed, as far as our present state of knowledge permits. Towards the end of the work I give my well-abused hypothesis of Pangenesis. An unverified hypothesis is of little or no value; but if any one should hereafter be led to make observations by which some such hypothesis could be established, I shall have done good service, as an astonishing number of isolated facts can be thus connected together and rendered intelligible. In 1875 a second and largely corrected edition, which cost me a good deal of labour, was brought out.

My *Descent of Man* was published in February 1871. As soon as I had become, in the year 1837 or 1838, convinced that species were mutable productions, I could not avoid the belief that man must come under the same law. Accordingly I collected notes on the subject for my own satisfaction, and not for a long time with any intention of publishing. Although in the *Origin of Species* the derivation of any particular species is never discussed, yet I thought it best, in order that no honourable man should accuse me of concealing my views, to add that by the work " light would be thrown on the origin of man and his history." It would have been useless, and injurious to the success of the book to have paraded, without giving any evidence, my conviction with respect to his origin.

But when I found that many naturalists fully accepted the doctrine of the evolution of species, it seemed to me advisable to work up such notes as I possessed, and to publish a special treatise on the origin of man. I was the more glad to do so, as it gave me an opportunity of fully discussing sexual selection—a subject which had always greatly interested me. This subject, and that of the variation of our domestic productions, together with the causes and laws of variation, inheritance, and the intercrossing of plants, are the sole subjects which I have been able to write about in full, so as to use all the materials which I have collected. The *Descent of Man* took me three years to write, but then as usual some of this time was lost by ill health, and some was consumed by preparing new editions and other minor works. A second and largely corrected edition of the *Descent* appeared in 1874.

My book on the *Expression of the Emotions in Men and Animals* was published in the autumn of 1872. I had intended to give only a chapter on the subject in the *Descent of Man*, but as soon as I began to put my notes together, I saw that it would require a separate treatise.

My first child was born on December 27th, 1839, and I at once commenced to make notes on the first dawn of the various expressions which he exhibited, for I felt convinced, even at this early period, that the most complex and fine shades of expression must all have had a gradual and natural origin. During the summer of the following year, 1840, I read Sir C. Bell's admirable work on expression, and this greatly increased the interest which I felt in the subject, though I could not at all agree with his belief that various muscles had been specially created for the sake of expression. From this time forward I occasionally attended to the subject, both with respect to man and our domesticated animals. My book sold largely; 5267 copies having been disposed of on the day of publication.

In the summer of 1860 I was idling and resting near Hartfield, where two species of [Sundew] abound; and I noticed that numerous insects had been entrapped by the leaves. I carried home some plants, and on giving them insects saw the movements of the tentacles, and this made me think it probable that the insects were caught for some special purpose. Fortunately a crucial test occurred to me, that of placing a large number of leaves in various nitrogenous and non-nitrogenous fluids of equal density; and as soon as I found that the former alone excited energetic movements, it was obvious that here was a fine new field for investigation.

During subsequent years, whenever I had leisure I pursued my experiments, and my book on *Insectivorous Plants* was published in July 1875—that is sixteen years after my first observations. The delay in this case, as with all my other books, has been a great advantage to me; for a man after a long interval can criticise his own work, almost as well as if it were that of another person. The fact that a plant should secrete, when properly excited, a fluid containing an acid and ferment, closely and analogous to the digestive fluid of an animal, was certainly a remarkable discovery.

During this autumn of 1876 I shall publish on the *Effects of Cross- and Self-Fertilisation in the Vegetable Kingdom.* This book will form a complement to that on the *Fertilisation of Orchids*, in which I showed how perfect were the means for cross-fertilisation, and here I shall show how important are the results. I was led to make, during eleven years, the numerous experiments recorded in this volume, by a mere accidental observation; and indeed it required the accident to be repeated before my attention was thoroughly aroused to the remarkable fact that seedlings of self-fertilised parentage are inferior, even in the first generation, in height and vigour to seedlings of cross-fertilised parentage. I hope also to republish a revised edition of my book on Orchids, and hereafter my papers on dimorphic and trimorphic plants, together with some additional observations on allied points which I never have had time to arrange. My strength will then probably be exhausted, and I shall be ready to exclaim " Nunc dimittis."

Written May 1st, 1881.—The Effects of Cross- and Self-Fertilisation was published in the autumn of 1876; and the results there arrived at explain, as I believe, the endless and wonderful contrivances for the transportal of pollen from one plant to another of the same species. I now believe, however, chiefly from the observations of Hermann Müller, that I ought to have insisted more strongly than I did on the many adaptations for self-fertilisation; though I was well aware of many such adaptations. A much enlarged edition of my *Fertilisation of Orchids* was published in 1877.

In this same year *The Different Forms of Flowers, &c.*, appeared, and in 1880 a second edition. This book consists chiefly of the several papers on Hetero-styled flowers originally published by the Linnean Society, corrected, with much new matter added, together with observations on some other cases in which the same plant bears two kinds of flowers. As before remarked, no little discovery of mine ever gave me so much pleasure as the making out the meaning of hetero-styled flowers. The results of crossing such flowers in an illegitimate manner, I believe to be very important, as bearing on the sterility of hybrids; although these results have been noticed by only a few persons.

In 1879, I had a translation of Dr. Ernst Krause's *Life of Erasmus Darwin* published, and I added a sketch of his character and habits from material in my possession.

Many persons have been much interested by this little life, and I am surprised that only 800 or 900 copies were sold.

In 1880 I published, with [my son] Frank's assistance our *Power of Movement in Plants*. This was a tough piece of work. The book bears somewhat the same relation to my little book on *Climbing Plants*, which *Cross-Fertilisation* did to the *Fertilisation of Orchids ;* for in accordance with the principle of evolution it was impossible to account for climbing plants having been developed in so many widely different groups unless all kinds of plants possess some slight power of movement of an analogous kind. This I proved to be the case ; and I was further led to a rather wide generalisation, viz., that the great and important classes of movements, excited by light, the attraction of gravity, &c., are all modified forms of the fundamental movement of circumnutation. It has always pleased me to exalt plants in the scale of organised beings; and I therefore felt an especial pleasure in showing how many and what admirably well adapted movements the tip of a root possesses.

I have now (May 1, 1881) sent to the printers the MS. of a little book on *The Formation of Vegetable Mould through the Action of Worms*. This is a subject of but small importance; and I know not whether it will interest any readers,* but it has interested me. It is the completion of a short paper read before the Geological Society more than forty years ago, and has revived old geological thoughts.

I have now mentioned all the books which I have published, and these have been the milestones in my life, so that little remains to be said. I am not conscious of any change in my mind during the last thirty years, excepting in one point presently to be mentioned ; nor, indeed, could any change have been expected unless one of general deterioration. But my father lived to his eighty-third year with his mind as lively as ever it was, and all his faculties undimmed ; and I hope that I may die before my mind fails to a sensible extent. I think that I have become a little more skilful in guessing right explanations and in devising experimental tests ; but this may probably be the result of mere practice, and of a larger store of knowledge. I have as much difficulty as ever in expressing myself clearly and concisely ; and

---

* Between November 1881 and February 1884, 8500 copies were sold.—F. D.

this difficulty has caused me a very great loss of time; but it has had the compensating advantage of forcing me to think long and intently about every sentence, and thus I have been led to see errors in reasoning and in my own observations or those of others.

There seems to be a sort of fatality in my mind leading me to put at first my statement or proposition in a wrong or awkward form. Formerly I used to think about my sentences before writing them down; but for several years I have found that it saves time to scribble in a vile hand, whole pages as quickly as I possibly can, contracting half the words; and then correct deliberately. Sentences thus scibbled down are often better ones than I could have written deliberately.

Having said thus much about my manner of writing, I will add that with my large books I spend a good deal of time over the general arrangement of the matter. I first make the rudest outline in two or three pages, and then a larger one in several pages, a few words or one word standing for a whole discussion or series of facts. Each one of these headings is again enlarged and often transferred before I begin to write *in extenso*. As in several of my books facts observed by others have been very extensively used, and as I have always had several quite distinct subjects in hand at the same time, I may mention that I keep from thirty to forty large portfolios, in cabinets with labelled shelves, into which I can at once put a detached reference or memorandum. I have bought many books, and at their ends I make an index of all the facts that concern my work; or, if the book is not my own, write out a separate abstract, and of such abstracts I have a large drawer full. Before beginning on any subject I look to all the short indexes and make a general and classified index, and by taking the one or more proper portfolios I have all the information collected during my life ready for use.

I have said that in one respect my mind has changed during the last twenty or thirty years. Up to the age of thirty, or beyond it, poetry of many kinds, such as the works of Milton, Gray, Byron, Wordsworth, Coleridge, and Shelley, gave me great pleasure, and even as a schoolboy I took intense delight in Shakespeare, especially in the historical plays. I have also said that formerly pictures gave me considerable, and music very great delight. But now for many years I cannot endure to read a line of poetry; I have tried lately to read Shakespeare, and found it so intolerably dull that it nauseated me. I have also almost lost my taste for pictures or music. Music generally sets me thinking too energetically on what I have been at work on, instead of giving me pleasure. I retain some taste for fine scenery, but it does not cause me the exquisite delight which it formerly did. On the other hand, novels, which are works of the imagination, though not of a very high order, have been for years a wonderful relief and pleasure to me, and I often bless all novelists. A surprising number have been read aloud to me, and I like all if moderately good, and if they do not end unhappily—against which a law ought to be passed. A novel, according to my taste, does not come into the first class unless it contains some person whom one can thoroughly love, and if a pretty woman all the better.

This curious and lamentable loss of the higher æsthetic tastes is all the odder, as books on history, biographies, and travels (independently of any scientific facts which they may contain), and essays on all sorts of subjects interest me as much as ever they did. My mind seems to have become a kind of machine for grinding general laws out of large collections of facts, but why this should have caused the atrophy of that part of the brain alone, on which the higher tastes depend, I cannot conceive. A man with a mind more highly organised or better constituted than mine, would not, I suppose, have thus suffered ; and if I had to live my life again, I would have made a rule to read some poetry and listen to some music at least once every week ; for perhaps the parts of my brain now atrophied would thus have been kept active through use. The loss of these tastes is a loss of happiness, and may possibly be injurious to the intellect, and more probably to the moral character, by enfeebling the emotional part of our nature.

My books have sold largely in England, have been translated into many languages, and passed through several editions in foreign countries. I have heard it said that the success of a work abroad is the best test of its enduring value. I doubt whether this is at all trustworthy ; but judged by this standard my name ought to last for a few years. Therefore it may be worth while to try to analyse the mental qualities and the conditions on which my success has depended ; though I am aware that no man can do this correctly.

I have no great quickness of apprehension or wit which

is so remarkable in some clever men, for instance, Huxley. I am therefore a poor critic: a paper or book, when first read, generally excites my admiration, and it is only after considerable reflection that I perceive the weak points. My power to follow a long and purely abstract train of thought is very limited; and therefore I could never have succeeded with metaphysics or mathematics. My memory is extensive, yet hazy: it suffices to make me cautious by vaguely telling me that I have observed or read something opposed to the conclusion which I am drawing, or on the other hand in favour of it; and after a time I can generally recollect where to search for my authority. So poor in one sense is my memory, that I have never been able to remember for more than a few days a single date or a line of poetry.

Some of my critics have said, " Oh, he is a good observer, but he has no power of reasoning!" I do not think that this can be true, for the *Origin of Species* is one long argument from the beginning to the end, and it has convinced not a few able men. No one could have written it without having some power of reasoning. I have a fair share of invention, and of common sense or judgment, such as every fairly successful lawyer or doctor must have, but not, I believe, in any higher degree.

On the favourable side of the balance, I think that I am superior to the common run of men in noticing things which easily escape attention, and in observing them carefully. My industry has been nearly as great as it could have been in the observation and collection of facts. What is far more important, my love of natural science has been steady and ardent.

This pure love has, however, been much aided by the ambition to be esteemed by my fellow naturalists. From my early youth I have had the strongest desire to understand or explain whatever I observed,—that is, to group all facts under some general laws. These causes combined have given me the patience to reflect or ponder for any number of years over any unexplained problem. As far as I can judge, I am not apt to follow blindly the lead of other men. I have steadily endeavoured to keep my mind free so as to give up any hypothesis, however much beloved (and I cannot resist forming one on every subject), as soon as facts are shown to be opposed to it. Indeed, I have had no choice but to act in this manner, for with the exception of the Coral Reefs, I cannot remember a single first-formed hypothesis which had not after a time to be given up or greatly modified. This has naturally led me to distrust greatly, deductive reasoning in the mixed sciences. On the other hand, I am not very sceptical,—a frame of mind which I believe to be injurious to the progress of science. A good deal of scepticism in a scientific man is advisable to avoid much loss of time, [but] I have met with not a few men, who, I feel sure, have often thus been deterred from experiment or observations, which would have proved directly or indirectly serviceable.

In illustration, I will give the oddest case which I have known. A gentleman (who, as I afterwards heard, is a good local botanist) wrote to me from the Eastern counties that the seeds or beans of the common field-bean had this year everywhere grown on the wrong side of the pod. I wrote back, asking for further information, as I did not understand what was meant; but I did not receive any answer for a very long time. I then saw in two newspapers, one published in Kent and the other in Yorkshire, paragraphs stating that it was a most remarkable fact that " the beans this year had all grown on the wrong side." So I thought there must be some foundation for so general a statement. Accordingly, I went to my gardener, an old Kentish man, and asked him whether he had heard anything about it, and he answered, " Oh, no, sir, it must be a mistake, for the beans grow on the wrong side only on leap-year." I then asked him how they grew in common years and how on leap-years, but soon found that he knew absolutely nothing of how they grew at any time, but he stuck to his belief.

After a time I heard from my first informant, who, with many apologies, said that he should not have written to me had he not heard the statement from several intelligent farmers; but that he had since spoken again to every one of them, and not one knew in the least what he had himself meant. So that here a belief—if indeed a statement with no definite idea attached to it can be called a belief—had spread over almost the whole of England without any vestige of evidence.

I have known in the course of my life only three intentionally falsified statements, and one of these may have been a hoax (and there have been several scientific hoaxes) which, however, took in an American Agricultural Journal. It related to the formation in Holland of a new breed of oxen by the crossing of distinct species of Bos (some of which I happen to know are sterile together), and the author had the impudence to state that he had corresponded with me, and that I had been deeply impressed with the importance of his result. The article was sent to me by the editor of an English Agricultural Journal, asking for my opinion before republishing it.

A second case was an account of several varieties raised by the author from several species of Primula, which had spontaneously yielded a full complement of seed, although the parent plants had been carefully protected from the access of insects. This account was published before I had discovered the meaning of heterostylism, and the whole statement must have been fraudulent, or there was neglect in excluding insects so gross as to be scarcely credible.

The third case was more curious: Mr. Huth published in his book on "Consanguineous Marriage" some long extracts from a Belgian author, who stated that he had interbred rabbits in the closest manner for very many generations, without the least injurious effects. The account was published in a most respectable Journal, that of the Royal Society of Belgium; but I could not avoid feeling doubts— I hardly know why, except that there were no accidents of any kind, and my experience in breeding animals made me think this improbable.

So with much hesitation I wrote to Professor Van Beneden, asking him whether the author was a trustworthy man. I soon heard in answer that the Society had been greatly shocked by discovering that the whole account was a fraud.* The writer had been publicly challenged in the journal to say where he had resided and kept his large stock of rabbits while carrying on his experiments, which must have consumed several years, and no answer could be extracted from him.

My habits are methodical, and this has been of not a little use for my particular line of work. Lastly, I have had ample leisure from not having to earn my own bread. Even ill-health, though it has annihilated several years of my life, has saved me from the distractions of society and amusement.

---

* The falseness of the published statements on which Mr. Huth relied were pointed out in a slip inserted in all the unsold copies of his book, *The Marriage of near Kin.*—F. D.

Therefore, my success as a man of science, whatever this may have amounted to, has been determined, as far as I can judge, by complex and diversified mental qualities and conditions. Of these, the most important have been—the love of science—unbounded patience in long reflecting over any subject—industry in observing and collecting facts—and a fair share of invention as well as of common-sense. With such moderate abilities as I possess, it is truly surprising that I should have influenced to a considerable extent the belief of scientific men on some important points.

CHAPTER III.

RELIGION.

MY father in his published works was reticent on the matter of religion, and what he has left on the subject was not written with a view to publication.*

I believe that his reticence arose from several causes. He felt strongly that a man's religion is an essentially private matter, and one concerning himself alone. This is indicated by the following extract from a letter of 1879 :—†

" What my own views may be is a question of no consequence to any one but myself. But, as you ask, I may state that my judgment often fluctuates . . . In my most extreme fluctuations I have never been an Atheist in the sense of denying the existence of a God. I think that generally (and more and more as I grow older), but not always, that an Agnostic would be the more correct description of my state of mind."

He naturally shrank from wounding the sensibilities of others in religious matters, and he was also influenced by the consciousness that a man ought not to publish on a subject to which he has not given special and continuous thought. That he felt this caution to apply to himself in the matter of religion is shown in a letter to Dr. F. E. Abbott, of Cambridge, U. S. (September 6, 1871). After explaining that the weakness arising from bad health prevented him from feeling " equal to deep reflection, on the deepest subject which can fill a man's mind," he goes on to say : " With respect to my former notes to you, I quite forget their contents. I have to write many letters, and

---

* As an exception, may be mentioned, a few words of concurrence with Dr. Abbott's *Truths for the Times*, which my father allowed to be published in the *Index*.

† Addressed to Mr. J. Fordyce, and published by him in his *Aspects of Scepticism*, 1883.

can reflect but little on what I write; but I fully believe
and hope that I have never written a word, which at the
time I did not think; but I think you will agree with me,
that anything which is to be given to the public ought to
be maturely weighed and cautiously put. It never occurred
to me that you would wish to print any extract from my
notes: if it had, I would have kept a copy. I put 'private'
from habit, only as yet partially acquired, from some hasty
notes of mine having been printed, which were not in the
least degree worth printing, though otherwise unobjectionable.
It is simply ridiculous to suppose that my former note to you
would be worth sending to me, with any part marked which
you desire to print; but if you like to do so, I will at once
say whether I should have any objection. I feel in some
degree unwilling to express myself publicly on religious
subjects, as I do not feel that I have thought deeply enough
to justify any publicity."

What follows is from another letter to Dr. Abbott (November 16, 1871), in which my father gives more fully his
reasons for not feeling competent to write on religious and
moral subjects :—

" I can say with entire truth that I feel honoured by
your request that I should become a contributor to the
Index, and am much obliged for the draft. I fully, also,
subscribe to the proposition that it is the duty of every one
to spread what he believes to be the truth; and I honour
you for doing so, with so much devotion and zeal. But I
cannot comply with your request for the following reasons;
and excuse me for giving them in some detail, as I should
be very sorry to appear in your eyes ungracious. My health
is very weak : I *never* pass 24 hours without many hours of
discomfort, when I can do nothing whatever. I have thus,
also, lost two whole consecutive months this season. Owing
to this weakness, and my head being often giddy, I am
unable to master new subjects requiring much thought, and
can deal only with old materials. At no time am I a quick
thinker or writer : whatever I have done in science has
solely been by long pondering, patience and industry.

" Now I have never systematically thought much on
religion in relation to science, or on morals in relation to
society; and without steadily keeping my mind on such
subjects for a long period, I am really incapable of writing
anything worth sending to the *Index*."

He was more than once asked to give his views on religion, and he had, as a rule, no objection to doing so in a private letter. Thus, in answer to a Dutch student, he wrote (April 2, 1873) :—

"I am sure you will excuse my writing at length, when I tell you that I have long been much out of health, and am now staying away from my home for rest.

"It is impossible to answer your question briefly; and I am not sure that I could do so, even if I wrote at some length. But I may say that the impossibility of conceiving that this grand and wondrous universe, with our conscious selves, arose through chance, seems to me the chief argument for the existence of God; but whether this is an argument of real value, I have never been able to decide. I am aware that if we admit a First Cause, the mind still craves to know whence it came and how it arose. Nor can I overlook the difficulty from the immense amount of suffering through the world. I am, also, induced to defer to a certain extent to the judgment of the many able men who have fully believed in God; but here again I see how poor an argument this is. The safest conclusion seems to me that the whole subject is beyond the scope of man's intellect; but man can do his duty."

Again in 1879 he was applied to by a German student, in a similar manner. The letter was answered by a member of my father's family, who wrote :—

"Mr. Darwin begs me to say that he receives so many letters, that he cannot answer them all.

"He considers that the theory of Evolution is quite compatible with the belief in a God; but that you must remember that different persons have different definitions of what they mean by God."

This, however, did not satisfy the German youth, who again wrote to my father, and received from him the following reply :—

"I am much engaged, an old man, and out of health, and I cannot spare time to answer your questions fully,— nor indeed can they be answered. Science has nothing to do with Christ, except in so far as the habit of scientific research makes a man cautious in admitting evidence. For myself, I do not believe that there ever has been any revelation. As for a future life, every man must judge for himself between conflicting vague probabilities."

The passages which here follow are extracts, somewhat abbreviated, from a part of the Autobiography, written in

1876, in which my father gives the history of his religious
views :—

"During these two years * I was led to think much
about religion. Whilst on board the *Beagle* I was quite
orthodox, and I remember being heartily laughed at by
several of the officers (though themselves orthodox) for
quoting the Bible as an unanswerable authority on some
point of morality. I supposed it was the novelty of the
argument that amused them. But I had gradually come
by this time, *i.e.* 1836 to 1839, to see that the Old Testament
was no more to be trusted than the sacred books of the
Hindoos. The question then continually rose before my
mind and would not be banished, is it credible that if God
were now to make a revelation to the Hindoos, he would
permit it to be connected with the belief in Vishnu, Siva,
&c., as Christianity is connected with the Old Testament?
This appeared to me utterly incredible.

" By further reflecting that the clearest evidence would
be requisite to make any sane man believe in the miracles
by which Christianity is supported,—and that the more we
know of the fixed laws of nature the more incredible do
miracles become,—that the men at that time were ignorant
and credulous to a degree almost incomprehensible by us,—
that the Gospels cannot be proved to have been written
simultaneously with the events,—that they differ in many
important details, far too important, as it seemed to me, to
be admitted as the usual inaccuracies of eye-witnesses;—by
such reflections as these, which I give not as having the
least novelty or value, but as they influenced me, I gradually
came to disbelieve in Christianity as a divine revelation.
The fact that many false religions have spread over large
portions of the earth like wildfire had some weight with me.

" But I was very unwilling to give up my belief; I feel
sure of this, for I can well remember often and often inventing day-dreams of old letters between distinguished
Romans, and manuscripts being discovered at Pompeii or
elsewhere, which confirmed in the most striking manner all
that was written in the Gospels. But I found it more and
more difficult, with free scope given to my imagination, to
invent evidence which would suffice to convince me. Thus
disbelief crept over me at a very slow rate, but was at last
complete. The rate was so slow that I felt no distress.

---

* October 1836 to January 1839.

" Although I did not think much about the existence of a personal God until a considerably later period my life, I will here give the vague conclusions to which I have been driven. The old argument from design in Nature, as given by Paley, which formerly seemed to me so conclusive, fails, now that the law of natural selection has been discovered. We can no longer argue that, for instance, the beautiful hinge of a bivalve shell must have been made by an intelligent being, like the hinge of a door by man. There seems to be no more design in the variability of organic beings, and in the action of natural selection, than in the course which the wind blows. But I have discussed this subject at the end of my book on the *Variation of Domesticated Animals and Plants*,* and the argument there given has never, as far as I can see, been answered.

" But passing over the endless beautiful adaptations which we everywhere meet with, it may be asked how can the generally beneficent arrangement of the world be accounted for? Some writers indeed are so much impressed with the amount of suffering in the world, that they doubt, if we look to all sentient beings, whether there is more of misery or of happiness; whether the world as a whole is a good or a bad one. According to my judgment happiness decidedly prevails, though this would be very difficult to prove. If the truth of this conclusion be granted, it harmonizes well with the effects which we might expect from natural selection. If all the individuals of any species were habitually to suffer to an extreme degree, they would neglect to propagate their kind; but we have no reason to believe that this has ever, or at least often occurred. Some other considerations, moreover, lead to the belief that all sentient beings have been formed so as to enjoy, as a general rule, happiness.

" Every one who believes, as I do, that all the corporeal and mental organs (excepting those which are neither ad-

---

* My father asks whether we are to believe that the forms are preordained of the broken fragments of rock which are fitted together by man to build his houses. If not, why should we believe that the variations of domestic animals or plants are preordained for the sake of the breeder? "But if we give up the principle in one case, . . . no shadow of reason can be assigned for the belief that variations, alike in nature and the result of the same general laws, which have been the groundwork through natural selection of the formation of the most perfectly adapted animals in the world, man included, were intentionally and specially guided."—*Variation of Animals and Plants*, 1st Edit. vol. ii. p. 431.—F. D.

vantageous nor disadvantageous to the possessor) of all beings have been developed through natural selection, or the survival of the fittest, together with use or habit, will admit that these organs have been formed so that their possessors may compete successfully with other beings, and thus increase in number. Now an animal may be led to pursue that course of action which is most beneficial to the species by suffering, such as pain, hunger, thirst, and fear; or by pleasure, as in eating and drinking, and in the propagation of the species, &c.; or by both means combined, as in the search for food. But pain or suffering of any kind, if long continued, causes depression and lessens the power of action, yet is well adapted to make a creature guard itself against any great or sudden evil. Pleasurable sensations, on the other hand, may be long continued without any depressing effect; on the contrary, they stimulate the whole system to increased action. Hence it has come to pass that most or all sentient beings have been developed in such a manner, through natural selection, that pleasurable sensations serve as their habitual guides. We see this in the pleasure from exertion, even occasionally from great exertion of the body or mind,—in the pleasure of our daily meals, and especially in the pleasure derived from sociability, and from loving our families. The sum of such pleasures as these, which are habitual or frequently recurrent, give, as I can hardly doubt, to most sentient beings an excess of happiness over misery, although many occasionally suffer much. Such suffering is quite compatible with the belief in Natural Selection, which is not perfect in its action, but tends only to render each species as successful as possible in the battle for life with other species, in wonderfully complex and changing circumstances.

" That there is much suffering in the world no one disputes. Some have attempted to explain this with reference to man by imagining that it serves for his moral improvement. But the number of men in the world is as nothing compared with that of all other sentient beings, and they often suffer greatly without any moral improvement. This very old argument from the existence of suffering against the existence of an intelligent First Cause seems to me a strong one; whereas, as just remarked, the presence of much suffering agrees well with the view that all organic beings have been developed through variation and natural selection.

"At the present day the most usual argument for the existence of an intelligent God is drawn from the deep inward conviction and feelings which are experienced by most persons.

"Formerly I was led by feelings such as those just referred to (although I do not think that the religious sentiment was ever strongly developed in me), to the firm conviction of the existence of God and of the immortality of the soul. In my Journal I wrote that whilst standing in the midst of the grandeur of a Brazilian forest, 'it is not possible to give an adequate idea of the higher feelings of wonder, admiration, and devotion which fill and elevate the mind.' I well remember my conviction that there is more in man than the mere breath of his body; but now the grandest scenes would not cause any such convictions and feelings to rise in my mind. It may be truly said that I am like a man who has become colour-blind, and the universal belief by men of the existence of redness makes my present loss of perception of not the least value as evidence. This argument would be a valid one if all men of all races had the same inward conviction of the existence of one God; but we know that this is very far from being the case. Therefore I cannot see that such inward convictions and feelings are of any weight as evidence of what really exists. The state of mind which grand scenes formerly excited in me, and which was intimately connected with a belief in God, did not essentially differ from that which is often called the sense of sublimity; and however difficult it may be to explain the genesis of this sense, it can hardly be advanced as an argument for the existence of God, any more than the powerful though vague and similar feelings excited by music.

"With respect to immortality, nothing shows me [so clearly] how strong and almost instinctive a belief it is as the consideration of the view now held by most physicists, namely, that the sun with all the planets will in time grow too cold for life, unless indeed some great body dashes into the sun and thus gives it fresh life. Believing as I do that man in the distant future will be a far more perfect creature than he now is, it is an intolerable thought that he and all other sentient beings are doomed to complete annihilation after such long-continued slow progress. To those who fully admit the immortality of the human soul, the destruction of our world will not appear so dreadful.

" Another source of conviction in the existence of God, connected with the reason and not with the feelings, impresses me as having much more weight. This follows from the extreme difficulty or rather impossibility of conceiving this immense and wonderful universe, including man with his capacity of looking far backwards and far into futurity, as the result of blind chance or necessity. When thus reflecting, I feel compelled to look to a First Cause having an intelligent mind in some degree analogous to that of man ; and I deserve to be called a Theist. This conclusion was strong in my mind about the time, as far as I can remember, when I wrote the *Origin of Species*, and it is since that time that it has very gradually, with many fluctuations, become weaker. But then arises the doubt— can the mind of man, which has, as I fully believe, been developed from a mind as low as that possessed by the lowest animals, be trusted when it draws such grand conclusions?

" I cannot pretend to throw the least light on such abstruse problems. The mystery of the beginning of all things is insoluble by us, and I for one must be content to remain an Agnostic."

The following letters repeat to some extent what is given above from the *Autobiography*. The first one refers to *The Boundaries of Science: a Dialogue*, published in *Macmillan's Magazine*, for July 1861.

C. D. to Miss Julia Wedgwood, July 11 [*1681*].

Some one has sent us *Macmillan*, and I must tell you how much I admire your Article, though at the same time I must confess that I could not clearly follow you in some parts, which probably is in main part due to my not being at all accustomed to metaphysical trains of thought. I think that you understand my book * perfectly, and that I find a very rare event with my critics. The ideas in the last page have several times vaguely crossed my mind. Owing to several correspondents, I have been led lately to think, or rather to try to think, over some of the chief points discussed by you. But the result has been with me a maze—something like thinking on the origin of evil, to which you allude. The mind refuses to look at this uni-

---

* The *Origin of Species*.

verse, being what it is, without having been designed; yet, where one would most expect design, viz. in the structure of a sentient being, the more I think on the subject, the less I can see proof of design. Asa Gray and some others look at each variation, or at least at each beneficial variation (which A. Gray would compare with the raindrops * which do not fall on the sea, but on to the land to fertilise it) as having been providentially designed. Yet when I ask him whether he looks at each variation in the rock-pigeon, by which man has made by accumulation a pouter or fantail pigeon, as providentially designed for man's amusement, he does not know what to answer; and if he, or any one, admits [that] these variations are accidental, as far as purpose is concerned (of course not accidental as to their cause or origin), then I can see no reason why he should rank the accumulated variations by which the beautifully adapted woodpecker has been formed as providentially designed. For it would be easy to imagine the enlarged crop of the pouter, or tail of the fantail, as of some use to birds, in a state of nature, having peculiar habits of life. These are the considerations which perplex me about design; but whether you will care to hear them, I know not.

On the subject of design, he wrote (July 1860) to Dr. Gray:

" One word more on 'designed laws' and 'undesigned results.' I see a bird which I want for food, take my gun, and kill it, I do this *designedly*. An innocent and good man stands under a tree and is killed by a flash of lightning Do you believe (and I really should like to hear) that God *designedly* killed this man? Many or most persons do believe this; I can't and don't. If you believe so, do you believe that when a swallow snaps up a gnat that God designed that that particular swallow should snap up that particular gnat at that particular instant? I believe that the man and the gnat are in the same predicament. If the death of neither man nor gnat are designed, I see no good reason to be-

---

* Dr. Gray's rain-drop metaphor occurs in the Essay, *Darwin and his Reviewers* (*Darwiniana*, p. 157): " The whole animate life of a country depends absolutely upon the vegetation, the vegetation upon the rain. The moisture is furnished by the ocean, is raised by the sun's heat from the ocean's surface, and is wafted inland by the winds. But what multitudes of rain-drops fall back into the ocean—are as much without a final cause as the incipient varieties which come to nothing! Does it therefore follow that the rains which are bestowed upon the soil with such rule and average regularity were not designed to support vegetable and animal life ? "

lieve that their *first* birth or production should be necessarily designed."

C. D. to W. Graham.   Down, July 3d, 1881.

DEAR SIR,—I hope you will not think it intrusive on my part to thank you heartily for the pleasure which I have derived from reading your admirably - written *Creed of Science*, though I have not yet quite finished it, as now that I am old I read very slowly. It is a very long time since any other book has interested me so much. The work must have cost you several years and much hard labour with full leisure for work. You would not probably expect any one fully to agree with you on so many abstruse subjects; and there are some points in your book which I cannot digest. The chief one is that the existence of so-called natural laws implies purpose. I cannot see this. Not to mention that many expect that the several great laws will some day be found to follow inevitably from some one single law, yet taking the laws as we now know them, and look at the moon, where the law of gravitation—and no doubt of the conservation of energy—of the atomic theory, &c., &c., hold good, and I cannot see that there is then necessarily any purpose. Would there be purpose if the lowest organisms alone, destitute of consciousness, existed in the moon? But I have had no practice in abstract reasoning, and I may be all astray. Nevertheless you have expressed my inward conviction, though far more vividly and clearly than I could have done, that the Universe is not the result of chance.* But then with me the horrid doubt always arises whether the convictions of man's mind which has been developed from the mind of the lower animals, are of any value or at all trustworthy. Would any one trust in the convictions of a monkey's mind, if there are any convictions in such a mind? Secondly, I think that I could make somewhat of a case against the enormous im-

---

* The Duke of Argyll (*Good Words*, April 1885, p. 244) has recorded a few words on this subject, spoken by my father in the last year of his life. ". . . in the course of that conversation I said to Mr. Darwin, with reference to some of his own remarkable works on the *Fertilisation of Orchids*, and upon *The Earthworms*, and various other observations he made of the wonderful contrivances for certain purposes in nature—I said it was impossible to look at these without seeing that they were the effect and the expression of mind. I shall never forget Mr. Darwin's answer. He looked at me very hard and said, ' Well, that often comes over me with overwhelming force; but at other times,' and he shook his head vaguely, adding, ' it seems to go away.'"

portance which you attribute to our greatest men; I have been accustomed to think second, third, and fourth-rate men of very high importance, at least in the case of Science. Lastly, I could show fight on natural selection having done and doing more for the progress of civilisation than you seem inclined to admit. Remember what risk the nations of Europe ran, not so many centuries ago, of being overwhelmed by the Turks, and how ridiculous such an idea now is! The more civilised so-called Caucasian races have beaten the Turkish hollow in the struggle for existence. Looking to the world at no very distant date, what an endless number of the lower races will have been eliminated by the higher civilised races throughout the world. But I will write no more, and not even mention the many points in your work which have much interested me. I have indeed cause to apologise for troubling you with my impressions, and my sole excuse is the excitement in my mind which your book has aroused.

I beg leave to remain, dear sir,

Yours faithfully and obliged.

Darwin spoke little on these subjects, and I can contribute nothing from my own recollection of his conversation which can add to the impression here given of his attitude towards Religion.* Some further idea of his views may, however, be gathered from occasional remarks in his letters.

---

* Dr. Aveling has published an account of a conversation with my father. I think that the readers of this pamphlet (*The Religious Views of Charles Darwin*, Free Thought Publishing Company, 1883) may be misled into seeing more resemblance than really existed between the positions of my father and Dr. Aveling: and I say this in spite of my conviction that Dr. Aveling gives quite fairly his impressions of my father's views. Dr. Aveling tried to show that the terms "Agnostic" and "Atheist" were practically equivalent—that an atheist is one who, without denying the existence of God, is without God, inasmuch as he is unconvinced of the existence of a Deity. My father's replies implied his preference for the unaggressive attitude of an Agnostic. Dr. Aveling seems (p. 5) to regard the absence of aggressiveness in my father's views as distinguishing them in an unessential manner from his own. But, in my judgment, it is precisely differences of this kind which distinguish him so completely from the class of thinkers to which Dr. Aveling belongs.

THE STUDY AT DOWN.*

CHAPTER IV.

REMINISCENCES OF MY FATHER'S EVERYDAY LIFE.

It is my wish in the present chapter to give some idea of
my father's everyday life. It has seemed to me that I might
carry out this object in the form of a rough sketch of a day's
life at Down, interspersed with such recollections as are
called up by the record. Many of these recollections, which
have a meaning for those who knew my father, will seem
colourless or trifling to strangers. Nevertheless, I give them
in the hope that they may help to preserve that impression
of his personality which remains on the minds of those who
knew and loved him—an impression at once so vivid and so
untranslatable into words.

---

* From the *Century Magazine*, January 1883.

Of his personal appearance (in these days of multiplied photographs) it is hardly necessary to say much. He was about six feet in height, but scarcely looked so tall, as he stooped a good deal; in later days he yielded to the stoop; but I can remember seeing him long ago swinging back his arms to open out his chest, and holding himself upright with a jerk. He gave one the idea that he had been active rather than strong; his shoulders were not broad for his height, though certainly not narrow. As a young man he must have had much endurance, for on one of the shore excursions from the *Beagle*, when all were suffering from want of water, he was one of the two who were better able than the rest to struggle on in search of it. As a boy he was active, and could jump a bar placed at the height of the " Adam's apple " in his neck.

He walked with a swinging action, using a stick heavily shod with iron, which he struck loudly against the ground, producing as he went round the " Sand-walk " at Down, a rhythmical click which is with all of us a very distinct remembrance. As he returned from the midday walk, often carrying the waterproof or cloak which had proved too hot, one could see that the swinging step was kept up by something of an effort. Indoors his step was often slow and laboured, and as he went upstairs in the afternoon he might be heard mounting the stairs with a heavy footfall, as if each step were an effort. When interested in his work he moved about quickly and easily enough, and often in the midst of dictating he went eagerly into the hall to get a pinch of snuff, leaving the study door open, and calling out the last words of his sentence as he left the room.

In spite of his activity, he had, I think, no natural grace or neatness of movement. He was awkward with his hands, and was unable to draw at all well.* This he always regretted, and he frequently urged the paramount necessity to a young naturalist of making himself a good draughtsman.

He could dissect well under the simple microscope, but I think it was by dint of his great patience and carefulness. It was characteristic of him that he thought any little bit of skilful dissection something almost superhuman. He used to speak with admiration of the skill with which he

---

* The figure in *Insectivorous Plants* representing the aggregated cell-contents was drawn by him.

saw Newport dissect a humble bee, getting out the nervous system with a few cuts of a pair of fine scissors. He used to consider cutting microscopic sections a great feat, and in the last year of his life, with wonderful energy, took the pains to learn to cut sections of roots and leaves. His hand was not steady enough to hold the object to be cut, and he employed a common microtome, in which the pith for holding the object was clamped, and the razor slid on a glass surface. He used to laugh at himself, and at his own skill in section-cutting, at which he would say he was " speechless with admiration." On the other hand, he must have had accuracy of eye and power of co-ordinating his movements, since he was a good shot with a gun as a young man, and as a boy was skilful in throwing. He once killed a hare sitting in the flower-garden at Shrewsbury by throwing a marble at it, and, as a man, he killed a cross-beak with a stone. He was so unhappy at having uselessly killed the cross-beak that he did not mention it for years, and then explained that he should never have thrown at it if he had not felt sure that his old skill had gone from him.

His beard was full and almost untrimmed, the hair being grey and white, fine rather than coarse, and wavy or frizzled. His moustache was somewhat disfigured by being cut short and square across. He became very bald, having only a fringe of dark hair behind.

His face was ruddy in colour, and this perhaps made people think him less of an invalid than he was. He wrote to Sir Joseph Hooker (June 13, 1849), " Every one tells me that I look quite blooming and beautiful; and most think I am shamming, but you have never been one of those." And it must be remembered that at this time he was miserably ill, far worse than in later years. His eyes were bluish grey under deep overhanging brows, with thick, bushy projecting eyebrows. His high forehead was deeply wrinkled, but otherwise his face was not much marked or lined. His expression showed no signs of the continual discomfort he suffered.

When he was excited with pleasant talk his whole manner was wonderfully bright and animated, and his face shared to the full in the general animation. His laugh was a free and sounding peal, like that of a man who gives himself sympathetically and with enjoyment to the person and the thing which have amused him. He often used some sort of gesture with his laugh, lifting up his hands or bring-

ing one down with a slap. I think, generally speaking, he
was given to gesture, and often used his hands in explaining
anything (e. g. the fertilisation of a flower) in a way that
seemed rather an aid to himself than to the listener. He
did this on occasions when most people would illustrate
their explanations by means of a rough pencil sketch.

He wore dark clothes, of a loose and easy fit. Of late
years he gave up the tall hat even in London, and wore a
soft black one in winter, and a big straw hat in summer.
His usual out-of-doors dress was the short cloak in which
Elliot and Fry's photograph* represents him, leaning
against the pillar of the verandah. Two peculiarities of
his indoor dress were that he almost always wore a shawl
over his shoulders, and that he had great loose cloth boots
lined with fur which he could slip on over his indoor
shoes.

He rose early, and took a short turn before breakfast, a
habit which began when he went for the first time to a
water-cure establishment, and was preserved till almost the
end of his life. I used, as a little boy, to like going out
with him, and I have a vague sense of the red of the winter
sunrise, and a recollection of the pleasant companionship,
and a certain honour and glory in it. He used to delight
me as a boy by telling me how, in still earlier walks, on
dark winter mornings, he had once or twice met foxes trot-
ting home at the dawning.

After breakfasting alone about 7.45, he went to work at
once, considering the 1½ hour between 8 and 9.30 one of
his best working times. At 9.30 he came in to the drawing-room for his letters—rejoicing if the post was a light
one and being sometimes much worried if it was not. He
would then hear any family letters read aloud as he lay on
the sofa.

The reading aloud, which also included part of a novel,
lasted till about half-past ten, when he went back to work
till twelve or a quarter past. By this time he considered
his day's work over, and would often say, in a satisfied voice,
"*I've* done a good day's work." He then went out of doors
whether it was wet or fine; Polly, his white terrier, went
with him in fair weather, but in rain she refused or might
be seen hesitating in the verandah, with a mixed expression
of disgust and shame at her own want of courage; gener-

---

* *Life and Letters*, vol. iii. frontispiece.

ally, however, her conscience carried the day, and as soon as
he was evidently gone she could not bear to stay behind.

My father was always fond of dogs, and as a young man
had the power of stealing away the affections of his sister's
pets; at Cambridge, he won the love of his cousin W. D.
Fox's dog, and this may perhaps have been the little beast
which used to creep down inside his bed and sleep at the
foot every night. My father had a surly dog, who was devoted to him, but unfriendly to every one else, and when he
came back from the *Beagle* voyage, the dog remembered
him, but in a curious way, which my father was fond of
telling. He went into the yard and shouted in his old manner; the dog rushed out and set off with him on his walk,
showing no more emotion or excitement than if the same
thing had happened the day before, instead of five years
ago. This story is made use of in the *Descent of Man*, 2nd
Edit. p. 74.

In my memory there were only two dogs which had
much connection with my father. One was a large black
and white half-bred retriever, called Bob, to which we, as
children, were much devoted. He was the dog of whom
the story of the "hot-house face" is told in the *Expression
of the Emotions.*

But the dog most closely associated with my father was
the above-mentioned Polly, a rough, white fox-terrier. She
was a sharp-witted, affectionate dog; when her master was
going away on a journey, she always discovered the fact by
the signs of packing going on in the study, and became low-
spirited accordingly. She began, too, to be excited by seeing the study prepared for his return home. She was a
cunning little creature, and used to tremble or put on an
air of misery when my father passed, while she was waiting
for dinner, just as if she knew that he would say (as he did
often say) that "she was famishing." My father used to
make her catch biscuits off her nose, and had an affectionate and mock-solemn way of explaining to her before-hand
that she must "be a very good girl." She had a mark on
her back where she had been burnt, and where the hair had
re-grown red instead of white, and my father used to com-
mend her for this tuft of hair as being in accordance with
his theory of pangenesis; her father had been a red bull-
terrier, thus the red hair appearing after the burn showed
the presence of latent red gemmules. He was delightfully
tender to Polly, and never showed any impatience at the

attentions she required, such as to be let in at the door, or
out at the verandah window, to bark at "naughty people,"
a self-imposed duty she much enjoyed. She died, or rather
had to be killed, a few days after his death.*

My father's mid-day walk generally began by a call at
the greenhouse, where he looked at any germinating seeds
or experimental plants which required a casual examination,
but he hardly ever did any serious observing at this time.
Then he went on for his constitutional—either round the
"Sand-walk," or outside his own grounds in the immediate
neighbourhood of the house. The "Sand-walk" was a narrow strip of land 1½ acre in extent, with a gravel-walk round
it. On one side of it was a broad old shaw with fair-sized
oaks in it, which made a sheltered shady walk; the other
side was separated from a neighbouring grass field by a low
quickset hedge, over which you could look at what view
there was, a quiet little valley losing itself in the upland
country towards the edge of the Westerham hill, with hazel
coppice and larch plantation, the remnants of what was
once a large wood, stretching away to the Westerham high
road. I have heard my father say that the charm of this
simple little valley was a decided factor in his choice of a
home.

The Sand-walk was planted by my father with a variety
of trees, such as hazel, alder, lime, hornbeam, birch, privet,
and dogwood, and with a long line of hollies all down the
exposed side. In earlier times he took a certain number of
turns every day, and used to count them by means of a heap
of flints, one of which he kicked out on the path each time
he passed. Of late years I think he did not keep to any
fixed number of turns, but took as many as he felt strength
for. The Sand-walk was our play-ground as children, and
here we continually saw my father as he walked round. He
liked to see what we were doing, and was ever ready to
sympathise in any fun that was going on. It is curious to
think how, with regard to the Sand-walk in connection with
my father, my earliest recollections coincide with my latest;
it shows the unvarying character of his habits.

Sometimes when alone he stood still or walked stealthily
to observe birds or beasts. It was on one of these occasions

---

* The basket in which she usually lay curled up near the fire in his study
is faithfully represented in Mr. Parson's drawing given at the head of the
chapter.

that some young squirrels ran up his back and legs, while their mother barked at them in an agony from the tree. He always found birds' nests even up to the last years of his life, and we, as children, considered that he had a special genius in this direction. In his quiet prowls he came across the less common birds, but I fancy he used to conceal it from me as a little boy, because he observed the agony of mind which I endured at not having seen the siskin or goldfinch, or some other of the less common birds. He used to tell us how, when he was creeping noiselessly along in the " Big-Woods," he came upon a fox asleep in the day-time, which was so much astonished that it took a good stare at him before it ran off. A Spitz dog which accompanied him showed no sign of excitement at the fox, and he used to end the story by wondering how the dog could have been so faint-hearted.

Another favourite place was " Orchis Bank," above the quiet Cudham valley, where fly- and musk-orchis grew among the junipers, and Cephalanthera and Neottia under the beech boughs; the little wood " Hangrove," just above this, he was also fond of, and here I remember his collecting grasses, when he took a fancy to make out the names of all the common kinds. He was fond of quoting the saying of one of his little boys, who, having found a grass that his father had not seen before, had it laid by his own plate during dinner, remarking, " I are an extraordinary grass-finder!"

My father much enjoyed wandering idly in the garden with my mother or some of his children, or making one of a party, sitting on a bench on the lawn; he generally sat, however, on the grass, and I remember him often lying under one of the big lime-trees, with his head on the green mound at its foot. In dry summer weather, when we often sat out, the fly-wheel of the well was commonly heard spinning round, and so the sound became associated with those pleasant days. He used to like to watch us playing at lawn-tennis, and often knocked up a stray ball for us with the curved handle of his stick.

Though he took no personal share in the management of the garden, he had great delight in the beauty of flowers —for instance, in the mass of Azaleas which generally stood in the drawing-room. I think he sometimes fused together his admiration of the structure of a flower and of its intrinsic beauty; for instance, in the case of the big pendulous pink and white flowers of Diclytra. In the same way he had an affection, half-artistic, half-botanical, for the little blue Lobelia. In admiring flowers, he would often laugh at the dingy high-art colours, and contrast them with the bright tints of nature. I used to like to hear him admire the beauty of a flower; it was a kind of gratitude to the flower itself, and a personal love for its delicate form and colour. I seem to remember him gently touching a flower he delighted in; it was the same simple admiration that a child might have.

He could not help personifying natural things. This feeling came out in abuse as well as in praise—*e. g.* of some seedlings—" The little beggars are doing just what I don't want them to." He would speak in a half-provoked, half-admiring way of the ingenuity of the leaf of a Sensitive Plant in screwing itself out of a basin of water in which he had tried to fix it. One might see the same spirit in his way of speaking of Sundew, earthworms, &c.*

Within my memory, his only outdoor recreation, besides walking, was riding; this was taken up at the recommendation of Dr. Bence Jones, and we had the luck to find for him the easiest and quietest cob in the world, named " Tommy." He enjoyed these rides extremely, and devised a series of short rounds which brought him home in time for lunch. Our country is good for this purpose, owing to the number of small valleys which give a variety to what in a flat country would be a dull loop of road. I think he felt surprised at himself, when he remembered how bold a rider he had been, and how utterly old age and bad health had taken away his nerve. He would say that riding prevented him thinking much more effectually than walking—that having to attend to the horse gave him occupation sufficient to prevent any really hard thinking. And the change of scene which it gave him was good for spirits and health.

If I go beyond my own experience, and recall what I have heard him say of his love for sport, &c., I can think of a good deal, but much of it would be a repetition of what is contained in his *Recollections*. He was fond of his gun as quite a boy, and became a good shot; he used to tell

* Cf. Leslie Stephen's *Swift*, 1882, p. 200, where Swift's inspection of the manners and customs of servants are compared to my father's observations on worms, " The difference is," says Mr. Stephen, " that Darwin had none but kindly feelings for worms."

how in South America he killed twenty-three snipe in twenty-four shots. In telling the story he was careful to add that he thought they were not quite so wild as English snipe.

Luncheon at Down came after his mid-day walk; and here I may say a word or two about his meals generally. He had a boy-like love of sweets, unluckily for himself, since he was constantly forbidden to take them. He was not particularly successful in keeping the "vows," as he called them, which he made against eating sweets, and never considered them binding unless he made them aloud.

He drank very little wine, but enjoyed and was revived by the little he did drink. He had a horror of drinking, and constantly warned his boys that any one might be led into drinking too much. I remember, in my innocence as a small boy, asking him if he had been ever tipsy; and he answered very gravely that he was ashamed to say he had once drunk too much at Cambridge. I was much impressed, so that I know now the place where the question was asked.

After his lunch he read the newspaper, lying on the sofa in the drawing-room. I think the paper was the only non-scientific matter which he read to himself. Everything else, novels, travels, history, was read aloud to him. He took so wide an interest in life, that there was much to occupy him in newspapers, though he laughed at the wordiness of the debates, reading them, I think, only in abstract. His interest in politics was considerable, but his opinion on these matter was formed rather by the way than with any serious amount of thought.

After he had read his paper, came his time for writing letters. These, as well as the MS. of his books, were written by him as he sat in a huge horse-hair chair by the fire, his paper supported on a board resting on the arms of the chair. When he had many or long letters to write, he would dictate them from a rough copy; these rough copies were written on the backs of manuscript or of proof-sheets, and were almost illegible, sometimes even to himself. He made a rule of keeping all letters that he received; this was a habit which he learnt from his father, and which he said had been of great use to him.

Many letters were addressed to him by foolish, unscrupulous people, and all of these received replies. He used to say that if he did not answer them, he had it on his conscience afterwards, and no doubt it was in great measure the courtesy with which he answered every one which produced the widespread sense of his kindness of nature which was so evident on his death.

He was considerate to his correspondents in other and lesser things—for instance, when dictating a letter to a foreigner, he hardly ever failed to say to me, " You'd better try and write well, as it's to a foreigner." His letters were generally written on the assumption that they would be carelessly read ; thus, when he was dictating, he was careful to tell me to make an important clause begin with an obvious paragraph, " to catch his eye," as he often said. How much he thought of the trouble he gave others by asking questions, will be well enough shown by his letters.

He had a printed form to be used in replying to troublesome correspondents, but he hardly ever used it ; I suppose he never found an occasion that seemed exactly suitable. I remember an occasion on which it might have been used with advantage. He received a letter from a stranger stating that the writer had undertaken to uphold Evolution at a debating society, and that being a busy young man, without time for reading, he wished to have a sketch of my father's views. Even this wonderful young man got a civil answer, though I think he did not get much material for his speech. His rule was to thank the donors of books, but not of pamphlets. He sometimes expressed surprise that so few thanked him for his books which he gave away liberally; the letters that he did receive gave him much pleasure, because he habitually formed so humble an estimate of the value of all his works, that he was genuinely surprised at the interest which they excited.

In money and business matters he was remarkably careful and exact. He kept accounts with great care, classifying them, and balancing at the end of the year like a merchant. I remember the quick way in which he would reach out for his account-book to enter each cheque paid, as though he were in a hurry to get it entered before he had forgotten it. His father must have allowed him to believe that he would be poorer than he really was, for some of the difficulty experienced over finding a house in the country must have arisen from the modest sum he felt prepared to give. Yet he knew, of course, that he would be in easy circumstances, for in his *Recollections* he mentions this as one of the reasons for his not having worked at medicine with so much zeal as he would have done if he had been obliged to gain his living.

He had a pet economy in paper, but it was rather a hobby than a real economy. All the blank sheets of letters received were kept in a portfolio to be used in making notes; it was his respect for paper that made him write so much on the backs of his old MS., and in this way, unfortunately, he destroyed large parts of the original MS. of his books. His feeling about paper extended to waste paper, and he objected, half in fun, to the habit of throwing a spill into the fire after it had been used for lighting a candle.

He had a great respect for pure business capacity, and often spoke with admiration of a relative who had doubled his fortune. And of himself would often say in fun that what he really *was* proud of was the money he had saved. He also felt satisfaction in the money he made by his books His anxiety to save came in great measure from his fears that his children would not have health enough to earn their own livings, a foreboding which fairly haunted him for many years. And I have a dim recollection of his saying, "Thank God, you'll have bread and cheese," when I was so young that I was inclined to take it literally.

When letters were finished, about three in the afternoon, he rested in his bedroom, lying on the sofa, smoking a cigarette, and listening to a novel or other book not scientific. He only smoked when resting, whereas snuff was a stimulant, and was taken during working hours. He took snuff for many years of his life, having learnt the habit at Edinburgh as a student. He had a nice silver snuff-box given him by Mrs. Wedgwood, of Maer, which he valued much—but he rarely carried it, because it tempted him to take too many pinches. In one of his early letters he speaks of having given up snuff for a month, and describes himself as feeling "most lethargic, stupid, and melancholy." Our former neighbour and clergyman, Mr. Brodie Innes, tells me that at one time my father made a resolve not to take snuff, except away from home, "a most satisfactory arrangement for me," he adds, "as I kept a box in my study to which there was access from the garden without summoning servants, and I had more frequently, than might have been otherwise the case, the privilege of a few minutes' conversation with my dear friend." He generally took snuff from a jar on the hall-table, because having to go this distance for a pinch was a slight check; the clink of the lid of the snuff-jar was a very familiar sound.   Sometimes when he was in the drawing-room, it would occur to him that the study fire must be burning low, and when one of us offered to see after it, it would turn out that he also wished to get a pinch of snuff.

Smoking he only took to permanently of late years, though on his Pampas rides he learned to smoke with the Gauchos, and I have heard him speak of the great comfort of a cup of *maté* and a cigarette when he halted after a long ride and was unable to get food for some time.

He came down at four o'clock to dress for his walk, and he was so regular that one might be quite certain it was within a few minutes of four when his descending steps were heard.

From about half-past four to half-past five he worked ; then he came to the drawing-room, and was idle till it was time (about six) to go up for another rest with novel-reading and a cigarette.

Latterly he gave up late dinner, and had a simple tea at half-past seven (while we had dinner), with an egg or a small piece of meat.   After dinner he never stayed in the room, and used to apologise by saying he was an old woman who must be allowed to leave with the ladies.   This was one of the many signs and results of his constant weakness and ill-health.   Half an hour more or less conversation would make to him the difference of a sleepless night and of the loss perhaps of half the next day's work.

After dinner he played backgammon with my mother, two games being played every night.   For many years a score of the games which each won was kept, and in this score he took the greatest interest.   He became extremely animated over these games, bitterly lamenting his bad luck and exploding with exaggerated mock-anger at my mother's good fortune.

After playing backgammon he read some scientific book to himself, either in the drawing-room, or, if much talking was going on, in the study.

In the evening—that is, after he had read as much as his strength would allow, and before the reading aloud began—he would often lie on the sofa and listen to my mother playing the piano.   He had not a good ear, yet in spite of this he had a true love of fine music.   He used to lament that his enjoyment of music had become dulled with age, yet within my recollection his love of a good tune was strong. I never heard him hum more than one tune, the Welsh song " Ar hyd y nos," which he went through correctly; he used also, I believe, to hum a little Otaheitan song. From his want of ear he was unable to recognise a tune when he heard it again, but he remained constant to what he liked, and would often say, when an old favourite was played, " That's a fine thing; what is it ? " He liked especially parts of Beethoven's symphonies and bits of Handel. He was sensitive to differences in style, and enjoyed the late Mrs. Vernon Lushington's playing intensely, and in June 1881, when Hans Richter paid a visit at Down, he was roused to strong enthusiasm by his magnificent performance on the piano. He enjoyed good singing, and was moved almost to tears by grand or pathetic songs. His niece Lady Farrer's singing of Sullivan's " Will he come " was a never-failing enjoyment to him. He was humble in the extreme about his own taste, and correspondingly pleased when he found that others agreed with him.

He became much tired in the evenings, especially of late years, and left the drawing-room about ten, going to bed at half-past ten. His nights were generally bad, and he often lay awake or sat up in bed for hours, suffering much discomfort. He was troubled at night by the activity of his thoughts, and would become exhausted by his mind working at some problem which he would willingly have dismissed. At night, too, anything which had vexed or troubled him in the day would haunt him, and I think it was then that he suffered if he had not answered some troublesome correspondent.

The regular readings, which I have mentioned, continued for so many years, enabled him to get through a great deal of the lighter kinds of literature. He was extremely fond of novels, and I remember well the way in which he would anticipate the pleasure of having a novel read to him as he lay down or lighted his cigarette. He took a vivid interest both in plot and characters, and would on no account know beforehand how a story finished; he considered looking at the end of a novel as a feminine vice. He could not enjoy any story with a tragical end; for this reason he did not keenly appreciate George Eliot, though he often spoke warmly in praise of *Silas Marner*. Walter Scott, Miss Austen, and Mrs. Gaskell were read and re-read till they could be read no more. He had two or three books in hand at the same time—a novel and perhaps a biography and a book of travels. He did not often read out-of-the-way or old standard books, but generally kept to the books of the day obtained from a circulating library.

His literary tastes and opinions were not on a level with the rest of his mind. He himself, though he was clear as to what he thought good, considered that in matters of literary tastes he was quite outside the pale, and often spoke of what those within it liked or disliked, as if they formed a class to which he had no claim to belong.

In all matters of art he was inclined to laugh at professed critics and say that their opinions were formed by fashion. Thus in painting, he would say how in his day every one admired masters who are now neglected. His love of pictures as a young man is almost a proof that he must have had an appreciation of a portrait as a work of art, not as a likeness. Yet he often talked laughingly of the small worth of portraits, and said that a photograph was worth any number of pictures, as if he were blind to the artistic quality in a painted portrait. But this was generally said in his attempts to persuade us to give up the idea of having his portrait painted, an operation very irksome to him.

This way of looking at himself as an ignoramus in all matters of art, was strengthened by the absence of pretence, which was part of his character. With regard to questions of taste, as well as to more serious things he had the courage of his opinions. I remember, however, an instance that sounds like a contradiction to this : when he was looking at the Turners in Mr. Ruskin's bedroom, he did not confess, as he did afterwards, that he could make out absolutely nothing of what Mr. Ruskin saw in them. But this little pretence was not for his own sake, but for the sake of courtesy to his host. He was pleased and amused when subsequently Mr. Ruskin brought him some photographs of pictures (I think Vandyke portraits), and courteously seemed to value my father's opinion about them.

Much of his scientific reading was in German, and this was a serious labour to him ; in reading a book after him, I was often struck at seeing, from the pencil-marks made each day where he left off, how little he could read at a time. He used to call German the " Verdammte," pronounced as if in English. He was especially indignant with Germans, because he was convinced that they could write simply if they chose, and often praised Professor

Hildebrand of Freiburg for writing German which was as clear as French. He sometimes gave a German sentence to a friend, a patriotic German lady, and used to laugh at her if she did not translate it fluently. He himself learnt German simply by hammering away with a dictionary; he would say that his only way was to read a sentence a great many times over, and at last the meaning occurred to him. When he began German long ago, he boasted of the fact (as he used to tell) to Sir J. Hooker, who replied, " Ah, my dear fellow, that's nothing; I've begun it many times."

In spite of his want of grammar, he managed to get on wonderfully with German, and the sentences that he failed to make out were generally difficult ones. He never attempted to speak German correctly, but pronounced the words as though they were English; and this made it not a little difficult to help him when he read out a German sentence and asked for a translation. He certainly had a bad ear for vocal sounds, so that he found it impossible to perceive small differences in pronunciation.

His wide interest in branches of science that were not specially his own was remarkable. In the biological sciences his doctrines make themselves felt so widely that there was something interesting to him in most departments. He read a good deal of many quite special works, and large parts of text books, such as Huxley's *Invertebrate Anatomy*, or such a book as Balfour's *Embryology*, where the detail, at any rate, was not specially in his own line. And in the case of elaborate books of the monograph type, though he did not make a study of them, yet he felt the strongest admiration for them.

In the non-biological sciences he felt keen sympathy with work of which he could not really judge. For instance, he used to read nearly the whole of *Nature*, though so much of it deals with mathematics and physics. I have often heard him say that he got a kind of satisfaction in reading articles which (according to himself) he could not understand. I wish I could reproduce the manner in which he would laugh at himself for it.

It was remarkable, too, how he kept up his interest in subjects at which he had formerly worked. This was strikingly the case with geology. In one of his letters to Mr. Judd he begs him to pay him a visit, saying that since Lyell's death he hardly ever gets a geological talk. His observations, made only a few years before his death, on the

upright pebbles in the drift at Southampton, and discussed
in a letter to Sir A. Geikie, afford another instance. Again,
in his letters to Dr. Dohrn, he shows how his interest in
barnacles remained alive. I think it was all due to the vi-
tality and persistence of his mind—a quality I have heard
him speak of as if he felt that he was strongly gifted in that
respect. Not that he used any such phrases as these about
himself, but he would say that he had the power of keeping
a subject or question more or less before him for a great
many years. The extent to which he possessed this power
appears when we consider the number of different problems
which he solved, and the early period at which some of
them began to occupy him.

It was a sure sign that he was not well when he was
idle at any times other than his regular resting hours ; for,
as long as he remained moderately well, there was no break
in the regularity of his life. Week-days and Sundays passed
by alike, each with their stated intervals of work and rest.
It is almost impossible, except for those who watched his
daily life, to realise how essential to his well-being was the
regular routine that I have sketched : and with what pain
and difficulty anything beyond it was attempted. Any public appearance, even of the most modest kind, was an effort
to him. In 1871 he went to the little village church for the
wedding of his elder daughter, but he could hardly bear the
fatigue of being present through the short service. The
same may be said of the few other occasions on which he
was present at similar ceremonies.

I remember him many years ago at a christening ; a
memory which has remained with me, because to us children his being at church was an extraordinary occurrence.
I remember his look most distinctly at his brother Erasmus's funeral, as he stood in the scattering of snow,
wrapped in a long black funeral cloak, with a grave look of
sad reverie.

When, after an absence of many years, he attended a
meeting of the Linnean Society, it was felt to be, and was
in fact, a serious undertaking ; one not to be determined on
without much sinking of heart, and hardly to be carried
into effect without paying a penalty of subsequent suffering.
In the same way a breakfast-party at Sir James Paget's,
with some of the distinguished visitors to the Medical Congress (1881), was to him a severe exertion.

The early morning was the only time at which he could

make any effort of the kind, with comparative impunity.
Thus it came about that the visits he paid to his scientific
friends in London were by preference made as early as ten
in the morning. For the same reason he started on his
journeys by the earliest possible train, and used to arrive at
the houses of relatives in London when they were beginning
their day.

He kept an accurate journal of the days on which he
worked and those on which his ill health prevented him
from working, so that it would be possible to tell how many
were idle days in any given year. In this journal—a little
yellow Letts's Diary, which lay open on his mantel-piece,
piled on the diaries of previous years—he also entered the
day on which he started for a holiday and that of his return.

The most frequent holidays were visits of a week to
London, either to his brother's house (6 Queen Anne Street),
or to his daughter's (4 Bryanston Street). He was generally
persuaded by my mother to take these short holidays, when
it became clear from the frequency of " bad days," or from
the swimming of his head, that he was being overworked.
He went unwillingly, and tried to drive hard bargains, stipu-
lating, for instance, that he should come home in five days
instead of six. The discomfort of a journey to him was, at
least latterly, chiefly in the anticipation, and in the miserable
sinking feeling from which he suffered immediately before
the start; even a fairly long journey, such as that to Coniston, tired him wonderfully little, considering how much an
invalid he was; and he certainly enjoyed it in an almost
boyish way, and to a curious degree.

Although, as he has said, some of his æsthetic tastes had
suffered a gradual decay, his love of scenery remained fresh
and strong. Every walk at Coniston was a fresh delight,
and he was never tired of praising the beauty of the broken
hilly country at the head of the lake.

Besides these longer holidays, there were shorter visits
to various relatives—to his brother-in-law's house, close to
Leith Hill, and to his son near Southampton He always
particularly enjoyed rambling over rough open country,
such as the commons near Leith Hill and Southampton, the
heath-covered wastes of Ashdown Forest, or the delightful
" Rough " near the house of his friend Sir Thomas Farrer.
He never was quite idle even on these holidays, and found
things to observe. At Hartfield he watched Drosera catch-
ing insects, etc.; at Torquay he observed the fertilisation of

an orchid (*Spiranthes*), and also made out the relations of
the sexes in Thyme.

He rejoiced at his return home after his holidays, and
greatly enjoyed the welcome he got from his dog Polly, who
would get wild with excitement, panting, squeaking, rushing round the room, and jumping on and off the chairs;
and he used to stoop down, pressing her face to his, letting
her lick him, and speaking to her with a peculiarly tender,
caressing voice.

My father had the power of giving to these summer
holidays a charm which was strongly felt by all his family.
The pressure of his work at home kept him at the utmost
stretch of his powers of endurance, and when released from
it, he entered on a holiday with a youthfulness of enjoyment
that made his companionship delightful; we felt that we
saw more of him in a week's holiday than in a month at
home.

Besides the holidays which I have mentioned, there were
his visits to water-cure establishments.   In 1849, when very
ill, suffering from constant sickness, he was urged by a
friend to try the water-cure, and at last agreed to go to Dr.
Gully's establishment at Malvern.   His letters to Mr. Fox
show how much good the treatment did him; he seems to
have thought that he had found a cure for his troubles, but,
like all other remedies, it had only a transient effect on him.
However, he found it, at first, so good for him, that when
he came home he built himself a douche-bath, and the
butler learnt to be his bathman.

He was, too, a frequent patient at Dr. Lane's water-cure
establishment, Moor Park, near Aldershot, visits to which
he always looked back with pleasure.

Some idea of his relation to his family and his friends may
be gathered from what has gone before; it would be impossible to attempt a complete account of these relationships,
but a slightly fuller outline may not be out of place.   Of
his married life I cannot speak, save in the briefest manner.
In his relationship towards my mother, his tender and sympathetic nature was shown in its most beautiful aspect.   In
her presence he found his happiness, and through her, his
life—which might have been overshadowed by gloom—became one of content and quiet gladness.

The *Expression of the Emotions* shows how closely he
watched his children; it was characteristic of him that (as
I have heard him tell), although he was so anxious to ob-

serve accurately the expression of a crying child, his sympathy with the grief spoiled his observation. His note-book,
in which are recorded sayings of his young children, shows
his pleasure in them. He seemed to retain a sort of regret-
ful memory of the childhoods which had faded away, and
thus he wrote in his *Recollections* :—" When you were very
young it was my delight to play with you all, and I think
with a sigh that such days can never return."

I quote, as showing the tenderness of his nature, some
sentences from an account of his little daughter Annie,
written a few days after her death :—

" Our poor child, Annie, was born in Gower Street, on
March 2, 1841, and expired at Malvern at mid-day on the
23rd of April, 1851.

" I write these few pages, as I think in after years, if we
live, the impressions now put down will recall more vividly
her chief characteristics. From whatever point I look back
at her, the main feature in her disposition which at once
rises before me, is her buoyant joyousness, tempered by two
other characteristics, namely, her sensitiveness, which might
easily have been overlooked by a stranger, and her strong
affection. Her joyousness and animal spirits radiated from
her whole countenance, and rendered every movement elas-
tic and full of life and vigour. It was delightful and cheerful to behold her. Her dear face now rises before me, as
she used sometimes to come running downstairs with a
stolen pinch of snuff for me, her whole form radiant with
the pleasure of giving pleasure. Even when playing with
her cousins, when her joyousness almost passed into bois-
terousness, a single glance of my eye, not of displeasure (for
I thank God I hardly ever cast one on her), but of want of
sympathy, would for some minutes alter her whole countenance.

" The other point in her character, which made her joyousness and spirits so delightful, was her strong affection,
which was of a most clinging, fondling nature. When quite
a baby, this showed itself in never being easy without touching her mother, when in bed with her ; and quite lately she
would, when poorly, fondle for any length of time one of
her mother's arms. When very unwell, her mother lying
down beside her, seemed to soothe her in a manner quite
different from what it would have done to any of our other
children. So, again, she would at almost any time spend
half-an-hour in arranging my hair, ' making it,' as she called

it, ' beautiful,' or in smoothing, the poor dear darling, my collar or cuffs—in short, in fondling me.

" Besides her joyousness thus tempered, she was in her manners remarkably cordial, frank, open, straightforward, natural, and without any shade of reserve. Her whole mind was pure and transparent. One felt one knew her thoroughly and could trust her. I always thought, that come what might, we should have had, in our old age, at least one loving soul, which nothing could have changed. All her movements were vigorous, active, and usually graceful. When going round the Sand-walk with me, although I walked fast, yet she often used to go before, pirouetting in the most elegant way, her dear face bright all the time with the sweetest smiles. Occasionally she had a pretty, coquettish manner towards me, the memory of which is charming. She often used exaggerated language, and when I quizzed her by exaggerating what she had said, how clearly can I now see the little toss of the head, and exclamation of ' Oh, papa, what a shame of you!' In the last short illness, her conduct in simple truth was angelic. She never once complained; never became fretful; was ever considerate of others, and was thankful in the most gentle, pathetic manner for everything done for her. When so exhausted that she could hardly speak, she praised everything that was given her, and said some tea ' was beautifully good.' When I gave her some water, she said, ' I quite thank you ; ' and these, I believe, were the last precious words ever addressed by her dear lips to me.

" We have lost the joy of the household, and the solace of our old age. She must have known how we loved her. Oh, that she could now know how deeply, how tenderly, we do still and shall ever love her dear joyous face ! Blessings on her !*

" April 30, 1851."

We, his children, all took especial pleasure in the games he played at with us, and in his stories, which, partly on account of their rarity, were considered specially delightful.

The way he brought us up is shown by a little story about my brother Leonard, which my father was fond of telling. He came into the drawing-room and found Leon-

---

* The words, " A good and dear child," form the descriptive part of the inscription on her gravestone. See the *Athenæum*, Nov. 26, 1887.

ard dancing about on the sofa, to the peril of the springs, and said, " Oh, Lenny, Lenny, that's against all rules," and received for answer, " Then I think you'd better go out of the room." I do not believe he ever spoke an angry word to any of his children in his life ; but I am certain that it never entered our heads to disobey him. I well remember one occasion when my father reproved me for a piece of carelessness ; and I can still recall the feeling of depression which came over me, and the care which he took to disperse it by speaking to me soon afterwards with especial kindness. He kept up his delightful, affectionate manner towards us all his life. I sometimes wonder that he could do so, with such an undemonstrative race as we are ; but I hope he knew how much we delighted in his loving words and manner. He allowed his grown-up children to laugh with and at him, and was generally speaking on terms of perfect equality with us.

He was always full of interest about each one's plans or successes. We used to laugh at him, and say he would not believe in his sons, because, for instance, he would be a little doubtful about their taking some bit of work for which he did not feel sure that they had knowledge enough. On the other hand, he was only too much inclined to take a favourable view of our work. When I thought he had set too high a value on anything that I had done, he used to be indignant and inclined to explode in mock anger. His doubts were part of his humility concerning what was in any way connected with himself ; his too favourable view of our work was due to his sympathetic nature, which made him lenient to every one.

He kept up towards his children his delightful manner of expressing his thanks ; and I never wrote a letter or read a page aloud to him, without receiving a few kind words of recognition. His love and goodness towards his little grandson Bernard were great ; and he often spoke of the pleasure it was to him to see " his little face opposite to him " at luncheon. He and Bernard used to compare their tastes ; e. g., in liking brown sugar better than white, &c. ; the result being, " We always agree, don't we ? "

My sister writes :—

" My first remembrances of my father are of the delights of his playing with us. He was passionately attached to his own children, although he was not an indiscriminate child-lover. To all of us he was the most delightful play-fellow, and the most perfect sympathiser. Indeed it is impossible adequately to describe how delightful a relation his was to his family, whether as children or in their later life.

"It is a proof of the terms on which we were, and also of how much he was valued as a play-fellow, that one of his sons when about four years old tried to bribe him with sixpence to come and play in working hours.

"He must have been the most patient and delightful of nurses. I remember the haven of peace and comfort it seemed to me when I was unwell, to be tucked up on the study sofa, idly considering the old geological map hung on the wall. This must have been in his working hours, for I always picture him sitting in the horse hair arm chair by the corner of the fire.

"Another mark of his unbounded patience was the way in which we were suffered to make raids into the study when we had an absolute need of sticking plaster, string, pins, scissors, stamps, foot rule, or hammer. These and other such necessaries were always to be found in the study, and it was the only place where this was a certainty. We used to feel it wrong to go in during work time; still, when the necessity was great, we did so. I remember his patient look when he said once, 'Don't you think you could not come in again, I have been interrupted very often.' We used to dread going in for sticking plaster, because he disliked to see that we had cut ourselves, both for our sakes and on account of his acute sensitiveness to the sight of blood. I well remember lurking about the passage till he was safe away, and then stealing in for the plaster.

"Life seems to me, as I look back upon it, to have been very regular in those early days, and except relations (and a few intimate friends), I do not think any one came to the house. After lessons, we were always free to go where we would, and that was chiefly in the drawing-room and about the garden, so that we were very much with both my father and mother. We used to think it most delightful when he told us any stories about the *Beagle*, or about early Shrewsbury days—little bits about school life and his boyish tastes.

"He cared for all our pursuits and interests, and lived our lives with us in a way that very few fathers do. But I am certain that none of us felt that this intimacy interfered the least with our respect and obedience. Whatever he said was absolute truth and law to us. He always put his whole mind into answering any of our questions. One trifling instance makes me feel how he cared for what we cared for. He had no special taste for cats, but yet he knew and remembered the individualities of my many cats, and would talk about the habits and characters of the more remarkable ones years after they had died.

"Another characteristic of his treatment of his children was his respect for their liberty, and for their personality. Even as quite a little girl, I remember rejoicing in this sense of freedom. Our father and mother would not even wish to know what we were doing or thinking unless we wished to tell. He always made us feel that we were each of us creatures whose opinions and thoughts were valuable to him, so that whatever there was best in us came out in the sunshine of his presence.

"I do not think his exaggerated sense of our good qualities, intellectual or moral, made us conceited, as might perhaps have been expected, but rather more humble and grateful to him. The reason being, no doubt, that the influence of his character, of his sincerity and greatness of nature, had a much deeper and more lasting effect than any small exaltation which his praises or admiration may have caused to our vanity." *

As head of a household he was much loved and respected ; he always spoke to servants with politeness, using the expression, " would you be so good," in asking for anything. He was hardly ever angry with his servants; it shows how seldom this occurred, that when, as a small boy, I overheard a servant being scolded, and my father speaking angrily, it impressed me as an appalling circumstance, and I remember running up stairs out of a general sense of awe. He did not trouble himself about the management of the garden, cows, &c. He considered the horses so little his concern, that he used to ask doubtfully whether he might have a horse and cart to send to Keston for Sundew, or to the Westerham nurseries for plants, or the like.

As a host my father had a peculiar charm : the presence of visitors excited him, and made him appear to his best advantage. At Shrewsbury, he used to say, it was his father's wish that the guests should be attended to constantly, and in one of the letters to Fox he speaks of the impos-

---

* Some pleasant recollections of my father's life at Down, written by our friend and former neighbour, Mrs. Wallis Nash, have been published in the *Overland Monthly* (San Francisco), October 1890.

sibility of writing a letter while the house was full of company. I think he always felt uneasy at not doing more for the entertainment of his guests, but the result was successful; and, to make up for any loss, there was the gain that the guests felt perfectly free to do as they liked. The most usual visitors were those who stayed from Saturday till Monday; those who remained longer were generally relatives, and were considered to be rather more my mother's affair than his.

Besides these visitors, there were foreigners and other strangers, who came down for luncheon and went away in the afternoon. He used conscientiously to represent to them the enormous distance of Down from London, and the labour it would be to come there, unconsciously taking for granted that they would find the journey as toilsome as he did himself. If, however, they were not deterred, he used to arrange their journeys for them, telling them when to come, and practically when to go. It was pleasant to see the way in which he shook hands with a guest who was being welcomed for the first time; his hand used to shoot out in a way that gave one the feeling that it was hastening to meet the guest's hands. With old friends his hand came down with a hearty swing into the other hand in a way I always had satisfaction in seeing. His good-bye was chiefly characterised by the pleasant way in which he thanked his guests, as he stood at the hall-door, for having come to see him.

These luncheons were successful entertainments, there was no drag or flagging about them, my father was bright and excited throughout the whole visit. Professor De Candolle has described a visit to Down, in his admirable and sympathetic sketch of my father.* He speaks of his manner as resembling that of a "savant" of Oxford or Cambridge. This does not strike me as quite a good comparison; in his ease and naturalness there was more of the manner of some soldiers; a manner arising from total absence of pretence or affectation It was this absence of pose, and the natural and simple way in which he began talking to his guests, so as to get them on their own lines, which made him so charming a host to a stranger. His happy choice of matter for talk seemed to flow out of his sympathetic nature, and humble, vivid interest in other people's work.

---

* *Darwin considéré au point de vue des causes de son succès* (Geneva, 1882).

To some, I think, he caused actual pain by his modesty;
I have seen the late Francis Balfour quite discomposed by
having knowledge ascribed to himself on a point about
which my father claimed to be utterly ignorant.

It is difficult to seize on the characteristics of my father's
conversation.

He had more dread than have most people of repeating
his stories, and continually said, " You must have heard me
tell," or " I daresay I've told you." One peculiarity he had,
which gave a curious effect to his conversation. The first
few words of a sentence would often remind him of some
exception to, or some reason against, what he was going to
say; and this again brought up some other point, so that
the sentence would become a system of parenthesis within
parenthesis, and it was often impossible to understand the
drift of what he was saying until he came to the end of his
sentence. He used to say of himself that he was not quick
enough to hold an argument with any one, and I think this
was true. Unless it was a subject on which he was just
then at work, he could not get the train of argument into
working order quickly enough. This is shown even in his
letters; thus, in the case of two letters to Professor Semper
about the effect of isolation, he did not recall the series of
facts he wanted until some days after the first letter had
been sent off.

When puzzled in talking, he had a peculiar stammer on
the first word of a sentence. I only recall this occurring
with words beginning with w; possibly he had a special
difficulty with this letter, for I have heard him say that as a
boy he could not pronounce w, and that sixpence was offered
him if he could say " white wine," which he pronounced
" rite rine." Possibly he may have inherited this tendency
from Erasmus Darwin who stammered.*

He sometimes combined his metaphors in a curious way,
using such a phrase as " holding on like life "—a mixture
of " holding on for his life," and " holding on like grim
death." It came from his eager way of putting emphasis
into what he was saying. This sometimes gave an air of
exaggeration where it was not intended; but it gave, too, a
noble air of strong and generous conviction; as, for in-

---

* My father related a Johnsonian answer of Erasmus Darwin's : " Don't
you find it very inconvenient stammering, Dr. Darwin ? "   " No, Sir, because
I have time to think before I speak, and don't ask impertinent questions."

stance, when he gave his evidence before the Royal Commission on vivisection, and came out with his words about
cruelty, " It deserves detestation and abhorrence." When
he felt strongly about any similar question, he could hardly
trust himself to speak, as he then easily became angry, a
thing which he disliked excessively. He was conscious that
his anger had a tendency to multiply itself in the utterance,
and for this reason dreaded (for example) having to reprove
a servant.

It was a proof of the modesty of his manner of talking,
that when, for instance, a number of visitors came over from
Sir John Lubbock's for a Sunday afternoon call, he never
seemed to be preaching or lecturing, although he had so
much of the talk to himself. He was particularly charming
when " chaffing " any one, and in high spirits over it. His
manner at such times was light-hearted and boyish, and his
refinement of nature came out most strongly. So, when he
was talking to a lady who pleased and amused him, the
combination of raillery and deference in his manner was
delightful to see. There was a personal dignity about him,
which the most familiar intercourse did not diminish. One
felt that he was the last person with whom anyone would
wish to take a liberty, nor do I remember an instance of such
a thing occurring to him.

When my father had several guests he managed them
well, getting a talk with each, or bringing two or three together round his chair. In these conversations there was
always a good deal of fun, and, speaking generally, there
was either a humourous turn in his talk, or a sunny geniali-
ty which served instead. Perhaps my recollection of a per-
vading element of humour is the more vivid, because the
best talks were with Mr. Huxley, in whom there is the apt-
ness which is akin to humour, even when humour itself is
not there. My father enjoyed Mr. Huxley's humour exceedingly, and would often say, " What splendid fun Huxley is ! " I think he probably had more scientific argument
(of the nature of a fight) with Lyell and Sir Joseph Hooker.

He used to say that it grieved him to find that for the
friends of his later life he had not the warm affection of his
youth. Certainly in his early letters from Cambridge he
gives proofs of strong friendship for Herbert and Fox; but
no one except himself would have said that his affection for
his friends was not, throughout life, of the warmest possible
kind. In serving a friend he would not spare himself, and

precious time and strength were willingly given. He undoubtedly had, to an unusual degree, the power of attaching his friends to him. He had many warm friendships, but to Sir Joseph Hooker he was bound by ties of affection stronger than we often see among men. He wrote in his *Recollections*, " I have known hardly any man more lovable than Hooker."

His relationship to the village people was a pleasant one; he treated them, one and all, with courtesy, when he came in contact with them, and took an interest in all relating to their welfare. Some time after he came to live at Down he helped to found a Friendly Club, and served as treasurer for thirty years. He took much trouble about the club, keeping its accounts with minute and scrupulous exactness, and taking pleasure in its prosperous condition. Every Whit-Monday the club marched round with a band and banner and paraded on the lawn in front of the house. There he met them, and explained to them their financial position in a little speech seasoned with a few well-worn jokes. He was often unwell enough to make even this little ceremony an exertion, but I think he never failed to meet them.

He was also treasurer of the Coal Club, which gave him a certain amount of work, and he acted for some years as a County Magistrate.

With regard to my father's interest in the affairs of the village, Mr. Brodie Innes has been so good as to give me his recollections :—

" On my becoming Vicar of Down in 1846, we became friends, and so continued till his death. His conduct towards me and my family was one of unvarying kindness, and we repaid it by warm affection.

" In all parish matters he was an active assistant; in matters connected with the schools, charities, and other business, his liberal contribution was ever ready, and in the differences which at times occurred in that, as in other parishes, I was always sure of his support. He held that where there was really no important objection, his assistance should be given to the clergyman, who ought to know the circumstances best, and was chiefly responsible."

His intercourse with strangers was marked with scrupulous and rather formal politeness, but in fact he had few opportunities of meeting strangers, and the quiet life he led at Down made him feel confused in a large gathering; for instance, at the Royal Society's *soirées* he felt oppressed by the numbers. The feeling that he ought to know people, and the difficulty he had in remembering faces in his latter years, also added to his discomfort on such occasions. He did not realise that he would be recognised from his photographs, and I remember his being uneasy at being obviously recognised by a stranger at the Crystal Palace Aquarium.

I must say something of his manner of working; a striking characteristic was his respect for time; he never forgot how precious it was. This was shown, for instance, in the way in which he tried to curtail his holidays; also, and more clearly, with respect to shorter periods. He would often say, that saving the minutes was the way to get work done; he showed this love of saving the minutes in the difference he felt between a quarter of an hour and ten minutes' work; he never wasted a few spare minutes from thinking that it was not worth while to set to work. I was often struck by his way of working up to the very limit of his strength, so that he suddenly stopped in dictating, with the words, " I believe I mustn't do any more." The same eager desire not to lose time was seen in his quick movements when at work. I particularly remember noticing this when he was making an experiment on the roots of beans, which required some care in manipulation; fastening the little bits of card upon the roots was done carefully and necessarily slowly, but the intermediate movements were all quick; taking a fresh bean, seeing that the root was healthy, impaling it on a pin, fixing it on a cork, and seeing that it was vertical, &c.; all these processes were performed with a kind of restrained eagerness. He gave one the impression of working with pleasure, and not with any drag. I have an image, too, of him as he recorded the result of some experiment, looking eagerly at each root, &c., and then writing with equal eagerness. I remember the quick movement of his head up and down as he looked from the object to the notes.

He saved a great deal of time through not having to do things twice. Although he would patiently go on repeating experiments where there was any good to be gained, he could not endure having to repeat an experiment which ought, if complete care had been taken, to have told its story at first—and this gave him a continual anxiety that the experiment should not be wasted; he felt the experiment to be sacred, however slight a one it was. He wished to learn as much as possible from an experiment, so that he did not confine himself to observing the single point to which the experiment was directed, and his power of seeing a number of other things was wonderful. I do not think he cared for preliminary or rough observations intended to serve as guides and to be repeated. Any experiment done was to be of some use, and in this connection I remember how strongly he urged the necessity of keeping the notes of experiments which failed, and to this rule he always adhered.

In the literary part of his work he had the same horror of losing time, and the same zeal in what he was doing at the moment, and this made him careful not to be obliged unnecessarily to read anything a second time.

His natural tendency was to use simple methods and few instruments. The use of the compound microscope has much increased since his youth, and this at the expense of the simple one. It strikes us nowadays as extraordinary that he should have had no compound microscope when he went his *Beagle* voyage; but in this he followed the advice of Robert Brown, who was an authority in such matters. He always had a great liking for the simple microscope, and maintained that nowadays it was too much neglected, and that one ought always to see as much as possible with the simple before taking to the compound microscope. In one of his letters he speaks on this point, and remarks that he suspects the work of a man who never uses the simple microscope.

His dissecting table was a thick board, let into a window of the study; it was lower than an ordinary table, so that he could not have worked at it standing; but this, from wishing to save his strength, he would not have done in any case. He sat at his dissecting-table on a curious low stool which had belonged to his father, with a seat revolving on a vertical spindle, and mounted on large castors, so that he could turn easily from side to side. His ordinary tools, &c., were lying about on the table, but besides these a number of odds and ends were kept in a round table full of radiating drawers, and turning on a vertical axis, which stood close by his left side, as he sat at his microscope-table. The drawers were labelled, " best tools," " rough tools," " specimens," " preparations for specimens," &c. The most marked peculiarity of the contents of these drawers was the care with which little scraps and almost useless things were preserved; he held the well-known belief, that if you throw a thing away you were sure to want it directly—and so things accumulated.

If any one had looked at his tools, &c., lying on the table, he would have been struck by an air of simpleness, make-shift, and oddity.

At his right hand were shelves, with a number of other odds and ends, glasses, saucers, tin biscuit boxes for germinating seeds, zinc labels, saucers full of sand, &c., &c. Considering how tidy and methodical he was in essential things, it is curious that he bore with so many make-shifts : for instance, instead of having a box made of a desired shape, and stained black inside, he would hunt up something like what he wanted and get it darkened inside with shoe-blacking; he did not care to have glass covers made for tumblers in which he germinated seeds, but used broken bits of irregular shape, with perhaps a narrow angle sticking uselessly out on one side. But so much of his experimenting was of a simple kind, that he had no need for any elaboration, and I think his habit in this respect was in great measure due to his desire to husband his strength and not waste it on inessential things.

His way of marking objects may here be mentioned. If he had a number of things to distinguish, such as leaves, flowers, &c., he tied threads of different colours round them. In particular he used this method when he had only two classes of objects to distinguish; thus in the case of crossed and self-fertilised flowers, one set would be marked with black and one with white thread, tied round the stalk of the flower. I remember well the look of two sets of capsules, gathered and waiting to be weighed, counted, &c., with pieces of black and of white thread to distinguish the trays in which they lay. When he had to compare two sets of seedlings, sowed in the same pot, he separated them by a partition of zinc-plate; and the zinc-label, which gave the necessary details about the experiment, was always placed on a certain side, so that it became instinctive with him to know without reading the label which were the "crossed" and which the "self-fertilised."

His love of each particular experiment, and his eager zeal not to lose the fruit of it, came out markedly in these crossing experiments—in the elaborate care he took not to make any confusion in putting capsules into wrong trays, &c., &c. I can recall his appearance as he counted seeds under the simple microscope with an alertness not usually characterising such mechanical work as counting. I think he personified each seed as a small demon trying to elude him by getting into the wrong heap, or jumping away altogether; and this gave to the work the excitement of a game. He had great faith in instruments, and I do not think it naturally occurred to him to doubt the accuracy of a scale, a measuring glass, &c. He was astonished when we found that one of his micrometers differed from the other. He did not require any great accuracy in most of his measurements, and had not good scales; he had an old three-foot rule, which was the common property of the household, and was constantly being borrowed, because it was the only one which was certain to be in its place—unless, indeed, the last borrower had forgotten to put it back. For measuring the height of plants, he had a seven-foot deal rod, graduated by the village carpenter. Latterly he took to using paper scales graduated to millimeters. I do not mean by this account of his instruments that any of his experiments suffered from want of accuracy in measurement, I give them as examples of his simple methods and faith in others—faith at least in instrument-makers, whose whole trade was a mystery to him.

A few of his mental characteristics, bearing especially on his mode of working, occur to me. There was one quality of mind which seemed to be of special and extreme advantage in leading him to make discoveries. It was the power of never letting exceptions pass unnoticed. Everybody notices a fact as an exception when it is striking or frequent, but he had a special instinct for arresting an exception. A point apparently slight and unconnected with his present work is passed over by many a man almost unconsciously with some half-considered explanation, which is in fact no explanation. It was just these things that he seized on to make a start from. In a certain sense there is nothing special in this procedure, many discoveries being made by means of it. I only mention it because, as I watched him at work, the value of this power to an experimenter was so strongly impressed upon me.

Another quality which was shown in his experimental work, was his power of sticking to a subject; he used almost to apologise for his patience, saying that he could not bear to be beaten, as if this were rather a sign of weakness on his part. He often quoted the saying, "It's dogged as does it;" and I think doggedness expresses his frame of mind almost better than perseverance. Perseverance seems hardly to express his almost fierce desire to force the truth to reveal itself. He often said that it was important that a man should know the right point at which to give up an inquiry. And I think it was his tendency to pass this point that inclined him to apologise for his perseverance, and gave the air of doggedness to his work.

He often said that no one could be a good observer unless he was an active theoriser. This brings me back to what I said about his instinct for arresting exceptions : it was as though he were charged with theorising power ready to flow into any channel on the slightest disturbance, so that no fact, however small, could avoid releasing a stream of theory, and thus the fact became magnified into importance. In this way it naturally happened that many untenable theories occurred to him ; but fortunately his richness of imagination was equalled by his power of judging and condemning the thoughts that occurred to him. He was just to his theories, and did not condemn them unheard ; and so it happened that he was willing to test what would seem to most people not at all worth testing. These rather wild trials he called " fool's experiments," and enjoyed extremely. As an example I may mention that finding the seed-leaves of a kind of sensitive plant, to be highly sensitive to vibrations of the table, he fancied that they might perceive the vibrations of sound, and therefore made me play my bassoon close to a plant.*

The love of experiment was very strong in him, and I can remember the way he would say, " I shan't be easy till I have tried it," as if an outside force were driving him. He enjoyed experimenting much more than work which only entailed reasoning, and when he was engaged on one of his books which required argument and the marshalling of facts, he felt experimental work to be a rest or holiday. Thus, while working upon the *Variations of Animals and Plants* in 1860–61, he made out the fertilisation of Orchids, and thought himself idle for giving so much time to them. It is interesting to think that so important a piece of research should have been undertaken and largely worked out as a pastime in place of more serious work. The letters to Hooker of this period contain expressions such as, " God

---

* This is not so much an example of superabundant theorising from a small cause as of his wish to test the most improbable ideas.

forgive me for being so idle; I am quite sillily interested in the work." The intense pleasure he took in understanding the adaptations for fertilisation is strongly shown in these letters. He speaks in one of his letters of his intention of working at Sundew as a rest from the *Descent of Man.* He has described in his *Recollections* the strong satisfaction he felt in solving the problem of heterostylism.* And I have heard him mention that the Geology of South America gave him almost more pleasure than anything else. It was perhaps this delight in work requiring keen observation that made him value praise given to his observing powers almost more than appreciation of his other qualities.

For books he had no respect, but merely considered them as tools to be worked with. Thus he did not bind them, and even when a paper book fell to pieces from use, as happened to Müller's *Befruchtung*, he preserved it from complete dissolution by putting a metal clip over its back. In the same way he would cut a heavy book in half, to make it more convenient to hold. He used to boast that he had made Lyell publish the second edition of one of his books in two volumes, instead of in one, by telling him how he had been obliged to cut it in half. Pamphlets were often treated even more severely than books, for he would tear out, for the sake of saving room, all the pages except the one that interested him. The consequence of all this was, that his library was not ornamental, but was striking from being so evidently a working collection of books.

He was methodical in his manner of reading books and pamphlets bearing on his own work. He had one shelf on which were piled up the books he had not yet read, and another to which they were transferred after having been read, and before being catalogued. He would often groan over his unread books, because there were so many which he knew he should never read. Many a book was at once transferred to the other heap, either marked with a cypher at the end, to show that it contained no marked passages, or inscribed, perhaps, " not read," or " only skimmed." The books accumulated in the " read " heap until the shelves overflowed, and then, with much lamenting, a day was given up to the cataloguing. He disliked this work, and as the necessity of undertaking the work became im-

---

* That is to say, the sexual relations in such plants as the cowslip.

perative, would often say, in a voice of despair, "We really must do these books soon."

In each book, as he read it, he marked passages bearing on his work. In reading a book or pamphlet, &c., he made pencil lines at the side of the page, often adding short remarks, and at the end made a list of the pages marked. When it was to be catalogued and put away, the marked pages were looked at, and so a rough abstract of the book was made. This abstract would perhaps be written under three or four headings on different sheets, the facts being sorted out and added to the previously collected facts in the different subjects. He had other sets of abstracts arranged, not according to subject, but according to the periodicals from which they were taken. When collecting facts on a large scale, in earlier years, he used to read through, and make abstracts, in this way, of whole series of journals.

In some of his early letters he speaks of filling several note-books with facts for his book on species; but it was certainly early that he adopted his plan of using portfolios, as described in the *Recollections*.* My father and M. de Candolle were mutually pleased to discover that they had adopted the same plan of classifying facts. De Candolle describes the method in his *Phytologie*, and in his sketch of my father mentions the satisfaction he felt in seeing it in action at Down.

Besides these portfolios, of which there are some dozens full of notes, there are large bundles of MS. marked "used" and put away. He felt the value of his notes, and had a horror of their destruction by fire. I remember, when some alarm of fire had happened, his begging me to be especially careful, adding very earnestly, that the rest of his life would be miserable if his notes and books were destroyed.

He shows the same feeling in writing about the loss of a manuscript, the purport of his words being, "I have a copy, or the loss would have killed me." In writing a book he would spend much time and labour in making a skeleton or plan of the whole, and in enlarging and sub-classing each heading, as described in his *Recollections*. I think this careful arrangement of the plan was not at all essential to the building up of his argument, but for its presentment,

---

* The racks in which the portfolios were placed are shown in the illustration at the head of the chapter, in the recess at the right-hand side of the fireplace.

and for the arrangement of his facts.　In his *Life of Eras-mus Darwin*, as it was first printed in slips, the growth of the book from a skeleton was plainly visible.　The arrange-ment was altered afterwards, because it was too formal and categorical, and seemed to give the character of his grand-father rather by means of a list of qualities than as a com-plete picture.

It was only within the last few years that he adopted a plan of writing which he was convinced suited him best, and which is described in the *Recollections;* namely, writing a rough copy straight off without the slighest attention to style.　It was characteristic of him that he felt unable to write with sufficient want of care if he used his best paper, and thus it was that he wrote on the backs of old proofs or manuscript.　The rough copy was then reconsidered, and a fair copy was made.　For this purpose he had foolscap paper ruled at wide intervals, the lines being needed to pre-vent him writing so closely that correction became difficult. The fair copy was then corrected, and was recopied before being sent to the printers.　The copying was done by Mr. E. Norman, who began this work many years ago when village schoolmaster at Down.　My father became so used to Mr. Norman's handwriting, that he could not correct manuscript, even when clearly written out by one of his children, until it had been recopied by Mr. Norman.　The MS., on returning from Mr. Norman, was once more cor-rected, and then sent off to the printers.　Then came the work of revising and correcting the proofs, which my father found especially wearisome.

When the book was passing through the "slip" stage he was glad to have corrections and suggestions from others. Thus my mother looked over the proofs of the *Origin*.　In some of the later works my sister, Mrs. Litchfield, did much of the correction.　After my sister's marriage perhaps most of the work fell to my share. ·

My sister, Mrs. Litchfield, writes :—

"This work was very interesting in itself, and it was inexpressibly exhilarating to work for him.　He was so ready to be convinced that any suggested alteration was an improvement, and so full of gratitude for the trouble taken. I do not think that he ever forgot to tell me what improve-ment he thought I had made, and he used almost to excuse himself if he did not agree with any correction.　I think I felt the singular modesty and graciousness of his nature through thus working for him in a way I never should otherwise have done."

Perhaps the commonest corrections needed were of obscurities due to the omission of a necessary link in the reasoning, evidently omitted through familiarity with the subject. Not that there was any fault in the sequence of the thoughts, but that from familiarity with his argument he did not notice when the words failed to reproduce his thought. He also frequently put too much matter into one sentence, so that it had to be cut up into two.

On the whole, I think the pains which my father took over the literary part of the work was very remarkable. He often laughed or grumbled at himself for the difficulty which he found in writing English, saying, for instance, that if a bad arrangement of a sentence was possible, he should be sure to adopt it. He once got much amusement and satisfaction out of the difficulty which one of the family found in writing a short circular. He had the pleasure of correcting and laughing at obscurities, involved sentences, and other defects, and thus took his revenge for all the criticism he had himself to bear with. He would quote with astonishment Miss Martineau's advice to young authors, to write straight off and send the MS. to the printer without correction. But in some cases he acted in a somewhat similar manner. When a sentence became hopelessly involved, he would ask himself, " now what *do* you want to say ?" and his answer written down, would often disentangle the confusion.

His style has been much praised ; on the other hand, at least one good judge has remarked to me that it is not a good style. It is, above all things, direct and clear ; and it is characteristic of himself in its simplicity bordering on naïveté, and in its absence of pretence. He had the strongest disbelief in the common idea that a classical scholar must write good English ; indeed, he thought that the contrary was the case. In writing, he sometimes showed the same tendency to strong expressions that he did in conversation. Thus in the *Origin*, p. 440, there is a description of a larval cirripede, " with six pairs of beautifully constructed natatory legs, a pair of magnificent compound eyes, and extremely complex antennæ." We used to laugh at him for this sentence, which we compared to an advertisement. This tendency to give himself up to the enthusiastic turn of his thought, without fear of being ludicrous appears elsewhere in his writings.

His courteous and conciliatory tone towards his reader is remarkable, and it must be partly this quality which revealed his personal sweetness of character to so many who had never seen him. I have always felt it to be a curious fact, that he who has altered the face of Biological Science, and is in this respect the chief of the moderns, should have written and worked in so essentially a non-modern spirit and manner. In reading his books one is reminded of the older naturalists rather than of any modern school of writers. He was a Naturalist in the old sense of the word, that is, a man who works at many branches of science, not merely a specialist in one. Thus it is, that, though he founded whole new divisions of special subjects—such as the fertilisation of flowers, insectivorous plants, &c.—yet even in treating these very subjects he does not strike the reader as a specialist. The reader feels like a friend who is being talked to by a courteous gentleman, not like a pupil being lectured by a professor. The tone of such a book as the *Origin* is charming, and almost pathetic; it is the tone of a man who, convinced of the truth of his own views, hardly expects to convince others; it is just the reverse of the style of a fanatic, who tries to force belief on his readers. The reader is never scorned for any amount of doubt which he may be imagined to feel, and his scepticism is treated with patient respect. A sceptical reader, or perhaps even an unreasonable reader, seems to have been generally present to his thoughts. It was in consequence of this feeling, perhaps, that he took much trouble over points which he imagined would strike the reader, or save him trouble, and so tempt him to read.

For the same reason he took much interest in the illustrations of his books, and I think rated rather too highly their value. The illustrations for his earlier books were drawn by professional artists. This was the case in *Animals and Plants*, the *Descent of Man*, and the *Expression of the Emotions*. On the other hand, *Climbing Plants, Insectivorous Plants*, the *Movements of Plants*, and *Forms of Flowers*, were, to a large extent, illustrated by some of his children—my brother George having drawn by far the most. It was delightful to draw for him, as he was enthusiastic in his praise of very moderate performances. I remember well his charming manner of receiving the drawings of one of his daughters-in-law, and how he would finish his words of praise by saying, " Tell A——, Michael

Angelo is nothing to it." Though he praised so generously, he always looked closely at the drawing, and easily detected mistakes or carelessness.

He had a horror of being lengthy, and seems to have
been really much annoyed and distressed when he found
how the *Variations of Animals and Plants* was growing
under his hands. I remember his cordially agreeing with
'Tristram Shandy's' words, "Let no man say, 'Come, I'll
write a duodecimo.'"

His consideration for other authors was as marked a
characteristic as his tone towards his reader. He speaks of
all other authors as persons deserving of respect. In cases
where, as in the case of ——'s experiments on Drosera, he
thought lightly of the author, he speaks of him in such a
way that no one would suspect it. In other cases he treats
the confused writings of ignorant persons as though the
fault lay with himself for not appreciating or understanding them. Besides this general tone of respect, he had a
pleasant way of expressing his opinion on the value of a
quoted work, or his obligation for a piece of private information.

His respectful feeling was not only admirable, but was I
think of practical use in making him ready to consider the
ideas and observations of all manner of people. He used
almost to apologise for this, and would say that he was at
first inclined to rate everything too highly.

It was a great merit in his mind that, in spite of having
so strong a respectful feeling towards what he read, he had
the keenest of instincts as to whether a man was trustworthy or not. He seemed to form a very definite opinion
as to the accuracy of the men whose books he read; and
employed this judgment in his choice of facts for use in
argument or as illustrations. I gained the impression that
he felt this power of judging of a man's trustworthiness to
be of much value.

He had a keen feeling of the sense of honour that ought
to reign among authors, and had a horror of any kind of
laxness in quoting. He had a contempt for the love of
honour and glory, and in his letters often blames himself for the pleasure he took in the success of his books, as
though he were departing from his ideal—a love of truth
and carelessness about fame. Often, when writing to Sir
J. Hooker what he calls a boasting letter, he laughs at
himself for his conceit and want of modesty. A wonder-

fully interesting letter is given in Chapter X. bequeathing to my mother, in case of his death, the care of publishing the manuscript of his first essay on evolution. This letter seems to me full of an intense desire that his theory should succeed as a contribution to knowledge, and apart from any desire for personal fame. He certainly had the healthy desire for success which a man of strong feelings ought to have. But at the time of the publication of the *Origin* it is evident that he was overwhelmingly satisfied with the adherence of such men as Lyell, Hooker, Huxley, and Asa Gray, and did not dream of or desire any such general fame as that to which he attained.

Connected with his contempt for the undue love of fame, was an equally strong dislike of all questions of priority. The letters to Lyell, at the time of the *Origin*, show the anger he felt with himself for not being able to repress a feeling of disappointment at what he thought was Mr. Wallace's forestalling of all his years of work. His sense of literary honour comes out strongly in these letters; and his feeling about priority is again shown in the admiration expressed in his *Recollections* of Mr. Wallace's self-annihilation.

His feeling about reclamations, including answers to attacks and all kinds of discussions, was strong. It is simply expressed in a letter to Falconer (1863) : " If I ever felt angry towards you, for whom I have a sincere friendship, I should begin to suspect that I was a little mad. I was very sorry about your reclamation, as I think it is in every case a mistake and should be left to others. Whether I should so act myself under provocation is a different question." It was a feeling partly dictated by instinctive delicacy, and partly by a strong sense of the waste of time, energy, and temper thus caused. He said that he owed his determination not to get into discussions * to the advice of Lyell,—advice which he transmitted to those among his friends who were given to paper warfare.

If the character of my father's working life is to be understood, the conditions of ill-health, under which he

---

* He departed from his rule in his " Note on the Habits of the Pampas Woodpecker, *Colaptes campestris*," *Proc. Zool. Soc.*, 1870, p. 705 : also in a letter published in the *Athenæum* (1863, p. 554), in which case he afterwards regretted that he had not remained silent. His replies to criticisms, in the latter editions of the *Origin*, can hardly be classed as infractions of his rule.

worked, must be constantly borne in mind. He bore his illness with such uncomplaining patience, that even his children can hardly, I believe, realise the extent of his habitual suffering. In their case the difficulty is heightened by the fact that, from the days of their earliest recollections, they saw him in constant ill-health,—and saw him, in spite of it, full of pleasure in what pleased them. Thus, in later life, their perception of what he endured had to be disentangled from the impression produced in childhood by constant genial kindness under conditions of unrecognised difficulty. No one indeed, except my mother, knows the full amount of suffering he endured, or the full amount of his wonderful patience. For all the latter years of his life she never left him for a night; and her days were so planned that all his resting hours might be shared with her. She shielded him from every avoidable annoyance, and omitted nothing that might save him trouble, or prevent him becoming overtired, or that might alleviate the many discomforts of his ill-health. I hesitate to speak thus freely of a thing so sacred as the life-long devotion which prompted all this constant and tender care. But it is, I repeat, a principal feature of his life, that for nearly forty years he never knew one day of the health of ordinary men, and that thus his life was one long struggle against the weariness and strain of sickness. And this cannot be told without speaking of the one condition which enabled him to bear the strain and fight out the struggle to the end.

CHAPTER V.

My father's Cambridge life comprises the time between the Lent Term, 1828, when he came up to Christ's College as a Freshman, and the end of the May Term, 1831, when he took his degree * and left the University.

He "kept" for a term or two in lodgings, over Bacon † the tobacconist's; not, however, over the shop in the Market Place, so well known to Cambridge men, but in Sydney Street. For the rest of his time he had pleasant rooms on the south side of the first court of Christ's.‡

What determined the choice of this college for his brother Erasmus and himself I have no means of knowing. Erasmus the elder, their grandfather, had been at St. John's, and this college might have been reasonably selected for them, being connected with Shrewsbury School. But the life of an undergraduate at St. John's seems, in those days, to have been a troubled one, if I may judge from the fact that a relative of mine migrated thence to Christ's to escape the harassing discipline of the place.

Darwin seems to have found no difficulty in living at peace with all men in and out of office at Lady Margaret's elder foundation. The impression of a contemporary of my father's is that Christ's in their day was a pleasant, fairly quiet college, with some tendency towards "horsiness"; many of the men made a custom of going to Newmarket during the races, though betting was not a regular practice. In this they were by no means discouraged by the Senior

---

* "On Tuesday last Charles Darwin, of Christ's College, was admitted B.A."—*Cambridge Chronicle*, Friday, April 29th, 1831.
† Readers of Calverley (another Christ's man) will remember his tobacco poem ending "Here's to thee, Bacon."
‡ The rooms are on the first floor, on the west side of the middle staircase. A medallion (given by my brother) has recently been let into the wall of the sitting-room.

Tutor, Mr. Shaw, who was himself generally to be seen on the Heath on these occasions.

Nor were the ecclesiastical authorities of the College over strict. I have heard my father tell how at evening chapel the Dean used to read alternate verses of the Psalms, without making even a pretence of waiting for the congregation to take their share. And when the Lesson was a lengthy one, he would rise and go on with the Canticles after the scholar had read fifteen or twenty verses.

It is curious that my father often spoke of his Cambridge life as if it had been so much time wasted,* forgetting that, although the set studies of the place were barren enough for him, he yet gained in the highest degree the best advantages of a University life—the contact with men and an opportunity for mental growth. It is true that he valued at its highest the advantages which he gained from associating with Professor Henslow and some others, but he seemed to consider this as a chance outcome of his life at Cambridge, not an advantage for which *Alma Mater* could claim any credit. One of my father's Cambridge friends was the late Mr. J. M. Herbert, County Court Judge for South Wales, from whom I was fortunate enough to obtain some notes which help us to gain an idea of how my father impressed his contemporaries. Mr. Herbert writes:—

" It would be idle for me to speak of his vast intellectual powers . . . but I cannot end this cursory and rambling sketch without testifying, and I doubt not all his surviving college friends would concur with me, that he was the most genial, warm-hearted, generous, and affectionate of friends; that his sympathies were with all that was good and true; and that he had a cordial hatred for everything false, or vile, or cruel, or mean, or dishonourable. He was not only great, but pre-eminently good, and just, and lovable."

Two anecdotes told by Mr. Herbert show that my father's feeling for suffering, whether of man or beast, was as strong in him as a young man as it was in later years : " Before he left Cambridge he told me that he had made up his mind not to shoot any more; that he had had two days' shooting at his friend's, Mr. Owen of Woodhouse; and that on the second day, when going over some of the ground they had

---

* For instance in a letter to Hooker (1847) :—" Many thanks for your welcome note from Cambridge, and I am glad you like my *Alma Mater*, which I despise heartily as a place of education, but love from many most pleasant recollections."

beaten on the day before, he picked up a bird not quite dead, but lingering from a shot it had received on the previous day; and that it had made and left such a painful impression on his mind, that he could not reconcile it to his conscience to continue to derive pleasure from a sport which inflicted such cruel suffering."

To realise the strength of the feeling that led to this resolve, we must remember how passionate was his love of sport. We must recall the boy shooting his first snipe,* and trembling with excitement so that he could hardly reload his gun. Or think of such a sentence as, "Upon my soul, it is only about a fortnight to the ' First,' then if there is a bliss on earth that is it."†

His old college friends agree in speaking with affectionate warmth of his pleasant, genial temper as a young man. From what they have been able to tell me, I gain the impression of a young man overflowing with animal spirits— leading a varied healthy life—not over-industrious in the set studies of the place, but full of other pursuits, which were followed with a rejoicing enthusiasm. Entomology, riding, shooting in the fens, suppers and card-playing, music at King's Chapel, engravings at the Fitzwilliam Museum, walks with Professor Henslow—all combined to fill up a happy life. He seems to have infected others with his enthusiasm. Mr. Herbert relates how, while on a reading-party at Barmouth, he was pressed into the service of "the science"—as my father called collecting beetles:—

"He armed me with a bottle of alcohol, in which I had to drop any beetle which struck me as not of a common kind. I performed this duty with some diligence in my constitutional walks; but, alas! my powers of discrimination seldom enabled me to secure a prize—the usual result, on his examining the contents of my bottle, being an exclamation, ' Well, old Cherbury ' ‡ (the nickname he gave me, and by which he usually addressed me), ' none of these will do.' "

Again, the Rev. T. Butler, who was one of the Barmouth reading-party in 1828, says: "He inoculated me with a taste for Botany which has stuck by me all my life."

Archdeacon Watkins, another old college friend of my father's, remembered him unearthing beetles in the willows

---

* Autobiography, p. 10.
† From a letter to W. D. Fox.
‡ No doubt in allusion to the title of Lord Herbert of Cherbury.

between Cambridge and Grantchester, and speaks of a
certain beetle the remembrance of whose name is " Crux
major." * How enthusiastically must my father have ex-
ulted over this beetle to have impressed its name on a companion so that he remembers it after half a century !

He became intimate with Henslow, the Professor of
Botany, and through him with some other older members
of the University. " But," Mr. Herbert writes, " he always
kept up the closest connection with the friends of his
own standing; and at our frequent social gatherings—at
breakfast, wine or supper parties—he was ever one of the
most cheerful, the most popular, and the most welcome."

My father formed one of a club for dining once a
week, called the Glutton Club, the members, besides himself and Mr. Herbert (from whom I quote), being Whitley
of St. John's, now Honorary Canon of Durham; † Heavi-
side of Sydney, now Canon of Norwich; Lovett Cameron
of Trinity, sometime vicar of Shoreham; R. Blane of Trinity,‡ who held a high post during the Crimean war; H.
Lowe ⁺ (afterwards Sherbrooke) of Trinity Hall; and F.
Watkins of Emmanuel, afterwards Archdeacon of York.
The origin of the club's name seems already to have become
involved in obscurity; it certainly implied no unusual luxury in the weekly gatherings.

At any rate, the meetings seemed to have been successful, and to have ended with " a game of mild vingt-
et-un."

Mr. Herbert speaks strongly of my father's love of
music, and adds, " What gave him the greatest delight was
some grand symphony or overture of Mozart's or Beethoven's, with their full harmonies." On one occasion Herbert remembers " accompanying him to the afternoon service at King's, when we heard a very beautiful anthem.
At the end of one of the parts, which was exceedingly
impressive, he turned round to me and said, with a deep
sigh, ' How's your backbone ? ' " He often spoke in later
years of a feeling of coldness or shivering in his back on
hearing beautiful music.

---

* *Panagæus crux-major.*
† Formerly Reader in Natural Philosophy at Durham University.
‡ Blane was afterwards, I believe, in the Life Guards; he was in the Crimean War, and afterwards Military Attaché at St. Petersburg. I am indebted to Mr. Hamilton for information about some of my father's contemporaries.
# Brother of Lord Sherbrooke.

Besides a love of music, he had certainly at this time a love of fine literature; and Mr. Cameron tells me that my father took much pleasure in Shakespeare readings carried on in his rooms at Christ's. He also speaks of Darwin's " great liking for first-class line engravings, especially those of Raphael Morghen and Müller; and he spent hours in the Fitzwilliam Museum in looking over the prints in that collection."

My father's letters to Fox show how sorely oppressed he felt by the reading for an examination. His despair over mathematics must have been profound, when he expresses a hope that Fox's silence is due to " your being ten fathoms deep in the Mathematics; and if you are, God help you, for so am I, only with this difference, I stick fast in the mud at the bottom, and there I shall remain." Mr. Herbert says : " He had, I imagine, no natural turn for mathematics, and he gave up his mathematical reading before he had mastered the first part of algebra, having had a special quarrel with Surds and the Binomial Theorem."

We get some evidence from my father's letters to Fox of his intention of going into the Church. " I am glad," he writes,* " to hear that you are reading divinity. I should like to know what books you are reading, and your opinions about them; you need not be afraid of preaching to me prematurely." Mr. Herbert's sketch shows how doubts arose in my father's mind as to the possibility of his taking Orders. He writes, " We had an earnest conversation about going into Holy Orders; and I remember his asking me, with reference to the question put by the Bishop in the Ordination Service, ' Do you trust that you are inwardly moved by the Holy Spirit, &c.,' whether I could answer in the affirmative, and on my saying I could not, he said, ' Neither can I, and therefore I cannot take orders.' " This conversation appears to have taken place in 1829, and if so, the doubts here expressed must have been quieted, for in May 1830, he speaks of having some thoughts of reading divinity with Henslow.

The greater number of his Cambridge letters are addressed by my father to his cousin, William Darwin Fox. My father's letters show clearly enough how genuine the friendship was. In after years, distance, large families, and ill-health on both sides, checked the intercourse; but a

---

* March 18, 1829.

warm feeling of friendship remained. The correspondence was never quite dropped and continued till Mr. Fox's death in 1880. Mr. Fox took orders, and worked as a country clergyman until forced by ill-health to leave his living in Delamere Forest. His love of natural history was strong, and he became a skilled fancier of many kinds of birds, &c. The index to *Animals and Plants*, and my father's later correspondence, show how much help he received from his old College friend.

C. D. to J. M. Herbert. September 14, 1828.*

MY DEAR OLD CHERBURY,—I am about to fulfil my promise of writing to you, but I am sorry to add there is a very selfish motive at the bottom. I am going to ask you a great favour, and you cannot imagine how much you will oblige me by procuring some more specimens of some insects which I dare say I can describe. In the first place, I must inform you that I have taken some of the rarest of the British Insects, and their being found near Barmouth, is quite unknown to the Entomological world: I think I shall write and inform some of the crack entomologists.

But now for business. *Several* more specimens, if you can procure them without much trouble, of the following insects:—The violet-black coloured beetle, found on Craig Storm,† under stones, also a large smooth black one very like it; a bluish metallic-coloured dung-beetle, which is *very* common on the hill-sides; also, if you *would* be so very kind as to cross the ferry, and you will find a great number under the stones on the waste land of a long, smooth, jet-black beetle (a great many of these); also, in the same situation, a very small pinkish insect, with black spots, with a curved thorax projecting beyond the head; also, upon the marshy land over the ferry, near the sea, under old sea weed, stones, &c., you will find a small yellowish transparent beetle, with two or four blackish marks on the back. Under these stones there are two sorts, one much darker than the other; the lighter coloured is that which I want. These last two insects are *excessively rare*, and you will really *extremely* oblige me by taking all this trouble

---

* The postmark being Derby seems to show that the letter was written from his cousin, W. D. Fox's house, Osmaston, near Derby.

† The top of the hill immediately behind Barmouth was called Craig-Storm, a hybrid Cambro-English word.

pretty soon. Remember me most kindly to Butler,* tell
him of my success, and I dare say both of you will easily
recognise these insects. I hope his caterpillars go on well.
I think many of the Chrysalises are well worth keeping. I
really am quite ashamed [of] so long a letter all about my
own concerns; but do return good for evil, and send me a
long account of all your proceedings.

In the first week I killed seventy-five head of game—a
very contemptible number—but there are very few birds. I
killed, however, a brace of black game. Since then I have
been staying at the Fox's, near Derby; it is a very pleasant
house, and the music meeting went off very well. I want
to hear how Yates likes his gun, and what use he has made
of it.

If the bottle is not large you can buy another for me,
and when you pass through Shrewsbury you can leave these
treasures, and I hope, if you possibly can, you will stay a
day or two with me, as I hope I need not say how glad I
shall be to see you again. Fox remarked what deuced good
natured fellows your friends at Barmouth must be; and if
I did not know that you and Butler were so, I would not
think of giving you so much trouble.

In the following January we find him looking forward
with pleasure to the beginning of another year of his Cambridge life: he writes to Fox, who had passed his examination :—

"I do so wish I were now in Cambridge (a very selfish
wish, however, as I was not with you in all your troubles
and misery), to join in all the glory and happiness, which
dangers gone by can give. How we would talk, walk, and
entomologise! Sappho should be the best of bitches, and
Dash, of dogs; then should be 'peace on earth, good will to
men,'—which, by the way, I always think the most perfect
description of happiness that words can give."

Later on in the Lent term he writes to Fox :—

"I am leading a quiet everyday sort of a life; a little of
Gibbon's History in the morning, and a good deal of *Van
John* in the evening; this, with an occasional ride with
Simcox and constitutional with Whitley, makes up the regu-
lar routine of my days. I see a good deal both of Herbert
and Whitley, and the more I see of them increases every

---

* Rev. T. Butler, a son of the former head master of Shrewsbury School.

day the respect I have for their excellent understandings and dispositions. They have been giving some very gay parties, nearly sixty men there both evenings."

C. D. to W. D. Fox. Christ's College, April 1 [1829].

MY DEAR Fox—In your letter to Holden you are pleased to observe " that of all the blackguards you ever met with I am the greatest." Upon this observation I shall make no remarks, excepting that I must give you all due credit for acting on it most rigidly. And now I should like to know in what one particular are you less of a blackguard than I am? You idle old wretch, why have you not answered my last letter, which I am sure I forwarded to Clifton nearly three weeks ago? If I was not really very anxious to hear what you are doing, I should have allowed you to remain till you thought it worth while to treat me like a gentleman. And now having vented my spleen in scolding you, and having told you, what you must know, how very much and how anxiously I want to hear how you and your family are getting on at Clifton, the purport of this letter is finished. If you did but know how often I think of you, and how often I regret your absence, I am sure I should have heard from you long enough ago.

I find Cambridge rather stupid, and as I know scarcely any one that walks, and this joined with my lips not being quite so well, has reduced me to a sort of hybernation . . I have caught Mr. Harbour * letting —— have the first pick of the beetles; accordingly we have made our final adieus, my part in the affecting scene consisted in telling him he was a d—d rascal, and signifying I should kick him down the stairs if ever he appeared in my rooms again. It seemed altogether mightily to surprise the young gentleman. I have no news to tell you; indeed, when a correspondence has been broken off like ours has been, it is difficult to make the first start again. Last night there was a terrible fire at Linton, eleven miles from Cambridge. Seeing the reflection so plainly in the sky, Hall, Woodyeare, Turner, and myself thought we would ride and see it. We set out at half-past nine, and rode like incarnate devils there, and did not return till two in the morning. Altogether it was a most awful sight. I cannot conclude with-

---

* No doubt a paid collector.

out telling you, that of all the blackguards I ever met with, you are the greatest and the best.

In July 1829 he had written to Fox :—

" I must read for my Little-go. Graham smiled and bowed so very civilly, when he told me that he was one of the six appointed to make the examination stricter, and that they were determined this would make it a very different thing from any previous examination, that from all this I am sure it will be the very devil to pay amongst all idle men and entomologists."

But things were not so bad as he feared, and in March 1830, he could write to the same correspondent :—

" I am through my Little-go ! ! ! I am too much exalted to humble myself by apologising for not having written before. But I assure you before I went in, and when my nerves were in a shattered and weak condition, your injured person often rose before my eyes and taunted me with my idleness. But I am through, through, through. I could write the whole sheet full with this delightful word. I went in yesterday, and have just heard the joyful news. I shall not know for a week which class I am in. The whole examination is carried on in a different system. It has one grand advantage—being over in one day. They are rather strict, and ask a wonderful number of questions.

And now I want to know something about your plans; of course you intend coming up here : what fun we will have together; what beetles we will catch; it will do my heart good to go once more together to some of our old haunts. I have two very promising pupils in Entomology, and we will make regular campaigns into the Fens. Heaven protect the beetles and Mr. Jenyns, for we won't leave him a pair in the whole country. My new Cabinet is come down, and a gay little affair it is."

In August he was diligently amusing himself in North Wales, finding no time to write to Fox, because :—

" This is literally the first idle day I have had to myself; for on the rainy days I go fishing, on the good ones entomologising."

November found him preparing for his degree, of which process he writes dolefully :—

" I have so little time at present, and am so disgusted by reading, that I have not the heart to write to anybody. I have only written once home since I came up. This must

excuse me for not having answered your three letters, for
which I am really very much obliged. . . .

"I have not stuck an insect this term, and scarcely
opened a case. If I had time I would have sent you the insects which I have so long promised; but really I had not
spirits or time to do anything. Reading makes me quite
desperate; the plague of getting up all my subjects is next
thing to intolerable. Henslow is my tutor, and a most *admirable* one he makes; the hour with him is the pleasantest
in the whole day. I think he is quite the most perfect man
I ever met with. I have been to some very pleasant parties
there this term. His good-nature is unbounded."

The new year brought relief, and on January 23, 1831,
he wrote to tell Fox that he was through his examination.

"I do not know why the degree should make one so
miserable, both before and afterwards. I recollect you were
sufficiently wretched before, and I can assure [you], I am
now; and what makes it the more ridiculous is, I know not
what about. I believe it is a beautiful provision of nature
to make one regret the less leaving so pleasant a place as
Cambridge; and amongst all its pleasures—I say it for
once and for all—none so great, as my friendship with you.
I sent you a newspaper yesterday, in which you will see
what a good place—tenth—I have got in the Poll. As for
Christ's, did you ever see such a college for producing Captains and Apostles?* There are no men either at Emmanuel or Christ's plucked. Cameron is gulfed,† together with
other three Trinity scholars! My plans are not at all settled. I think I shall keep this term, and then go and
economise at Shrewsbury, return and take my degree.

"A man may be excused for writing so much about himself when he has just passed the examination; so you must
excuse [me]. And on the same principle do you write a
letter brimful of yourself and plans."

THE APPOINTMENT TO THE 'BEAGLE.'

In a letter addressed to Captain Fitz-Roy, before the
Beagle sailed, my father wrote, "What a glorious day the

---

* The "Captain" is at the head of the "Poll": the "Apostles" are the last
twelve in the Mathematical Tripos.
† For an explanation of the word "gulfed" or "gulphed," see Mr. W. W.
Rouse Balls' interesting *History of the Study of Mathematics at Cambridge*
(1889), p. 160.

4th of November * will be to me—my second life will then commence, and it shall be as a birthday for the rest of my life."

Foremost in the chain of circumstances which led to his appointment to the *Beagle*, was his friendship with Professor Henslow, of which the autobiography gives a sufficient account.†

An extract from a pocket-book, in which Darwin briefly recorded the chief events of his life, gives the history of his introduction to that science which was so soon to be his chief occupation—geology.

"1831. *Christmas.*—Passed my examination for B.A. degree and kept the two following terms. During these months lived much with Professor Henslow, often dining with him and walking with him; became slightly acquainted with several of the learned men in Cambridge, which much quickened the zeal which dinner parties and hunting had not destroyed. In the spring Henslow persuaded me to think of Geology, and introduced me to Sedgwick. During Midsummer geologized a little in Shropshire."

This geological work was doubtless of importance as giving him some practical experience, and perhaps of more importance in helping to give him some confidence in himself. In July of the same year, 1831, he was "working like a tiger" at Geology, and trying to make a map of Shropshire, but not finding it "as easy as I expected."

In writing to Henslow about the same time, he gives some account of his work :—

"I have been working at so many things that I have not got on much with geology. I suspect the first expedition I take, clinometer and hammer in hand, will send me back very little wiser and a good deal more puzzled than when I started. As yet I have only indulged in hypotheses, but they are such powerful ones that I suppose, if they were put into action but for one day, the world would come to an end."

He was evidently most keen to get to work with Sedgwick, who had promised to take him on a geological tour in North Wales, for he wrote to Henslow : "I have not heard from Professor Sedgwick, so I am afraid he will not pay the

---

* The *Beagle* should have started on Nov. 4, but was delayed until Dec. 27.
† See, too, a sketch by my father of his old master, in the Rev. L. Blomefield's *Memoir of Professor Henslow.*

Severn formations a visit. I hope and trust you did your best to urge him."

My father has given in his *Recollections* some account of this Tour; there too we read of the projected excursion to the Canaries.

In April 1831, he writes to Fox: "At present I talk, think, and dream of a scheme I have almost hatched of going to the Canary Islands. I have long had a wish of seeing tropical scenery and vegetation, and, according to Humboldt, Teneriffe is a very pretty specimen." And again in May: "As for my Canary scheme, it is rash of you to ask questions; my other friends most sincerely wish me there, I plague them so with talking about tropical scenery, &c. Eyton will go next summer, and I am learning Spanish."

Later on in the summer the scheme took more definite form, and the date seems to have been fixed for June 1832. He got information in London about passage-money, and in July was working at Spanish and calling Fox "un grandísimo lebron," in proof of his knowledge of the language. But even then he seems to have had some doubts about his companions' zeal, for he writes to Henslow (July 27, 1831): "I hope you continue to fan your Canary ardour. I read and re-read Humboldt;* do you do the same. I am sure nothing will prevent us seeing the Great Dragon Tree."

Geological work and Teneriffe dreams carried him through the summer, till on returning from Barmouth for the sacred 1st of September, he received the offer of appointment as Naturalist to the *Beagle*.

The following extract from the pocket-book will be a help in reading the letters:—

"Returned to Shrewsbury at end of August. Refused offer of voyage.

"*September*.—Went to Maer, returned with Uncle Jos. to Shrewsbury, thence to Cambridge. London.

"*11th*.—Went with Captain Fitz-Roy in steamer to Plymouth to see the *Beagle*.

"*22nd*.—Returned to Shrewsbury, passing through Cambridge.

"*October 2nd*.—Took leave of my home. Stayed in London.

---

* The copy of Humboldt given by Henslow to my father, which is in my possession, is a double memento of the two men—the author and the donor, who so greatly influenced his life.

" *24th.*—Reached Plymouth.

" *October and November.*—These months very miserable.

" *December 10th.*—Sailed, but were obliged to put back.

" *21st.*—Put to sea again, and were driven back.

" *27th.*—Sailed from England on our Circumnavigation."

George Peacock * *to J. S. Henslow* [1831].

MY DEAR HENSLOW—Captain Fitz-Roy is going out to survey the southern coast of Tierra del Fuego, and afterwards to visit many of the South Sea Islands, and to return by the Indian Archipelago. The vessel is fitted out expressly for scientific purposes, combined with the survey; it will furnish, therefore, a rare opportunity for a naturalist, and it would be a great misfortune that it should be lost.

An offer has been made to me to recommend a proper person to go out as a naturalist with this expedition; he will be treated with every consideration. The Captain is a young man of very pleasing manners (a nephew of the Duke of Grafton), of great zeal in his profession, and who is very highly spoken of; if Leonard Jenyns could go, what treasures he might bring home with him, as the ship would be placed at his disposal whenever his inquiries made it necessary or desirable. In the absence of so accomplished a naturalist, is there any person whom you could strongly recommend? he must be such a person as would do credit to our recommendation. Do think of this subject; it would be a serious loss to the cause of natural science if this fine opportunity was lost.

The contents of the foregoing letter were communicated to Darwin by Henslow (August 24th, 1831) :—

" I have been asked by Peacock, who will read and forward this to you from London, to recommend him a Naturalist as companion to Captain Fitz-Roy, employed by Government to survey the southern extremity of America. I have stated that I consider you to be the best qualified person I know of who is likely to undertake such a situation. I state this not in the supposition of your being a *finished* naturalist, but as amply qualified for collecting, observing,

---

* Formerly Dean of Ely, and Lowndean Professor of Astronomy at Cambridge.

and noting anything worthy to be noted in Natural History.
Peacock has the appointment at his disposal, and if he cannot find a man willing to take the office, the opportunity
will probably be lost.   Captain Fitz-Roy wants a man (I
understand) more as a companion than a mere collector,
and would not take any one, however good a naturalist, who
was not recommended to him likewise as a *gentleman.*   Particulars of salary, &c., I know nothing.   The voyage is to
last two years, and if you take plenty of books with you,
anything you please may be done.   You will have ample
opportunities at command.   In short, I suppose there never
was a finer chance for a man of zeal and spirit; Captain
Fitz-Roy is a young man.   What I wish you to do is instantly to come and consult with Peacock (at No. 7 Suffolk
Street, Pall Mall East, or else at the University Club), and
learn further particulars.   Don't put on any modest doubts
or fears about your disqualifications, for I assure you I think
you are the very man they are in search of; so conceive
yourself to be tapped on the shoulder by your bum-bailiff
and affectionate friend, J. S. HENSLOW."

On the strength of Henslow's recommendation, Peacock
offered the post to Darwin, who wrote from Shrewsbury to
Henslow (August 30, 1831) :

" Mr. Peacock's letter arrived on Saturday, and I received it late yesterday evening.   As far as my own mind is
concerned, I should, I think *certainly*, most gladly have accepted the opportunity which you so kindly have offered me.
But my father, although he does not decidedly refuse me,
gives such strong advice against going, that I should not be
comfortable if I did not follow it.

" My father's objections are these : the unfitting me to
settle down as a Clergyman, my little habit of seafaring, *the
shortness of the time*, and the chance of my not suiting
Captain Fitz-Roy.   It is certainly a very serious objection,
the very short time for all my preparations, as not only body
but mind wants making up for such an undertaking.   But
if it had not been for my father I would have taken all risks.
What was the reason that a Naturalist was not long ago
fixed upon?   I am very much obliged for the trouble you
have had about it; there certainly could not have been a
better opportunity . . . .

" Even if I was to go, my father disliking would take

away all energy, and I should want a good stock of that. Again I must thank you, it adds a little to the heavy but pleasant load of gratitude which I owe to you."

The following letter was written by Darwin from Maer, the house of his uncle Josiah Wedgwood the younger. It is plain that at first he intended to await a written reply from Dr. Darwin, and that the expedition to Shrewsbury, mentioned in the *Autobiography*, was an afterthought.

[Maer] August 31 [1831].

MY DEAR FATHER—I am afraid I am going to make you again very uncomfortable. But, upon consideration, I think you will excuse me once again stating my opinions on the offer of the voyage. My excuse and reason is the different way all the Wedgwoods view the subject from what you and my sisters do.

I have given Uncle Jos* what I fervently trust is an accurate and full list of your objections, and he is kind enough to give his opinions on all. The list and his answers will be enclosed. But may I beg of you one favour, it will be doing me the greatest kindness, if you will send me a decided answer, yes or no? If the latter, I should be most ungrateful if I did not implicitly yield to your better judgment, and to the kindest indulgence you have shown me all through my life; and you may rely upon it I will never mention the subject again. If your answer should be yes; I will go directly to Henslow and consult deliberately with him, and then come to Shrewsbury.

The danger appears to me and all the Wedgwoods not great. The expense can not be serious, and the time I do not think, anyhow, would be more thrown away than if I stayed at home. But pray do not consider that I am so bent on going that I would for one *single moment* hesitate, if you thought that after a short period you should continue uncomfortable.

I must again state I cannot think it would unfit me hereafter for a steady life. I do hope this letter will not give you much uneasiness. I send it by the car to-morrow morning; if you make up your mind directly will you send me an answer on the following day by the same means? If this letter should not find you at home, I hope you will answer as soon as you conveniently can.

---

* Josiah Wedgwood.

I do not know what to say about Uncle Jos' kindness; I never can forget how he interests himself about me.

Believe me, my dear father, your affectionate son,
CHARLES DARWIN.

Here follow the objections above referred to :—

" (1.) Disreputable to my character as a Clergyman hereafter.

" (2.) A wild scheme.

" (3.) That they must have offered to many others before me the place of Naturalist.

" (4.) And from its not being accepted there must be some serious objection to the vessel or expedition.

" (5.) That I should never settle down to a steady life hereafter.

" (6.) That my accommodations would be most uncomfortable.

" (7.) That you [*i.e.* Dr. Darwin] should consider it as again changing my profession.

" (8.) That it would be a useless undertaking.

Josiah Wedgwood having demolished this curious array of argument, and the Doctor having been converted, Darwin left home for Cambridge. On his arrival at the Red Lion he sent a messenger to Henslow with the following note (September 2nd) :—

" I am just arrived; you will guess the reason. My father has changed his mind. I trust the place is not given away.

" I am very much fatigued, and am going to bed.

" I dare say you have not yet got my second letter.

" How soon shall I come to you in the morning? Send a verbal answer."

C. D. to Miss Susan Darwin. Cambridge [September 4, 1831].

. . . . . The whole of yesterday I spent with Henslow, thinking of what is to be done, and that I find is a great deal. By great good luck I know a man of the name of Wood, nephew of Lord Londonderry. He is a great friend of Captain Fitz-Roy, and has written to him about me. I heard a part of Captain Fitz-Roy's letter, dated some time ago, in which he says: 'I have a right good set of officers, and most of my men have been there before.' It seems he has been there for the last few years; he was then second in command with the same vessel that he has now chosen. He is only twenty-three years old, but [has] seen a deal of service, and won the gold medal at Portsmouth. The Admiralty say his maps are most perfect. He had choice of two vessels, and he chose the smallest. Henslow will give me letters to all travellers in town whom he thinks may assist me.

. . . . . I write as if it was settled, but Henslow tells me *by no means* to make up my mind till I have had long conversations with Captains Beaufort and Fitz-Roy. Good-bye. You will hear from me constantly. Direct 17 Spring Gardens. *Tell nobody* in Shropshire yet. Be sure not.

I was so tired that evening I was in Shrewsbury that I thanked none of you for your kindness half so much as I felt. Love to my father.

The reason I don't want people told in Shropshire: in case I should not go, it will make it more flat.

At this stage of the transaction, a hitch occurred. Captain Fitz-Roy, it seems, wished to take a friend (Mr. Chester) as companion on the voyage, and accordingly wrote to Cambridge in such a discouraging strain, that Darwin gave up hope and hardly thought it worth his while to go to London (September 5). Fortunately, however, he did go, and found that Mr. Chester could not leave England. When the physiognomical, or nose-difficulty (Autobiography, p. 26) occurred, I have no means of knowing: for at this interview Fitz-Roy was evidently well-disposed towards him.

My father wrote:—

"He offers me to go shares in everything in his cabin if I like to come, and every sort of accommodation I can have, but they will not be numerous. He says nothing would be so miserable for him as having me with him if I was uncomfortable, as in a small vessel we must be thrown together, and thought it his duty to state everything in the worst point of view. I think I shall go on Sunday to Plymouth to see the vessel.

"There is something most extremely attractive in his manners and way of coming straight to the point. If I live with him, he says I must live poorly—no wine, and the plainest dinners. The scheme is not certainly so good as Peacock describes. Captain Fitz-Roy advises me not [to] make up my mind quite yet, but that, seriously, he thinks it will have much more pleasure than pain for me. . . .

" The want of room is decidedly the most serious objection; but Captain Fitz-Roy (probably owing to Wood's letter) seems determined to make me [as] comfortable as he possibly can. I like his manner of proceeding. He asked me at once, ' Shall you bear being told that I want the cabin to myself—when I want to be alone ? If we treat each other this way, I hope we shall suit; if not, probably we should wish each other at the devil.' "

C. D. to Miss Susan Darwin. London [September 6, 1831].

MY DEAR SUSAN—Again I am going to trouble you. I suspect, if I keep on at this rate, you will sincerely wish me at Tierra del Fuego, or any other Terra, but England. First, I will give my commissions. Tell Nancy to make me some twelve instead of eight shirts. Tell Edward to send me up in my carpet-bag (he can slip the key in the bag tied to some string), my slippers, a pair of lightish walking-shoes, my Spanish books, my new microscope (about six inches long and three or four deep), which must have cotton stuffed inside; my geological compass; my father knows that; a little book, if I have got it in my bedroom—*Taxidermy.* Ask my father if he thinks there would be any objection to my taking arsenic for a little time, as my hands are not quite well, and I have always observed that if I once get them well and change my manner of living about the same time, they will generally remain well. What is the dose ? Tell Edward my gun is dirty. What is Erasmus's direction ? Tell me if you think there is time to write and to receive an answer before I start, as I should like particularly to know what he thinks about it. I suppose you do not know Sir J. Mackintosh's direction ?

I write all this as if it was settled, but it is not more than it was, excepting that from Captain Fitz-Roy wishing me so much to go, and, from his kindness, I feel a predistination I shall start. I spent a very pleasant evening with him yesterday. He must be more than twenty-three years old ; he is of a slight figure, and a dark but handsome edition of Mr. Kynaston, and, according to my notions, pre-eminently good manners. He is all for economy, excepting on one point —viz., fire arms. He recommends me strongly to get a case of pistols like his, which cost £60!! and never to go on shore anywhere without loaded ones, and he is doubting about a rifle; he says I cannot appreciate the luxury of fresh meat here. Of course I shall buy nothing till everything is settled; but I work all day long at my lists, putting in and striking out articles. This is the first really cheerful day I have spent since I received the letter, and it all is owing to the sort of involuntary confidence I place in my *beau ideal* of a Captain.

We stop at Teneriffe. His object is to stop at as many places as possible. He takes out twenty chronometers, and it will be a "sin" not to settle the longitude. He tells me to get it down in writing at the Admiralty that I have the free choice to leave as soon and whenever I like. I daresay you expect I shall turn back at the Madeira; if I have a morsel of stomach left, I won't give up. Excuse my so often troubling and writing: the one is of great utility, the other a great amusement to me. Most likely I shall write tomorrow. Answer by return of post. Love to my father, dearest Susan.

C. D. to J. S. Henslow.　　Devonport [November 15, 1831].

MY DEAR HENSLOW—The orders are come down from the Admiralty, and everything is finally settled. We positively sail the last day of this month, and I think before that time the vessel will be ready. She looks most beautiful, even a landsman must admire her. *We* all think her the most perfect vessel ever turned out of the Dockyard. One thing is certain, no vessel has been fitted out so expensively, and with so much care. Everything that can be made so is of mahogany, and nothing can exceed the neatness and beauty of all the accommodations. The instructions are very general, and leave a great deal to the Captain's discretion and judgment, paying a substantial as well as a verbal compliment to him . . . . .

No vessel ever left England with such a set of Chronometers, viz. twenty-four, all very good ones. In short, everything is well, and I have only now to pray for the sickness to moderate its fierceness, and I shall do very well. Yet I should not call it one of the very best opportunities for natural history that has ever occurred. The absolute want of room is an evil that nothing can surmount. I think L. Jenyns did very wisely in not coming, that is, judging from my own feelings, for I am sure if I had left college some few years, or been those years older I *never* could have endured it. The officers (excepting the Captain) are like the freshest freshmen, that is in their manners, in everything else widely different. Remember me most kindly to him, and tell him if ever he dreams in the night of palm-trees, he may in the morning comfort himself with the assurance that the voyage would not have suited him.

I am much obliged for your advice, *de Mathematicis.* I suspect when I am struggling with a triangle, I shall often wish myself in your room, and as for those wicked sulky surds, I do not know what I shall do without you to conjure them. My time passes away very pleasantly. I know one or two pleasant people, foremost of whom is Mr. Thunder-and-lightning Harris,* whom I daresay you have heard of. My chief employment is to go on board the *Beagle,* and try to look as much like a sailor as I can. I have no evidence of having taken in man, woman or child.

I am going to ask you to do one more commission, and I trust it will be the last. When I was in Cambridge, I wrote to Mr. Ash, asking him to send my College account to my father, after having subtracted about £30 for my furniture. This he has forgotten to do, and my father has paid the bill, and I want to have the furniture-money transmitted to my father. Perhaps you would be kind enough to speak to Mr. Ash. I have cost my father so much money, I am quite ashamed of myself.

I will write once again before sailing, and perhaps you will write to me before then.

Believe me, yours affectionately.

C. D. to J. S. Henslow. Devonport [December 3, 1831].

MY DEAR HENSLOW—It is now late in the evening, and to-night I am going to sleep on board. On Monday we most certainly sail, so you may guess in what a desperate state of confusion we are all in. If you were to hear the various exclamations of the officers, you would suppose we had scarcely had a week's notice. I am just in the same way taken all *aback*, and in such a bustle I hardly know what to do. The number of things to be done is infinite. I look forward even to sea-sickness with something like sat-

---

* William Snow Harris, the Electrician.

isfaction, anything must be better than this state of anxiety. I am very much obliged for your last kind and affectionate letter. I always like advice from you, and no one whom I have the luck to know is more capable of giving it than yourself. Recollect, when you write, that I am a sort of *protégé* of yours, and that it is your bounden duty to lecture me.

I will now give you my direction : it is at first, Rio ; but if you will send me a letter on the first Tuesday (when the packet sails) in February, directed to Monte Video, it will give me very great pleasure : I shall so much enjoy hearing a little Cambridge news. Poor dear old *Alma Mater!* I am a very worthy son in as far as affection goes. I have little more to write about . . . I cannot end this without telling you how cordially I feel grateful for the kindness you have shown me during my Cambridge life. Much of the pleasure and utility which I may have derived from it is owing to you. I long for the time when we shall again meet, and till then believe me, my dear Henslow,

Your affectionate and obliged friend,

CH. DARWIN.

THE 'BEAGLE' LAID ASHORE, RIVER SANTA CRUZ.

CHAPTER VI.

THE VOYAGE.

"There is a natural good-humoured energy in his letters just like himself."—From a letter of Dr. R. W. Darwin's to Professor Henslow.

THE object of the *Beagle* voyage is briefly described in my father's *Journal of Researches*, p. 1, as being "to complete the Survey of Patagonia and Tierra del Fuego, commenced under Captain King in 1826 to 1830; to survey the shores of Chile, Peru, and some islands in the Pacific; and to carry a chain of chronometrical measurements round the world."

The *Beagle* is described* as a well-built little vessel, of 235 tons, rigged as a barque, and carrying six guns. She belonged to the old class of ten-gun brigs, which were nick-named "coffins," from their liability to go down in severe weather. They were very "deep-waisted," that is, their bulwarks were high in proportion to their size, so that a heavy sea breaking over them might be highly dangerous.

---

* *Voyages of the Adventure and Beagle*, vol. i. introduction xii. The illustration at the head of the chapter is from vol. ii. of the same work.

Nevertheless, she had already lived through five years' work, in the most stormy regions in the world, under Commanders Stokes and Fitz-Roy without a serious accident. When re-commissioned in 1831 for her second voyage, she was found (as I learned from the late Admiral Sir James Sulivan) to be so rotten that she had practically to be rebuilt, and it was this that caused the long delay in refitting.

She was fitted out for the expedition with all possible care : to quote my father's description, written from Devonport, November 17, 1831: " Everybody, who can judge, says it is one of the grandest voyages that has almost ever been sent out. Everything is on a grand scale. . . . In short, everything is as prosperous as human means can make it." The twenty-four chronometers and the mahogany fittings seem to have been especially admired, and are more than once alluded to.

Owing to the smallness of the vessel, every one on board was cramped for room, and my father's accommodation seems to have been narrow enough.

Yet of this confined space he wrote enthusiastically, September 17, 1831 :—" When I wrote last, I was in great alarm about my cabin. The cabins were not then marked out, but when I left they were, and mine is a capital one, certainly next best to the Captain's and remarkably light. My companion most luckily, I think, will turn out to be the officer whom I shall like best. Captain Fitz-Roy says he will take care that one corner is so fitted up that I shall be comfortable in it and shall consider it my home, but that also I shall have the run of his. My cabin is the drawing one ; and in the middle is a large table, on which we two sleep in hammocks. But for the first two months there will be no drawing to be done, so that it will be quite a luxurious room, and a good deal larger than the Captain's cabin."

My father used to say that it was the absolute necessity of tidiness in the cramped space on the Beagle that helped " to give him his methodical habits of working." On the Beagle, too, he would say, that he learned what he considered the golden rule for saving time ; i.e., taking care of the minutes.

In a letter to his sister (July 1832), he writes contentedly of his manner of life at sea :—" I do not think I have ever given you an account of how the day passes. We breakfast at eight o'clock. The invariable maxim is to throw away all politeness—that is, never to wait for each other, and bolt off the minute one has done eating, &c. At sea, when the weather is calm, I work at marine animals, with which the whole ocean abounds. If there is any sea up I am either sick or contrive to read some voyage or travels. At one we dine. You shore-going people are lamentably mistaken about the manner of living on board. We have never yet (nor shall we) dined off salt meat. Rice and peas and *calavanses* are excellent vegetables, and, with good bread, who could want more? Judge Alderson could not be more temperate, as nothing but water comes on the table. At five we have tea."

The crew of the *Beagle* consisted of Captain Fitz-Roy, "Commander and Surveyor," two lieutenants, one of whom (the first lieutenant) was the late Captain Wickham, Governor of Queensland; the late Admiral Sir James Sulivan, K.C.B., was the second lieutenant. Besides the master and two mates, there was an assistant-surveyor, the late Admiral Lort Stokes. There were also a surgeon, assistant-surgeon, two midshipmen, master's mate, a volunteer (1st class), purser, carpenter, clerks, boatswain, eight marines, thirty-four seamen, and six boys.

There are not now (1892) many survivors of my father's old ship-mates. Admiral Mellersh, and Mr. Philip King, of the Legislative Council of Sydney, are among the number. Admiral Johnson died almost at the same time as my father.

My father retained to the last a most pleasant recollection of the voyage of the *Beagle*, and of the friends he made on board her. To his children their names were familiar, from his many stories of the voyage, and we caught his feeling of friendship for many who were to us nothing more than names.

It is pleasant to know how affectionately his old companions remember him.

Sir James Sulivan remained, throughout my father's lifetime, one of his best and truest friends. He writes:— " I can confidently express my belief that during the five years in the *Beagle*, he was never known to be out of temper, or to say one unkind or hasty word *of* or *to* any one. You will therefore readily understand how this, combined with the admiration of his energy and ability, led to our giving him the name of ' the dear old Philosopher.' " *

His other nickname was " The Flycatcher." I have heard my father tell

Admiral Mellersh writes to me:—" Your father is as vividly in my mind's eye as if it was only a week ago that I was in the *Beagle* with him; his genial smile and conversation can never be forgotten by any who saw them and heard them. I was sent on two or three occasions away in a boat with him on some of his scientific excursions, and always looked forward to these trips with great pleasure, an anticipation that, unlike many others, was always realised. I think he was the only man I ever knew against whom I never heard a word said; and as people when shut up in a ship for five years are apt to get cross with each other, that is saying a good deal."

Admiral Stokes, Mr. King, Mr. Usborne, and Mr. Hamond, all speak of their friendship with him in the same warm-hearted way.

Captain Fitz-Roy was a strict officer, and made himself thoroughly respected both by officers and men. The occasional severity of his manner was borne with because every one on board knew that his first thought was his duty, and that he would sacrifice anything to the real welfare of the ship. My father writes, July 1834: " We all jog on very well together, there is no quarrelling on board, which is something to say. The Captain keeps all smooth by rowing every one in turn."

My father speaks of the officers as a fine determined set of men, and especially of Wickham, the first lieutenant, as a " glorious fellow." The latter being responsible for the smartness and appearance of the ship strongly objected to Darwin littering the decks, and spoke of specimens as " d—d beastly devilment," and used to add, " If I were skipper, I would soon have you and all your d—d mess out of the place."

A sort of halo of sanctity was given to my father by the fact of his dining in the Captain's cabin, so that the midshipmen used at first to call him " Sir," a formality, however, which did not prevent his becoming fast friends with the younger officers. He wrote about the year 1861 or 1862 to Mr. P. G. King, M.L.C., Sydney, who, as before stated, was a midshipman on board the *Beagle*:—" The remembrance of old days, when we used to sit and talk on the

---

how he overheard the boatswain of the *Beagle* showing another boatswain over the ship, and pointing out the officers: " That's our first lieutenant; that's our doctor; that's our flycatcher."

booms of the *Beagle*, will always, to the day of my death, make me glad to hear of your happiness and prosperity." Mr. King describes the pleasure my father seemed to take " in pointing out to me as a youngster the delights of the tropical nights, with their balmy breezes eddying out of the sails above us, and the sea lighted up by the passage of the ship through the never-ending streams of phosphorescent animalculæ."

It has been assumed that his ill-health in later years was due to his having suffered so much from sea-sickness. This he did not himself believe, but rather ascribed his bad health to the hereditary fault which took shape as gout in some of the past generations. I am not quite clear as to how much he actually suffered from sea-sickness ; my impression is distinct that, according to his own memory, he was not actually ill after the first three weeks, but constantly uncomfortable when the vessel pitched at all heavily. But, judging from his letters, and from the evidence of some of the officers, it would seem that in later years he forgot the extent of the discomfort. Writing June 3, 1836, from the Cape of Good Hope, he says : " It is a lucky thing for me that the voyage is drawing to its close, for I positively suffer more from sea-sickness now than three years ago."

C. D. to R. W. Darwin. [February 8, 1832.] Bahia, or San Salvador, Brazil.

> I find after the first page I have been writing to my sisters.

MY DEAR FATHER—I am writing this on the 8th of February, one day's sail past St. Jago (Cape de Verd), and intend taking the chance of meeting with a homeward-bound vessel somewhere about the equator. The date, however, will tell this whenever the opportunity occurs. I will now begin from the day of leaving England, and give a short account of our progress. We sailed, as you know, on the 27th of December, and have been fortunate enough to have had from that time to the present a fair and moderate breeze. It afterwards proved that we had escaped a heavy gale in the Channel, another at Madeira, and another on [the] Coast of Africa. But in escaping the gale, we felt its consequence—a heavy sea. In the Bay of Biscay there was a long and continuous swell, and the misery I endured from sea-sickness is far beyond what I ever guessed at. I believe you are curious about it. I will give you all my dear-bought experience. Nobody who has only been to sea for twenty-four hours has a right to say that sea-sickness is even uncomfortable. The real misery only begins when you are so exhausted that a little exertion makes a feeling of faintness come on. I found nothing but lying in my hammock did me any good. I must especially except your receipt of raisins, which is the only food that the stomach will bear.

On the 4th of January we were not many miles from Madeira, but as there was a heavy sea running, and the island lay to windward, it was not thought worth while to beat up to it. It afterwards has turned out it was lucky we saved ourselves the trouble. I was much too sick even to get up to see the distant outline. On the 6th, in the evening, we sailed into the harbour of Santa Cruz. I now first felt even moderately well, and I was picturing to myself all the delights of fresh fruit growing in beautiful valleys, and reading Humboldt's description of the island's glorious views, when perhaps you may nearly guess at our disappointment, when a small pale man informed us we must perform a strict quarantine of twelve days. There was a death-like stillness in the ship till the Captain cried "up jib," and we left this long wished-for place.

We were becalmed for a day between Teneriffe and the Grand Canary, and here I first experienced any enjoyment. The view was glorious. The Peak of Teneriffe was seen amongst the clouds like another world. Our only drawback was the extreme wish of visiting this glorious island. From Teneriffe to St. Jago the voyage was extremely pleasant. I had a net astern the vessel which caught great numbers of curious animals, and fully occupied my time in my cabin, and on deck the weather was so delightful and clear, that the sky and water together made a picture. On the 16th we arrived at Port Praya, the capital of the Cape de Verds, and there we remained twenty-three days, viz. till yesterday, the 7th of February. The time has flown away most delightfully, indeed nothing can be pleasanter; exceedingly busy, and that business both a duty and a great delight. I do not believe I have spent one half-hour idly since leaving Teneriffe. St. Jago has afforded me an exceedingly rich harvest in several branches of Natural History. I find the descriptions scarcely worth anything of many of the commoner animals that inhabit the Tropics. I allude, of course, to those of the lower classes.

Geologising in a volcanic country is most delightful; besides the interest attacked to itself, it leads you into most beautiful and retired spots. Nobody but a person fond of Natural History can imagine the pleasure of strolling under cocoa-nuts in a thicket of bananas and coffee-plants, and an endless number of wild flowers. And this island, that has given me so much instruction and delight, is reckoned the most interesting place that we perhaps shall touch at during our voyage. It certainly is generally very barren, but the valleys are more exquisitely beautiful, from the very contrast. It is utterly useless to say anything about the scenery; it would be as profitable to explain to a blind man colours, as to a person who has not been out of Europe, the total dissimilarity of a tropical view. Whenever I enjoy anything, I always either look forward to writing it down, either in my log-book (which increases in bulk), or in a letter; so you must excuse raptures, and those raptures badly expressed. I find my collections are increasing wonderfully, and from Rio I think I shall be obliged to send a cargo home.

All the endless delays which we experienced at Plymouth have been most fortunate, as I verily believe no person ever went out better provided for collecting and observing in the different branches of Natural History. In a multitude of counsellors I certainly found good. I find to my great surprise that a ship is singularly comfortable for all sorts of work. Everything is so close at hand, and being cramped makes one so methodical, that in the end I have been a gainer. I already have got to look at going to sea as a regular quiet place, like going back to home after staying away from it. In short, I find a ship a very comfortable house, with everything you want, and if it was not for sea-sickness the whole world would be sailors. I do not think there is much danger of Erasmus setting the example, but in case there should be, he may rely upon it he does not know one-tenth of the sufferings of sea-sickness.

I like the officers much more than I did at first, especially Wickham, and young King and Stokes, and indeed all of them. The Captain continues steadily very kind, and does everything in his power to assist me. We see very little of each other when in harbour, our pursuits lead us in such different tracks. I never in my life met with a man who could endure nearly so great a share of fatigue. He works incessantly, and when apparently not employed, he is thinking. If he does not kill himself, he will during this voyage do a wonderful quantity of work. . . .

February 26th.—About 280 miles from Bahia. We have been singularly unlucky in not meeting with any homeward-bound vessels, but I suppose [at] Bahia we certainly shall be able to write to England. Since writing the first part of [this] letter nothing has occurred except crossing the Equator, and being shaved. This most disagreeable operation consists in having your face rubbed with paint and tar, which forms a lather for a saw, which represents the razor, and then being half drowned in a sail filled with salt water. About 50 miles north of the line we touched at the rocks of St. Paul; this little speck (about ¼ of a mile across) in the Atlantic has seldom been visited. It is totally barren, but is covered by hosts of birds; they were so unused to men that we found we could kill plenty with stones and sticks. After remaining some hours on the island, we returned on board with the boat loaded with our prey.* From this we went to Fernando Noronha, a small island where the [Brazilians] send their exiles. The landing there was attended with so much difficulty owing [to] a heavy surf that the Captain determined to sail the next day after arriving. My one day on shore was exceedingly interesting, the whole island is one single wood so matted together by creepers that it is very difficult to move out of the beaten path. I find the Natural History of all these unfrequented spots most exceedingly interesting, especially the geology. I have written this much in order to save time at Bahia.

Decidedly the most striking thing in the Tropics is the novelty of the vegetable forms. Cocoa-nuts could well be imagined from drawings, if you add to them a graceful lightness which no European tree partakes of. Bananas and plantains are exactly the same as those in hothouses, the acacias or tamarinds are striking from the blueness of their foliage; but of the glorious orange trees, no description, no drawings, will give any just idea; instead of the sickly green of our oranges, the native ones exceed the Portugal laurel in the darkness of their tint, and infinitely exceed it in beauty of form. Cocoa-nuts, papaws, the light-

---

* There was such a scene here. Wickham (1st Lieutenant) and I were the only two who landed with guns and geological hammers, &c. The birds by myriads were too close to shoot; we then tried stones, but at last, *proh pudor!* my geological hammer was the instrument of death. We soon loaded the boat with birds and eggs. Whilst we were so engaged, the men in the boat were fairly fighting with the sharks for such magnificent fish as you could not see in the London market. Our boat would have made a fine subject for Snyders, such a medley of game it contained."—From a letter to Herbert.

green bananas, and oranges, loaded with fruit, generally surround the more luxuriant villages. Whilst viewing such scenes, one feels the impossibility that any description should come near the mark, much less be overdrawn.

March 1st.—Bahia, or San Salvador. I arrived at this place on the 28th of February, and am now writing this letter after having in real earnest strolled in the forests of the new world. No person could imagine anything so beautiful as the ancient town of Bahia, it is fairly embosomed in a luxuriant wood of beautiful trees, and situated on a steep bank, and overlooks the calm waters of the great bay of All Saints. The houses are white and lofty, and, from the windows being narrow and long, have a very light and elegant appearance. Convents, porticos, and public buildings, vary the uniformity of the houses; the bay is scattered over with large ships; in short, and what can be said more, it is one of the finest views in the Brazils. But the exquisite glorious pleasure of walking amongst such flowers, and such trees, cannot be comprehended but by those who have experienced it.* Although in so low a latitude the locality is not disagreeably hot, but at present it is very damp, for it is the rainy season. I find the climate as yet agrees admirably with me; it makes me long to live quietly for some time in such a country. If you really want to have [an idea] of tropical countries, study Humboldt. Skip the scientific parts, and commence after leaving Teneriffe. My feelings amount to admiration the more I read him. . . .

This letter will go on the 5th, and I am afraid will be some time before it reaches you; it must be a warning how in other parts of the world you may be a long time without hearing. A year might by accident thus pass. About the 12th we start for Rio, but we remain some time on the way in sounding the Albrolhos shoals. . . .

We have beat all the ships in manœuvring, so much so that the commanding officer says we need not follow his example; because we do everything better than his great ship. I begin to take great interest in naval points, more especially now, as I find they all say we are the No. 1 in South America. I suppose the Captain is a most excellent officer. It was quite glorious to-day how we beat the *Samarang* in furling sails. It is quite a new thing for a

---

* " My mind has been, since leaving England, in a perfect hurricane of delight and astonishment."—*C. D. to Fox*, May 1832, from Botofogo Bay.

" sounding ship " to beat a regular man-of-war ; and yet the *Beagle* is not at all a particular ship. Erasmus will clearly perceive it when he hears that in the night I have actually sat down in the sacred precincts of the quarter deck. You must excuse these queer letters, and recollect they are generally written in the evening after my day's work. I take more pains over my log-book, so that eventually you will have a good account of all the places I visit. Hitherto the voyage has answered *admirably* to me, and yet I am now more fully aware of your wisdom in throwing cold water on the whole scheme; the chances are so numerous of [its] turning out quite the reverse; to such an extent do I feel this, that if my advice was asked by any person on a similar occasion, I should be very cautious in encouraging him. I have not time to write to anybody else, so send to Maer to let them know, that in the midst of the glorious tropical scenery, I do not forget how instrumental they were in placing me there. I will not rapturise again, but I give myself great credit in not being crazy out of pure delight.

Give my love to every soul at home, and to the Owens.

I think one's affections, like other good things, flourish and increase in these tropical regions.

The conviction that I am walking in the New World is even yet marvellous in my own eyes, and I daresay it is little less so to you, the receiving a letter from a son of yours in such a quarter.

Believe me, my dear father, your most affectionate son.

The *Beagle* letters give ample proof of his strong love of home, and all connected with it, from his father down to Nancy, his old nurse, to whom he sometimes sends his love.

His delight in home-letters is shown in such passages as : " But if you knew the glowing, unspeakable delight, which I felt at being certain that my father and all of you were well, only four months ago, you would not grudge the labour lost in keeping up the regular series of letters."

" You would be surprised to know how entirely the pleasure in arriving at a new place depends on letters."

" I saw the other day a vessel sail for England ; it was quite dangerous to know how easily I might turn deserter. As for an English lady, I have almost forgotten what she is —something very angelic and good."

" I have just received a bundle more letters. I do not know how to thank you all sufficiently. One from Cathe-

rine, February 8th, another from Susan, March 3d, together
with notes from Caroline and from my father; give my
best love to my father. I almost cried for pleasure at receiving it; it was very kind thinking of writing to me. My
letters are both few, short, and stupid in return for all yours;
but I always ease my conscience by considering the Journal
as a long letter."

Or again—his longing to return in words like these :—
" It is too delightful to think that I shall see the leaves fall
and hear the robin sing next autumn at Shrewsbury. My
feelings are those of a schoolboy to the smallest point; I
doubt whether ever boy longed for his holidays as much as
I do to see you all again. I am at present, although nearly
half the world is between me and home, beginning to arrange what I shall do, where I shall go during the first
week."

" No schoolboys ever sung the half-sentimental and half-
jovial strain of ' dulce domum ' with more fervour than we
all feel inclined to do. But the whole subject of ' dulce
domum,' and the delight of seeing one's friends, is most
dangerous, it must infallibly make one very prosy or very
boisterous. Oh, the degree to which I long to be once again
living quietly with not one single novel object near me!
No one can imagine it till he has been whirled round the
world during five long years in a ten-gun brig."

The following extracts may serve to give an idea of the
impressions now crowding on him, as well as of the vigorous
delight with which he plunged into scientific work.

May 18, 1832, to Henslow :—
" Here [Rio], I first saw a tropical forest in all its sublime grandeur—nothing but the reality can give any idea
how wonderful, how magnificent the scene is. If I was to
specify any one thing I should give the pre-eminence to the
host of parasitical plants. Your engraving is exactly true,
but underrates rather than exaggerates the luxuriance. I
never experienced such intense delight. I formerly admired
Humboldt, I now almost adore him; he alone gives any
notion of the feelings which are raised in the mind on first
entering the Tropics. I am now collecting fresh-water and
land animals; if what was told me in London is true, viz.,
that there are no small insects in the collections from the
Tropics, I tell Entomologists to look out and have their
pens ready for describing. I have taken as minute (if not

more so) as in England, Hydropori, Hygroti, Hydrobii, Pselaphi, Stapyhlini, Curculio, &c., &c. It is exceedingly interesting observing the difference of genera and species from those which I know; it is however much less than I had expected. I am at present red-hot with spiders; they are very interesting, and if I am not mistaken I have already taken some new genera. I shall have a large box to send very soon to Cambridge, and with that I will mention some more natural history particulars."

" One great source of perplexity to me is an utter ignorance whether I note the right facts, and whether they are of sufficient importance to interest others. In the one thing collecting I cannot go wrong."

" Geology carries the day: it is like the pleasure of gambling. Speculating, on first arriving, what the rocks may be, I often mentally cry out 3 to 1 tertiary against primitive; but the latter have hitherto won all the bets. So much for the grand end of my voyage : in other respects things are equally flourishing. My life, when at sea, is so quiet, that to a person who can employ himself, nothing can be pleasanter; the beauty of the sky and brilliancy of the ocean together make a picture. But when on shore, and wandering in the sublime forests, surrounded by views more gorgeous than even Claude ever imagined, I enjoy a delight which none but those who have experienced it can understand. At our ancient snug breakfasts, at Cambridge, I little thought that the wide Atlantic would ever separate us; but it is a rare privilege that with the body, the feelings and memory are not divided. On the contrary, the pleasantest scenes in my life, many of which have been in Cambridge, rise from the contrast of the present, the more vividly in my imagination. Do you think any diamond beetle will ever give me so much pleasure as our old friend *crux-major?* . . . . It is one of my most constant amusements to draw pictures of the past ; and in them I often see you and poor little Fan. Oh, Lord, and then old Dash poor thing ! Do you recollect how you all tormented me about his beautiful tail ? "—[From a letter to Fox.]

To his sister, June 1833 :—

" I am quite delighted to find the hide of the Megatherium has given you all some little interest in my employments. These fragments are not, however, by any means the most valuable of the geological relics. I trust and believe that the time spent in this voyage, if thrown away for all other respects, will produce its full worth in Natural History; and it appears to me the doing what *little* we can to increase the general stock of knowledge is as respectable an object of life as one can in any likelihood pursue. It is more the result of such reflections (as I have already said) than much immediate pleasure which now makes me continue the voyage, together with the glorious prospect of the future, when passing the Straits of Magellan, we have in truth the world before us."

To Fox, July 1835 :—

" I am glad to hear you have some thoughts of beginning Geology. I hope you will; there is so much larger a field for thought than in the other branches of Natural History. I am become a zealous disciple of Mr. Lyell's views, as known in his admirable book. Geologising in South America, I am tempted to carry parts to a greater extent even than he does. Geology is a capital science to begin, as it requires nothing but a little reading, thinking, and hammering. I have a considerable body of notes together; but it is a constant subject of perplexity to me, whether they are of sufficient value for all the time I have spent about them, or whether animals would not have been of more certain value."

In the following letter to his sister Susan he gives an account,—adapted to the non-geological mind,—of his South American work :—

Valparaiso, April 23, 1835.

MY DEAR SUSAN,—I received, a few days since, your letter of November; the three letters which I before mentioned are yet missing, but I do not doubt they will come to life. I returned a week ago from my excursion across the Andes to Mendoza. Since leaving England I have never made so successful a journey; it has, however, been very expensive. I am sure my father would not regret it, if he could know how deeply I have enjoyed it : it was something more than enjoyment; I cannot express the delight which I felt at such a famous winding-up of all my geology in South America. I literally could hardly sleep at nights for thinking over my day's work. The scenery was so new, and so majestic; everything at an elevation of 12,000 feet bears so different an aspect from that in a lower country. I have seen many views more beautiful, but none with so strongly marked a character. To a geologist, also, there are such manifest proofs of excessive violence; the strata of the highest pinnacles are tossed about like the crust of a broken pie.

I do not suppose any of you can be much interested in geological details, but I will just mention my principal results:—Besides understanding to a certain extent the description and manner of the force which has elevated this great line of mountains, I can clearly demonstrate that one part of the double line is of an age long posterior to the other. In the more ancient line, which is the true chain of the Andes, I can describe the sort and order of the rocks which compose it. These are chiefly remarkable by containing a bed of gypsum nearly 2000 feet thick—a quantity of this substance I should think unparalleled in the world. What is of much greater consequence, I have procured fossil shells (from an elevation of 12,000 feet). I think an examination of these will give an approximate age to these mountains, as compared to the strata of Europe. In the other line of the Cordilleras there is a strong presumption (in my own mind, conviction) that the enormous mass of mountains, the peaks of which rise to 13,000 and 14,000 feet, are so very modern as to be contemporaneous with the plains of Patagonia (or about with the *upper* strata of the Isle of Wight). If this result shall be considered as proved,* it is a very important fact in the theory of the formation of the world; because, if such wonderful changes have taken place so recently in the crust of the globe, there can be no reason for supposing former epochs of excessive violence. . . . .

Another feature in his letters is the surprise and delight with which he hears of his collections and observations being of some use. It seems only to have gradually occurred to him that he would ever be more than a collector of specimens and facts, of which the great men were to make use. And even as to the value of his collections he seems to have had much doubt, for he wrote to Henslow in 1834: " I really began to think that my collections were so poor that you were puzzled what to say; the case is now quite on the opposite tack, for you are guilty of exciting all my vain feelings to a most comfortable pitch; if hard work will atone for these thoughts, I vow it shall not be spared."

---

* The importance of these results has been fully recognized by geologists.

Again, to his sister Susan in August, 1836 :—

" Both your letters were full of good news; especially
the expressions which you tell me Professor Sedgwick* used
about my collections. I confess they are deeply gratifying
—I trust one part at least will turn out true, and that I
shall act as I now think—as a man who dares to waste one
hour of time has not discovered the value of life. Professor
Sedgwick mentioning my name at all gives me hopes that
he will assist me with his advice, of which, in my geological
questions, I stand much in need."

Occasional allusions to slavery show us that his feeling
on this subject was at this time as strong as in later life. †:—

" The Captain does everything in his power to assist me,
and we get on very well, but I thank my better fortune he
has not made me a renegade to Whig principles. I would
not be a Tory, if it was merely on account of their cold
hearts about that scandal to Christian nations—Slavery."

" I have watched how steadily the general feeling, as
shown at elections, has been rising against Slavery. What
a proud thing for England if she is the first European
nation which utterly abolishes it ! I was told before leaving
England that after living in slave countries all my opinions
would be altered : the only alteration I am aware of is forming a much higher estimate of the negro character. It is
impossible to see a negro and not feel kindly towards him;
such cheerful, open, honest expressions and such fine mus-

---

* Sedgwick wrote (November 7, 1835) to Dr. Butler, the head master of
Shrewsbury School :—" He is doing admirable work in South America, and
has already sent home a collection above all price. It was the best thing in
the world for him that he went out on the voyage of discovery. There was
some risk of his turning out an idle man, but his character will now be fixed,
and if God spares his life he will have a great name among the naturalists of
Europe. . ."—I am indebted to my friend Mr. J. W. Clark, the biographer of
Sedgwick, for the above extract.

† Compare the following passage from a letter (Aug. 25, 1845) addressed to
Lyell, who had touched on slavery in his *Travels in North America.* " I was
delighted with your letter in which you touch on Slavery ; I wish the same
feelings had been apparent in your published discussion. But I will not write
on this subject, I should perhaps annoy you, and most certainly myself. I
have exhaled myself with a paragraph or two in my Journal on the sin of
Brazilian slavery ; you perhaps will think that it is in answer to you; but
such is not the case. I have remarked on nothing which I did not hear on
the coast of South America. My few sentences, however, are merely an explosion of feeling. How could you relate so placidly that atrocious sentiment
about separating children from their parents ; and in the next page speak of
being distressed at the whites not having prospered ; I assure you the contrast
made me exclaim out. But I have broken my intention, and so no more on
this odious deadly subject." It is fair to add that the " atrocious sentiments "
were not Lyell's but those of a planter.

cular bodies. I never saw any of the diminutive Portuguese, with their murderous countenances, without almost wishing for Brazil to follow the example of Hayti; and, considering the enormous healthy-looking black population, it will be wonderful if, at some future day, it does not take place. There is at Rio a man (I know not his title) who has a large salary to prevent (I believe) the landing of slaves; he lives at Botofogo, and yet that was the bay where, during my residence, the greater number of smuggled slaves were landed.   Some of the Anti-Slavery people ought to question about his office; it was the subject of conversation at Rio amongst the lower English. . .

C. D. to J. S. Henslow.   Sydney [January, 1836].

MY DEAR HENSLOW—This is the last opportunity of communicating with you before that joyful day when I shall reach Cambridge. I have very little to say: but I must write if it is only to express my joy that the last year is concluded, and that the present one, in which the *Beagle* will return, is gliding onward. We have all been disappointed here in not finding even a single letter; we are, indeed, rather before our expected time, otherwise, I dare say, I should have seen your handwriting. I must feed upon the future, and it is beyond bounds delightful to feel the certainty that within eight months I shall be residing once again most quietly in Cambridge. Certainly, I never was intended for a traveller; my thoughts are always rambling over past or future scenes; I cannot enjoy the present happiness for anticipating the future, which is about as foolish as the dog who dropped the real bone for its shadow. . . .

I must return to my old resource and think of the future, but that I may not become more prosy, I will say farewell till the day arrives, when I shall see my Master in Natural History, and can tell him how grateful I feel for his kindness and friendship.

Believe me, dear Henslow, ever yours most faithfully.

C. D. to J. S. Henslow.   Shrewsbury [October 6, 1836].

MY DEAR HENSLOW—I am sure you will congratulate me on the delight of once again being home. The *Beagle* arrived at Falmouth on Sunday evening, and I reached Shrewsbury yesterday morning. I am exceedingly anxious to see you, and as it will be necessary in four or five days to return to London to get my goods and chattels out of the *Beagle*, it appears to me my best plan to pass through Cambridge. I want your advice on many points; indeed I am in the clouds, and neither know what to do or where to go. My chief puzzle is about the geological specimens—who will have the charity to help me in describing the mineralogical nature? Will you be kind enough to write to me one line by *return of post*, saying whether you are now at Cambridge? I am doubtful till I hear from Captain Fitz-Roy whether I shall not be obliged to start before the answer can arrive, but pray try the chance. My dear Henslow, I do long to see you; you have been the kindest friend to me that ever man possessed. I can write no more, for I am giddy with joy and confusion.

Farewell for the present,
Yours most truly obliged.

After his return and settlement in London, he began to realise the value of what he had done, and wrote to Captain Fitz-Roy—" However others may look back to the *Beagle's* voyage, now that the small disagreeable parts are well-nigh forgotten, I think it far the *most fortunate circumstance in my life* that the chance afforded by your offer of taking a Naturalist fell on me. I often have the most vivid and delightful pictures of what I saw on board the *Beagle* * pass before my eyes. These recollections, and what I learnt on Natural History, I would not exchange for twice ten thousand a year."

---

* According to the *Japan Weekly Mail*, as quoted in *Nature*, March 8, 1888, the *Beagle* is in use as a training ship at Yokosuka in Japan. Part of the old ship is, I am glad to think, in my possession, in the form of a box (which I owe to the kindness of Admiral Mellersh) made out of her main cross-tree.

CHAPTER VII.

1836–1842.

THE period illustrated in the present chapter includes the years between Darwin's return from the voyage of the *Beagle* and his settling at Down. It is marked by the gradual appearance of that weakness of health which ultimately forced him to leave London and take up his abode for the rest of his life in a quiet country house.

There is no evidence of any intention of entering a profession after his return from the voyage, and early in 1840 he wrote to Fitz-Roy: " I have nothing to wish for excepting stronger health to go on with the subjects to which I have joyfully determined to devote my life."

These two conditions—permanent ill-health and a passionate love of scientific work for its own sake—determined thus early in his career, the character of his whole future life. They impelled him to lead a retired life of constant labour, carried on to the utmost limits of his physical power, a life which signally falsified his melancholy prophecy:—" It has been a bitter mortification for me to digest the conclusion that the ' race is for the strong,' and that I shall probably do little more, but be content to admire the strides others make in science."

The end of the last chapter saw my father safely arrived at Shrewsbury on October 4, 1836, " after an absence of five years and two days." He wrote to Fox: " You cannot imagine how gloriously delightful my first visit was at home ; it was worth the banishment." But it was a pleasure that he could not long enjoy, for in the last days of October he was at Greenwich unpacking specimens from the *Beagle.* As to the destination of the collections he writes, somewhat despondingly, to Henslow :—

"I have not made much progress with the great men. I find, as you told me, that they are all overwhelmed with their own business. Mr. Lyell has entered, in the *most* good-natured manner, and almost without being asked, into all my plans. He tells me, however, the same story, that I must do all myself. Mr. Owen seems anxious to dissect some of the animals in spirits, and, besides these two, I have scarcely met any one who seems to wish to possess any of my specimens. I must except Dr. Grant, who is willing to examine some of the corallines. I see it is quite unreasonable to hope for a minute that any man will undertake the examination of a whole order. It is clear the collectors so much outnumber the real naturalists that the latter have no time to spare.

"I do not even find that the Collections care for receiving the unnamed specimens. The Zoological Museum * is nearly full, and upwards of a thousand specimens remain unmounted. I dare say the British Museum would receive them, but I cannot feel, from all I hear, any great respect even for the present state of that establishment. Your plan will be not only the best, but the only one, namely, to come down to Cambridge, arrange and group together the different families, and then wait till people, who are already working in different branches, may want specimens. . . .

"I have forgotten to mention Mr. Lonsdale,† who gave me a most cordial reception, and with whom I had much most interesting conversation. If I was not much more inclined for geology than the other branches of Natural History, I am sure Mr. Lyell's and Lonsdale's kindness ought to fix me. You cannot conceive anything more thoroughly good-natured than the heart-and-soul manner in which he put himself in my place and thought what would be best to do."

A few days later he writes more cheerfully : "I became acquainted with Mr. Bell,‡ who, to my surprise, expressed a good deal of interest about my crustacea and reptiles, and

---

* The Museum of the Zoological Society, then at 33 Bruton Street. The collection was some years later broken up and dispersed.

† William Lonsdale, b. 1794, d. 1871, was originally in the army, and served at the battles of Salamanca and Waterloo. After the war he left the service and gave himself up to science. He acted as assistant-secretary to the Geological Society from 1829–42, when he resigned, owing to ill-health.

‡ T. Bell, F.R.S., formerly Professor of Zoology in King's College, London, and sometime secretary to the Royal Society. He afterwards described the reptiles for the *Zoology of the Voyage of the Beagle.*

seems willing to work at them. I also heard that Mr.
Broderip would be glad to look over the South American
shells, so that things flourish well with me."

Again, on November 6 :—

"All my affairs, indeed, are most prosperous; I find
there are plenty who will undertake the description of
whole tribes of animals, of which I know nothing."

As to his Geological Collection he was soon able to write :
" I [have] disposed of the most important part [of] my collections, by giving all the fossil bones to the College of Sur-
geons, casts of them will be distributed, and descriptions
published.   They are very curious and valuable; one head
belongèd to some gnawing animal, but of the size of a Hip-
popotamus !   Another to an ant-eater of the size of a
horse ! "

My father's specimens included (besides the above-mentioned Toxodon and Scelidotherium) the remains of Mylo-
don, Glossotherium, another gigantic animal allied to the
ant-eater, and Macrauchenia. His discovery of these remains
is a matter of interest in itself, but it has a special importance as a point in his own life, his speculation on the extinction of these extraordinary creatures * and on their relationship to living forms having formed one of the chief
starting-points of his views on the origin of species. This
is shown in the following extract from his Pocket Book
for this year (1837): " In July opened first note-book
on Transmutation of Species.   Had been greatly struck
from about the month of previous March on character
of South American fossils, and species on Galapagos
Arhipelago. These facts (especially latter), origin of all
my views."

His affairs being thus far so prosperously managed he
was able to put into execution his plan of living at Cambridge, where he settled on December 10th, 1836.

"Cambridge," he writes, "·yet continues a very pleasant
but not half so merry a place as before.   To walk through
the courts of Christ's College, and not know an inhabitant
of a single room, gave one a feeling half melancholy.   The
only evil I found in Cambridge was its being too pleasant:
there was some agreeable party or another every evening,

---

* I have often heard him speak of the despair with which he had to break
off the projecting extremity of a huge, partly excavated bone, when the boat
waiting for him would wait no longer.

and one cannot say one is engaged with so much impunity there as in this great city." *

Early in the spring of 1837 he left Cambridge for London, and a week later he was settled in lodgings at 36 Great Marlborough Street; and except for a "short visit to Shrewsbury" in June, he worked on till September, being almost entirely employed on his *Journal*, of which he wrote (March) :—

"In your last letter you urge me to get ready *the* book. I am now hard at work and give up everything else for it. Our plan is as follows: Capt. Fitz-Roy writes two volumes out of the materials collected during the last voyage under Capt. King to Tierra del Fuego, and during our circumnavigation. I am to have the third volume, in which I intend giving a kind of journal of a naturalist, not following, however, always the order of time, but rather the order of position."

A letter to Fox (July) gives an account of the progress of his work :—

"I gave myself a holiday and a visit to Shrewsbury [in June], as I had finished my Journal. I shall now be very busy in filling up gaps and getting it quite ready for the press by the first of August. I shall always feel respect for every one who has written a book, let it be what it may, for I had no idea of the trouble which trying to write common English could cost one. And, alas, there yet remains the worst part of all, correcting the press. As soon as ever that is done I must put my shoulder to the wheel and commence at the Geology. I have read some short papers to the Geological Society, and they were favourably received by the great guns, and this gives me much confidence, and I hope not a very great deal of vanity, though I confess I feel too often like a peacock admiring his tail. I never expected that my Geology would ever have been worth the consid-

---

* A trifling record of my father's presence in Cambridge occurs in the book kept in Christ's College Combination-room, in which fines and bets are recorded, the earlier entries giving a curious impression of the after-dinner frame of mind of the Fellows. The bets are not allowed to be made in money, but are, like the fines, paid in wine. The bet which my father made and lost is thus recorded :

"*Feb.* 23, 1837.—Mr. Darwin *v.* Mr. Baines, that the combination-room measures from the ceiling to the floor more than x feet.

"1 Bottle paid same day."

The bets are usually recorded in such a way as not to preclude future speculation on a subject which has proved itself capable of supplying a discussion (and a bottle) to the Room, hence the x in the above quotation.

eration of such men as Lyell, who has been to me, since my return, a most active friend. My life is a very busy one at present, and I hope may ever remain so; though Heaven knows there are many serious drawbacks to such a life, and chief amongst them is the little time it allows one for seeing one's natural friends. For the last three years, I have been longing and longing to be living at Shrewsbury, and after all now in the course of several months, I see my good dear people at Shrewsbury for a week. Susan and Catherine have, however, been staying with my brother here for some weeks, but they had returned home before my visit."

In August he writes to Henslow to announce the success of the scheme for the publication of the *Zoology of the Voyage of the Beagle*, through the promise of a grant of £1000 from the Treasury : " I had an interview with the Chancellor of the Exchequer.* He appointed to see me this morning, and I had a long conversation with him, Mr. Peacock being present. Nothing could be more thoroughly obliging and kind than his whole manner. He made no sort of restriction, but only told me to make the most of the money, which of course I am right willing to do.

" I expected rather an awful interview, but I never found anything less so in my life. It will be my fault if I do not make a good work; but I sometimes take an awful fright that I have not materials enough. It will be excessively satisfactory at the end of some two years to find all materials made the most they were capable of."

Later in the autumn he wrote to Henslow : " I have not been very well of late, with an uncomfortable palpitation of the heart, and my doctors urge me *strongly* to knock off all work, and go and live in the country for a few weeks." He accordingly took a holiday of about a month at Shrewsbury and Maer, and paid Fox a visit in the Isle of Wight. It was, I believe, during this visit, at Mr. Wedgwood's house at Maer, that he made his first observations on the work done by earth-worms, and late in the autumn he read a paper on the subject at the Geological Society.

Here he was already beginning to make his mark. Lyell wrote to Sedgwick (April 21, 1837) :—

" Darwin is a glorious addition to any society of geologists, and is working hard and making way both in his book and in our discussions. I really never saw that bore

---

* Spring Rice.

Dr. Mitchell so successfully silenced, or such a bucket of cold water so dexterously poured down his back, as when Darwin answered some impertinent and irrelevant questions about South America. We escaped fifteen minutes of Dr. M.'s vulgar harangue in consequence. . . ."

Early in the following year (1838), he was, much against his will, elected Secretary of the Geological Society, an office he held for three years. A chief motive for his hesitation in accepting the post was the condition of his health, the doctors having urged "me to give up entirely all writing and even correcting press for some weeks. Of late anything which flurries me completely knocks me up afterwards, and brings on a violent palpitation of the heart."

In the summer of 1838 he started on his expedition to Glen Roy, where he spent "eight good days" over the Parallel Roads. His Essay on this subject was written out during the same summer, and published by the Royal Society.* He wrote in his Pocket Book : "September 6 (1838). Finished the paper on 'Glen Roy,' one of the most difficult and instructive tasks I was ever engaged on." It will be remembered that in his *Autobiography* he speaks of this paper as a failure, of which he was ashamed.†

C. D. to Lyell. [August 9th, 1838.]

36 Great Marlborough Street.

MY DEAR LYELL—I did not write to you at Norwich, for I thought I should have more to say, if I waited a few more days. Very many thanks for the present of your *Elements*, which I received (and I believe the *very first* copy distributed) together with your note. I have read it through every word, and am full of admiration of it, and, as I now see no geologist, I must talk to you about it. There is no pleasure in reading a book if one cannot have a good talk over it; I repeat, I am full of admiration of it, it is as

---

* *Phil. Trans.*, 1839, pp. 39–82.

† Sir Archibald Geikie has been so good as to allow me to quote a passage from a letter addressed to me (Nov. 19, 1884) :—" Had the idea of transient barriers of glacier-ice occurred to him, he would have found the difficulties vanish from the lake-theory which he opposed, and he would not have been unconsciously led to minimise the altogether overwhelming objections to the supposition that the terraces are of marine origin."
It may be added that the idea of the barriers being formed by glaciers could hardly have occurred to him, considering the state of knowledge at the time, and bearing in mind his want of opportunities of observing glacial action on a large scale.

clear as daylight, in fact I felt in many parts some mortification at thinking how geologists have laboured and strug-
gled at proving what seems, as you have put it, so evidently
probable.   I read with much interest your sketch of the
secondary deposits; you have contrived to make it quite
"juicy," as we used to say as children of a good story.
There was also much new to me, and I have to copy out
some fifty notes and references.   It must do good, the here-
tics against common-sense must yield. . . . By the way, do
you recollect my telling you how much I disliked the manner X. referred to his other works, as much as to say, " You
must, ought, and shall buy everything I have written."
To my mind, you have somehow quite avoided this; your
references only seem to say, "I can't tell you all in this
work, else I would, so you must go to the *Principles;*" and
many a one, I trust, you will send there, and make them,
like me, adorers of the good science of rock-breaking.*
You will see I am in a fit of enthusiasm, and good cause I
have to be, when I find you have made such infinitely
more use of my Journal than I could have anticipated.   I
will say no more about the book, for it is all praise.   I must,
however, admire the elaborate honesty with which you quote
the words of all living and dead geologists.

My Scotch expedition answered brilliantly; my trip in
the steam-packet was absolutely pleasant, and I enjoyed the
spectacle, wretch that I am, of two ladies, and some small
children quite sea-sick, I being well.   Moreover, on my return from Glasgow to Liverpool, I triumphed in a similar
manner over some full-grown men.   I stayed one whole
day in Edinburgh, or more truly on Salisbury Craigs; I
want to hear some day what you think about that classical
ground,—the structure was to me new and rather curious,—
that is, if I understand it right.   I crossed from Edinburgh
in gigs and carts (and carts without springs, as I never shall
forget) to Loch Leven.   I was disappointed in the scenery,
and reached Glen Roy on Saturday evening, one week after
leaving Marlborough Street.   Here I enjoyed five [?] days
of the most beautiful weather with gorgeous sunsets, and

---

* In a letter of Sept. 13 he wrote :—" It will be a curious point to geologists
hereafter to note how long a man's name will support a theory so completely
exposed as that of De Beaumont has been by you; you say you ' begin to
hope that the great principles there insisted on will stand the test of time.'
Begin to hope: why, the *possibility* of a doubt has never crossed my mind for
many a day.   This may be very unphilosophical, but my geological salvation
is staked on it."

all nature looking as happy as I felt.   I wandered over the
mountains in all directions, and examined that most extraordinary district.   I think, without any exceptions, not even
the first volcanic island, the first elevated beach, or the
passage of the Cordillera, was so interesting to me as this
week.   It is far the most remarkable area I ever examined.
I have fully convinced myself (after some doubting at first)
that the shelves are sea-beaches, although I could not find
a trace of a shell; and I think I can explain away most, if
not all, the difficulties.   I found a piece of a road in another
valley, not hitherto observed, which is important; and I
have some curious facts about erratic blocks, one of which
was perched up on a peak 2200 feet above the sea.   I am
now employed in writing a paper on the subject, which I
find very amusing work, excepting that I cannot anyhow
condense it into reasonable limits.   At some future day I
hope to talk over some of the conclusions with you, which
the examination of Glen Roy has led me to.   Now I have
had my talk out, I am much easier, for I can assure you
Glen Roy has astonished me.

I am living very quietly, and therefore pleasantly, and
am crawling on slowly but steadily with my work.   I have
come to one conclusion, which you will think proves me
to be a very sensible man, namely, that whatever you say
proves right; and as a proof of this, I am coming into your
way of only working about two hours at a spell; I then go
out and do my business in the streets, return and set to work
again, and thus make two separate days out of one.   The
new plan answers capitally; after the second half day is
finished, I go and dine at the Athenæum like a gentleman,
or rather like a lord, for I am sure the first evening I sat in
that great drawing-room, all on a sofa by myself, I felt just
like a duke.   I am full of admiration at the Athenæum, one
meets so many people there that one likes to see. . . .

I have heard from more than one quarter that quarrelling is expected at Newcastle *; I am sorry to hear it.   I
met old —— this evening at the Athenæum, and he mut-
tered something about writing to you or some one on the
subject; I am, however, all in the dark.   I suppose, however, I shall be illuminated, for I am going to dine with him
in a few days, as my inventive powers failed in making any
excuse.   A friend of mine dined with him the other day, a

---

* At the meeting of the British Association.

party of four, and they finished ten bottles of wine—a pleasant prospect for me; but I am determined not even to taste
his wine, partly for the fun of seeing his infinite disgust and
surprise. . . .

I pity you the infliction of this most unmerciful letter.
Pray remember me most kindly to Mrs. Lyell when you ar-
rive at Kinnordy. Tell Mrs. Lyell to read the second series
of 'Mr. Slick of Slickville's Sayings.' . . . He almost beats
'Samivel,' that prince of heroes. Good night, my dear
Lyell; you will think I have been drinking some strong
drink to write so much nonsense, but I did not even taste
Minerva's small beer to-day. . . .

A record of what he wrote during the year 1838 would
not give a true index of the most important work that was
in progress—the laying of the foundation-stones of what
was to be the achievement of his life. This is shown in the
following passages from a letter to Lyell (September), and
from a letter to Fox, written in June :—

" I wish with all my heart that my Geological book was
out. I have every motive to work hard, and will, following
your steps, work just that degree of hardness to keep well.
I should like my volume to be out before your new edition
of the *Principles* appears. Besides the Coral theory, the
volcanic chapters will, I think, contain some new facts. I
have lately been sadly tempted to be idle—that is, as far as
pure geology is concerned—by the delightful number of new
views which have been coming in thickly and steadily—on
the classification and affinities and instincts of animals—
bearing on the question of species. Note-book after note-
book has been filled with facts which begin to group themselves *clearly* under sub-laws."

" I am delighted to hear you are such a good man as not
to have forgotten my questions about the crossing of animals. It is my prime hobby, and I really think some day I
shall be able to do something in that most intricate subject,
species and varieties."

In the winter of 1839 (Jan. 29) my father was married
to his cousin, Emma Wedgwood.* The house in which
they lived for the first few years of their married life, No.
12 Upper Gower Street, was a small common-place London

---

* Daughter of Josiah Wedgwood of Maer, and grand-daughter of the
founder of the Etruria Pottery Works.

house, with a drawing-room in front, and a small room behind, in which they lived for the sake of quietness. In later years my father used to laugh over the surpassing ugliness of the furniture, carpets, &c., of the Gower Street house. The only redeeming feature was a better garden than most London houses have, a strip as wide as the house, and thirty yards long. Even this small space of dingy grass made their London house more tolerable to its two country-bred inhabitants.

Of his life in London he writes to Fox (October 1839) : "We are living a life of extreme quietness; Delamere itself, which you describe as so secluded a spot, is, I will answer for it, quite dissipated compared with Gower Street. We have given up all parties, for they agree with neither of us ; and if one is quiet in London, there is nothing like its quietness—there is a grandeur about its smoky fogs, and the dull distant sounds of cabs and coaches ; in fact you may perceive I am becoming a thorough-paced Cockney, and I glory in the thought that I shall be here for the next six months."

The entries of ill health in the Diary increase in number during these years, and as a consequence the holidays become longer and more frequent.

The entry under August 1839 is : "Read a little, was much unwell and scandalously idle. I have derived this much good, that *nothing* is so intolerable as idleness."

At the end of 1839 his first child was born, and it was then that he began his observations ultimately published in the *Expression of the Emotions.* His book on this subject, and the short paper published in *Mind*,* show how closely he observed his child. He seems to have been surprised at his own feeling for a young baby, for he wrote to Fox (July 1840) : "He [*i.e.* the baby] is so charming that I cannot pretend to any modesty. I defy anybody to flatter us on our baby, for I defy anyone to say anything in its praise of which we are not fully conscious. . . . I had not the smallest conception there was so much in a five-month baby. You will perceive by this that I have a fine degree of paternal fervour."

In 1841 some improvement in his health became apparent; he wrote in September :—

"I have steadily been gaining ground, and really believe now I shall some day be quite strong. I write daily for a

---

* July 1877.

couple of hours on my Coral volume, and take a little walk or ride every day. I grow very tired in the evenings, and am not able to go out at that time, or hardly to receive my nearest relations ; but my life ceases to be burdensome now that I can do something."

The manuscript of *Coral Reefs* was at last sent to the printers in January 1842, and the last proof corrected in May. He thus writes of the work in his diary :—

"I commenced this work three years and seven months ago. Out of this period about twenty months (besides work during *Beagle's* voyage) has been spent on it, and besides it, I have only compiled the Bird part of Zoology ; Appendix to Journal, paper on Boulders, and corrected papers on Glen Roy and earthquakes, reading on species, and rest all lost by illness."

The latter part of this year belongs to the period including the settlement at Down, and is therefore dealt with in another chapter.

CHAPTER VIII.

LIFE AT DOWN.

1842–1854.

" My life goes on like clockwork, and I am fixed on the spot where I shall end it."

Letter to Captain Fitz-Roy, October, 1846.

CERTAIN letters which, chronologically considered, belong to the period 1845–54 have been utilised in a later chapter where the growth of the *Origin of Species* is described. In the present chapter we only get occasional hints of the growth of my father's views, and we may suppose ourselves to be seeing his life, as it might have appeared to those who had no knowledge of the quiet development of his theory of evolution during this period.

On September 14, 1842, my father left London with his family and settled at Down.* In the Autobiographical chapter, his motives for moving into the country are briefly given. He speaks of the attendance at scientific societies and ordinary social duties as suiting his health so " badly that we resolved to live in the country, which we both preferred and have never repented of." His intention of keeping up with scientific life in London is expressed in a letter to Fox (Dec., 1842) :—

" I hope by going up to town for a night every fortnight or three weeks, to keep up my communication with scientific men and my own zeal, and so not to turn into a complete Kentish hog."

Visits to London of this kind were kept up for some years at the cost of much exertion on his part. I have

---

* I must not omit to mention a member of the household who accompanied him. This was his butler, Joseph Parslow, who remained in the family, a valued friend and servant, for forty years, and became, as Sir Joseph Hooker once remarked to me, " an integral part of the family, and felt to be such by all visitors at the house."

often heard him speak of the wearisome drives of ten miles
to or from Croydon or Sydenham—the nearest stations—
with an old gardener acting as coachman, who drove with
great caution and slowness up and down the many hills. In
later years, regular scientific intercourse with London became, as before mentioned, an impossibility.

The choice of Down was rather the result of despair than
of actual preference : my father and mother were weary of
house-hunting, and the attractive points about the place
thus seemed to them to counterbalance its somewhat more
obvious faults. It had at least one desideratum, namely,
quietness. Indeed it would have been difficult to find a
more retired place so near to London. In 1842 a coach
drive of some twenty miles was the usual means of access to
Down ; and even now that railways have crept closer to it,
it is singularly out of the world, with nothing to suggest
the neighbourhood of London, unless it be the dull haze of
smoke that sometimes clouds the sky. The village stands
in an angle between two of the larger high-roads of the
country, one leading to Tunbridge and the other to Westerham and Edenbridge. It is cut off from the Weald by a
line of steep chalk hills on the south, and an abrupt hill,
now smoothed down by a cutting and embankment, must
formerly have been something of a barrier against enroach-
ments from the side of London. In such a situation, a village, communicating with the main lines of traffic, only by
stony tortuous lanes, may well have preserved its retired
character. Nor is it hard to believe in the smugglers and
their strings of pack-horses making their way up from the
lawless old villages of the Weald, of which the memory still
existed when my father settled in Down. The village stands
on solitary upland country, 500 to 600 feet above the sea—a
country with little natural beauty, but possessing a certain
charm in the shaws, or straggling strips of wood, capping
the chalky banks and looking down upon the quiet ploughed
lands of the valleys. The village, of three or four hundred
inhabitants, consists of three small streets of cottages, meeting in front of the little flint-built church. It is a place
where new-comers are seldom seen, and the names occurring
far back in the old church registers are still known in the
village. The smock-frock is not yet quite extinct, though
chiefly used as a ceremonial dress by the " bearers " at
funerals ; but as a boy I remember the purple or green
smocks of the men at church.

The house stands a quarter of a mile from the village, and is built, like so many houses of the last century, as near as possible to the road—a narrow lane winding away to the Westerham high-road. In 1842, it was dull and unattractive enough: a square brick building of three storeys, covered with shabby whitewash, and hanging tiles. The garden had none of the shrubberies or walls that now give shelter; it was overlooked from the lane, and was open, bleak, and desolate. One of my father's first undertakings was to lower the lane by about two feet, and to build a flint wall along that part of it which bordered the garden. The earth thus excavated was used in making banks and mounds round the lawn: these were planted with evergreens, which now give to the garden its retired and sheltered character.

The house was made to look neater by being covered with stucco, but the chief improvement effected was the building of a large bow extending up through three storeys. This bow became covered with a tangle of creepers, and pleasantly varied the south side of the house. The drawing-room, with its verandah opening into the garden, as well as the study in which my father worked during the later years of his life, were added at subsequent dates.

Eighteen acres of land were sold with the house, of which twelve acres on the south side of the house form a pleasant field, scattered with fair-sized oaks and ashes. From this field a strip was cut off and converted into a kitchen garden in which the experimental plot of ground was situated, and where the greenhouses were ultimately put up.

During the whole of 1843 he was occupied with geological work, the result of which was published in the spring of the following year. It was entitled *Geological Observations on the Volcanic Islands, visited during the Voyage of H.M.S. Beagle, together with some brief notices on the geology of Australia and the Cape of Good Hope;* it formed the second part of the *Geology of the Voyage of the Beagle*, published "with the Approval of the Lords Commissioners of Her Majesty's Treasury." The volume on *Coral Reefs* forms Part I. of the series, and was published, as we have seen, in 1842. For the sake of the non-geological reader, I may here quote Sir A. Geikie's words * on these two volumes —which were up to this time my father's chief geological works. Speaking of the *Coral Reefs*, he says (p. 17): "This

---

* Charles Darwin, *Nature* Series, 1882.

well-known treatise, the most original of all its author's geological memoirs, has become one of the classics of geological literature. The origin of those remarkable rings of coral-rock in mid-ocean has given rise to much speculation, but no satisfactory solution of the problem had been proposed. After visiting many of them, and examining also coral reefs that fringe islands and continents, he offered a theory which for simplicity and grandeur, strikes every reader with astonishment. It is pleasant, after the lapse of many years, to recall the delight with which one first read the *Coral Reefs*, how one watched the facts being marshalled into their places, nothing being ignored or passed lightly over; and how, step by step, one was led to the grand conclusion of wide oceanic subsidence. No more admirable example of scientific method was ever given to the world, and even if he had written nothing else, the treatise alone would have placed Darwin in the very front of investigators of nature."

It is interesting to see in the following extract from one of Lyell's letters * how warmly and readily he embraced the theory. The extract also gives incidentally some idea of the theory itself.

"I am very full of Darwin's new theory of Coral Islands, and have urged Whewell to make him read it at our next meeting. I must give up my volcanic crater theory for ever, though it cost me a pang at first, for it accounted for so much, the annular form, the central lagoon, the sudden rising of an isolated mountain in a deep sea; all went so well with the notion of submerged, crateriform, and conical volcanoes, . . . and then the fact that in the South Pacific we had scarcely any rocks in the regions of coral islands, save two kinds, coral limestone and volcanic! Yet in spite of all this, the whole theory is knocked on the head, and the annular shape and central lagoon have nothing to do with volcanoes, nor even with a crateriform bottom. Perhaps Darwin told you when at the Cape what he considers the true cause? Let any mountain be submerged gradually, and coral grow in the sea in which it is sinking, and there will be a ring of coral, and finally only a lagoon in the centre. . . . Coral islands are the last efforts of drowning continents to lift their heads above water. Regions of elevation

---

* To Sir John Herschel, May 24, 1837. *Life of Sir Charles Lyell*, vol. ii. p. 12.

and subsidence in the ocean may be traced by the state of the coral reefs."

The second part of the *Geology of the Voyage of the Beagle*, *i.e.* the volume on Volcanic Islands, which specially concerns us now, cannot be better described than by again quoting from Sir A. Geikie (p. 18) :—

" Full of detailed observations, this work still remains the best authority on the general geological structure of most of the regions it describes. At the time it was written the ' crater of elevation theory,' though opposed by Constant Prévost, Scrope, and Lyell, was generally accepted, at least on the Continent. Darwin, however, could not receive it as a valid explanation of the facts; and though he did not share the view of its chief opponents, but ventured to propose a hypothesis of his own, the observations impartially made and described by him in this volume must be regarded as having contributed towards the final solution of the difficulty." Geikie continues (p. 21) : " He is one of the earliest writers to recognize the magnitude of the denudation to which even recent geological accumulations have been subjected. One of the most impressive lessons to be learnt from his account of ' Volcanic Islands ' is the prodigious extent to which they have been denuded. . . . He was disposed to attribute more of this work to the sea than most geologists would now admit; but he lived himself to modify his original views, and on this subject his latest utterances are quite abreast of the time."

An extract from a letter of my father's to Lyell shows his estimate of his own work. " You have pleased me much by saying that you intend looking through my *Volcanic Islands* : it cost me eighteen months ! ! ! and I have heard of very few who have read it.* Now I shall feel, whatever little (and little it is) there is confirmatory of old work, or new, will work its effect and not be lost."

The second edition of the *Journal of Researches* † was

---

* He wrote to Herbert :—" I have long discovered that geologists never read each other's works, and that the only object in writing a book is a proof of earnestness, and that you do not form your opinions without undergoing labour of some kind. Geology is at present very oral, and what I here say is to a great extent quite true." And to Fitz-Roy, on the same subject, he wrote : "¶I have sent my *South American Geology* to Dover Street, and you will get it, no doubt, in the course of time. You do not know what you threaten when you propose to read it—it is purely geological. I said to my brother, ' You will of course read it,' and his answer was, ' Upon my life, I would sooner even buy it.' "

† The first edition was published in 1839, as vol. iii. of the *Voyages of the* '*Adventure*' *and* '*Beagle.*'

completed in 1845. It was published by Mr. Murray in the *Colonial and Home Library*, and in this more accessible form soon had a large sale.

C. D. to Lyell. Down [July, 1845].

MY DEAR LYELL—I send you the first part * of the new edition, which I so entirely owe to you. You will see that I have ventured to dedicate it to you, and I trust that this cannot be disagreeable. I have long wished, not so much for your sake, as for my own feelings of honesty, to acknowledge more plainly than by mere reference, how much I geologically owe you. Those authors, however, who, like you, educate people's minds as well as teach them special facts, can never, I should think, have full justice done them except by posterity, for the mind thus insensibly improved can hardly perceive its own upward ascent. I had intended putting in the present acknowledgment in the third part of my Geology, but its sale is so exceedingly small that I should not have had the satisfaction of thinking that as far as lay in my power I had owned, though imperfectly, my debt. Pray do not think that I am so silly, as to suppose that my dedication can any ways gratify you, except so far as I trust you will receive it, as a most sincere mark of my gratitude and friendship. I think I have improved this edition, especially the second part, which I have just finished. I have added a good deal about the Fuegians, and cut down into half the mercilessly long discussion on climate and glaciers, &c. I do not recollect anything added to the first part, long enough to call your attention to; there is a page of description of a very curious breed of oxen in Banda Oriental. I should like you to read the few last pages; there is a little discussion on extinction, which will not perhaps strike you as new, though it has so struck me, and has placed in my mind all the difficulties with respect to the causes of extinction, in the same class with other difficulties which are generally quite overlooked and undervalued by naturalists; I ought, however, to have made my discussion longer and shown by facts, as I easily could, how steadily every species must be checked in its numbers.

A pleasant notice of the *Journal* occurs in a letter from Humboldt to Mrs. Austin, dated June 7, 1844 †:—

---

* No doubt proof-sheets.
† *Three Generations of Englishwomen*, by Janet Ross (1888), vol. i. p. 195.

" Alas! you have got some one in England whom you do not read—young Darwin, who went with the expedition to the Straits of Magellan. He has succeeded far better than myself with the subject I took up. There are admirable descriptions of tropical nature in his journal, which you do not read because the author is a zoologist, which you imagine to be synonymous with bore. Mr. Darwin has another merit, a very rare one in your country—he has praised me."

October 1846 to October 1854.

The time between October 1846, and October 1854, was practically given up to working at the Cirripedia (Barnacles); the results were published in two volumes by the Ray Society in 1851 and 1854. His volumes on the Fossil Cirripedes were published by the Palæontographical Society in 1851 and 1854.

Writing to Sir J. D. Hooker in 1845, my father says: " I hope this next summer to finish my South American Geology,* then to get out a little Zoology, and hurrah for my species work. . . ." This passage serves to show that he had at this time no intention of making an exhaustive study of the Cirripedes. Indeed it would seem that his original intention was, as I learn from Sir J. D. Hooker, merely to work out one special problem. This is quite in keeping with the following passage in the *Autobiography:* " When on the coast of Chile, I found a most curious form, which burrowed into the shells of Concholepas, and which differed so much from all other Cirripedes that I had to form a new sub-order for its sole reception. . . . To understand the structure of my new Cirripede I had to examine and dissect many of the common forms; and this gradually led me on to take up the whole group." In later years he seems to have felt some doubt as to the value of these eight years of work—for instance when he wrote in his *Autobiography*— " My work was of considerable use to me, when I had to discuss in the *Origin of Species* the principles of a natural

---

* This refers to the third and last of his geological books, *Geological Observation on South America*, which was published in 1846. A sentence from a letter of Dec. 11, 1860, may be quoted here—" David Forbes has been carefully working the Geology of Chile, and as I value praise for accurate observation far higher than for any other quality, forgive (if you can) the *insufferable* vanity of my copying the last sentence in his note : ' I regard your Monograph on Chile as, without exception, one of the finest specimens of Geological inquiry.' I feel inclined to strut like a turkey-cock ! "

classification. Nevertheless I doubt whether the work was
worth the consumption of so much time." Yet I learn
from Sir J. D. Hooker that he certainly recognized at the
time its value to himself as systematic training. Sir Joseph
writes to me : " Your father recognized three stages in his
career as a biologist : the mere collector at Cambridge ; the
collector and observer in the *Beagle*, and for some years
afterwards ; and the trained naturalist after, and only after
the Cirripede work. That he was a thinker all along is true
enough, and there is a vast deal in his writings previous to
the Cirripedes that a trained naturalist could but emulate.
. . . He often alluded to it as a valued discipline, and added
that even the 'hateful' work of digging out synonyms, and
of describing, not only improved his methods but opened
his eyes to the difficulties and merits of the works of the
dullest of cataloguers. One result was that he would never
allow a depreciatory remark to pass unchallenged on the
poorest class of scientific workers, provided that their work
was honest, and good of its kind. I have always regarded
it as one of the finest traits of his character,—this generous
appreciation of the hod-men of science, and of their labours
. . . and it was monographing the Barnacles that brought
it about."

Mr. Huxley allows me to quote his opinion as to the
value of the eight years given to the Cirripedes :—

" In my opinion your sagacious father never did a wiser
thing than when he devoted himself to the years of patient
toil which the Cirripede-book cost him.

" Like the rest of us, he had no proper training in biological science, and it has always struck me as a remarkable
instance of his scientific insight, that he saw the necessity
of giving himself such training, and of his courage, that he
did not shirk the labour of obtaining it.

" The great danger which besets all men of large speculative faculty, is the temptation to deal with the accepted
statements of fact in natural science, as if they were not
only correct, but exhaustive ; as if they might be dealt with
deductively, in the same way as propositions in Euclid may
be dealt with. In reality, every such statement, however
true it may be, is true only relatively to the means of observation and the point of view of those who have enunciated
it. So far it may be depended upon. But whether it will
bear every speculative conclusion that may be logically de-
duced from it, is quite another question.

" Your father was building a vast superstructure upon
the foundations furnished by the recognised facts of geological and biological science. In Physical Geography, in
Geology proper, in Geographical Distribution, and in Palæ-
ontology, he had acquired an extensive practical training
during the voyage of the *Beagle*. He knew of his own knowledge the way in which the raw materials of these branches
of science are acquired, and was therefore a most competent
judge of the speculative strain they would bear. That
which he needed, after his return to England, was a corresponding acquaintance with Anatomy and Development,
and their relation to Taxonomy—and he acquired this by
his Cirripede work."

Though he became excessively weary of the work before
the end of the eight years, he had much keen enjoyment in
the course of it. Thus he wrote to Sir J. D. Hooker
(1847?) :—" As you say, there is an extraordinary pleasure
in pure observation; not but what I suspect the pleasure in
this case is rather derived from comparisons forming in
one's mind with allied structures. After having been so
long employed in writing my old geological observations, it
is delightful to use one's eyes and fingers again." It was,
in fact, a return to the work which occupied so much of
his time when at sea during his voyage. Most of his work
was done with the simple dissecting microscope—and it was
the need which he found for higher powers that induced
him, in 1846, to buy a compound microscope. He wrote to
Hooker :—" When I was drawing with L., I was so delighted with the appearance of the objects, especially with
their perspective, as seen through the weak powers of a
good compound microscope, that I am going to order one;
indeed, I often have structures in which the $\frac{1}{30}$ is not power
enough."

During part of the time covered by the present chapter,
my father suffered perhaps more from ill-health than at
any other period of his life. He felt severely the depressing influence of these long years of illness; thus as early as
1840 he wrote to Fox: " I am grown a dull, old, spiritless
dog to what I used to be. One gets stupider as one grows
older I think." It is not wonderful that he should so have
written, it is rather to be wondered at that his spirit with-
stood so great and constant a strain. He wrote to Sir
Joseph Hooker in 1845 : " You are very kind in your
inquiries about my health ; I have nothing to say about it,

being always much the same, some days better and some
worse. I believe I have not had one whole day, or rather
night, without my stomach having been greatly disordered,
during the last three years, and most days great prostration
of strength : thank you for your kindness ; many of my
friends, I believe, think me a hypochondriac."

During the whole of the period now under consideration,
he was in constant correspondence with Sir Joseph Hooker.
The following characteristic letter on Sigillaria (a gigantic
fossil plant found in the Coal Measures) was afterwards
characterised by himself as not being " reasoning, or even
speculation, but simply as mental rioting."

[Down, 1847 ?]

" . . . I am delighted to hear that Brongniart thought
Sigillaria aquatic, and that Binny considers coal a sort of
submarine peat. I would bet 5 to 1 that in twenty years
this will be generally admitted ; * and I do not care for
whatever the botanical difficulties or impossibilities may be.
If I could but persuade myself that Sigillaria and Co. had a
good range of depth, *i.e.* could live from 5 to 10 fathoms
under water, all difficulties of nearly all kinds would be removed (for the simple fact of muddy ordinary shallow sea
implies proximity of land). [N.B.—I am chuckling to
think how you are sneering all this time.] It is not much
of a difficulty, there not being shells with the coal, considering how unfavourable deep mud is for most Mollusca,
and that shells would probably decay from the humic acid,
as seems to take place in peat and in the *black* moulds (as
Lyell tells me) of the Mississippi. So coal question settled
—Q. E. D. Sneer away ! "

The two following extracts give the continuation and
conclusion of the coal battle.

" By the way, as submarine coal made you so wrath, I
thought I would experimentise on Falconer and Bunbury †
together, and it made [them] even more savage ; ' such infernal nonsense ought to be thrashed out of me.' Bunbury
was more polite and contemptuous. So I now know how to
stir up and show off any Botanist. I wonder whether Zoologists and Geologists have got their tender points ; I wish
I could find out."

" I cannot resist thanking you for your most kind note.

---

* An unfulfilled prophecy.
† The late Sir C. Bunbury, well known as a palæobotanist.

Pray do not think that I was annoyed by your letter : I perceived that you had been thinking with animation, and accordingly expressed yourself strongly, and so I understood it. Forfend me from a man who weighs every expression with Scotch prudence. I heartily wish you all success in your noble problem, and I shall be very curious to have some talk with you and hear your ultimatum."

He also corresponded with the late Hugh Strickland,—a well-known ornithologist, on the need of reform in the principle of nomenclature. The following extract (1849) gives an idea of my father's view :—

" I feel sure as long as species-mongers have their vanity tickled by seeing their own names appended to a species, because they miserably described it in two or three lines, we shall have the same *vast* amount of bad work as at present, and which is enough to dishearten any man who is willing to work out any branch with care and time. I find every genus of Cirripedia has half-a-dozen names, and not one careful description of any one species in any one genus. I do not believe that this would have been the case if each man knew that the memory of his own name depended on his doing his work well, and not upon merely appending a name with a few wretched lines indicating only a few prominent external characters."

In 1848 Dr. R. W. Darwin died, and Charles Darwin wrote to Hooker, from Malvern :—

" On the 13th of November, my poor dear father died, and no one who did not know him would believe that a man above eighty-three years old could have retained so tender and affectionate a disposition, with all his sagacity unclouded to the last. I was at the time so unwell, that I was unable to travel, which added to my misery.

" All this winter I have been bad enough . . . and my nervous system began to be affected, so that my hands trembled, and head was often swimming. I was not able to do anything one day out of three, and was altogether too dispirited to write to you, or to do anything but what I was compelled. I thought I was rapidly going the way of all flesh. Having heard, accidentally, of two persons who had received much benefit from the water-cure, I got Dr. Gully's book, and made further inquiries, and at last started here, with wife, children, and all our servants. We have taken a house for two months, and have been here a fortnight. I am already a little stronger . . . Dr. Gully feels pretty sure he can do me good, which most certainly the regular doctors could not. . . . . . I feel certain that the water-cure is no quackery.

"How I shall enjoy getting back to Down with renovated health, if such is to be my good fortune, and resuming the beloved Barnacles. Now I hope that you will forgive me for my negligence in not having sooner answered your letter. I was uncommonly interested by the sketch you give of your intended grand expedition, from which I suppose you will soon be returning. How earnestly I hope that it may prove in every way successful. . . ."

C. D. to W. D. Fox.   [March 7, 1852.]

Our long silence occurred to me a few weeks since, and I had then thought of writing, but was idle. I congratulate and condole with you on your *tenth* child; but please to observe when I have a tenth, send only condolences to me. We have now seven children, all well, thank God, as well as their mother; of these seven, five are boys; and my father used to say that it was certain that a boy gave as much trouble as three girls; so that *bona fide* we have seventeen children. It makes me sick whenever I think of professions; all seem hopelessly bad, and as yet I cannot see a ray of light. I should very much like to talk over this (by the way, my three bugbears are Californian and Australian gold, beggaring me by making my money on mortgage worth nothing; the French coming by the Westerham and Sevenoaks roads, and therefore enclosing Down; and thirdly, professions for my boys), and I should like to talk about education, on which you ask me what we are doing. No one can more truly despise the old stereotyped stupid classical education than I do; but yet I have not had courage to break through the trammels. After many doubts we have just sent our eldest boy to Rugby, where for his age he has been very well placed. . . I honour, admire, and envy you for educating your boys at home. What on earth shall you do with your boys? Very many thanks for your most kind and large invitation to Delamere, but I fear we can hardly compass it. I dread going anywhere, on account of my stomach so easily failing under any excitement. I rarely even now go to London; not that I am at all worse, perhaps rather better, and lead a very comfortable life with my three hours of daily work, but it is the life of a hermit.

My nights are *always* bad, and that stops my becoming vigorous. You ask about water-cure. I take at intervals of
two or three months, five or six weeks of *moderately* severe
treatment, and always with good effect. Do you come here,
I pray and beg whenever you can find time ; you cannot
tell how much pleasure it would give me and E. What
pleasant times we had in drinking coffee in your rooms at
Christ's College, and think of the glories of Crux-major.*
Ah, in those days there were no professions for sons, no ill-
health to fear for them, no Californian gold, no French in-
vasions. How paramount the future is to the present when
one is surrounded by children. My dread is hereditary ill-
health. Even death is better for them.

> My dear Fox, your sincere friend.

P.S.—Susan † has lately been working in a way which
I think truly heroic about the scandalous violation of the
Act against children climbing chimneys. We have set up
a little Society in Shrewsbury to prosecute those who break
the law. It is all Susan's doing. She has had very nice
letters from Lord Shaftesbury and the Duke of Sutherland,
but the brutal Shropshire squires are as hard as stones to
move. The Act out of London seems most commonly vio-
lated. It makes one shudder to fancy one of one's own
children at seven years old being forced up a chimney—to
say nothing of the consequent loathsome disease and ulcer-
ated limbs, and utter moral degradation. If you think
strongly on this subject, do make some enquiries; add to
your many good works, this other one, and try to stir up
the magistrates. . . .

The following letter refers to the Royal Medal, which
was awarded to him in November, 1853 :

C. D. to J. D. Hooker. Down [November 1853].

MY DEAR HOOKER—Amongst my letters received this
morning, I opened first one from Colonel Sabine; the contents certainly surprised me very much, but, though the
letter was a *very kind one*, somehow, I cared very little indeed for the announcement it contained. I then opened
yours, and such is the effect of warmth, friendship, and

---

* The beetle *Panagæus crux-major.*   † His sister.

kindness from one that is loved, that the very same fact, told as you told it, made me glow with pleasure till my very heart throbbed. Believe me, I shall not soon forget the pleasure of your letter. Such hearty, affectionate sympathy is worth more than all the medals that ever were or will be coined. Again, my dear Hooker, I thank you. I hope Lindley * will never hear that he was a competitor against me; for really it is almost *ridiculous* (of course you would never repeat that I said this, for it would be thought by others, though not, I believe by you, to be affectation) his not having the medal long before me; I must feel *sure* that you did quite right to propose him; and what a good, dear, kind fellow you are, nevertheless, to rejoice in this honour being bestowed on me.

What *pleasure* I have felt on the occasion, I owe almost entirely to you.†

Farewell, my dear Hooker, yours affectionately.

The following series of extracts, must, for want of space, serve as a sketch of his feeling with regard to his seven years' work at Barnacles ‡ :—

September 1849.—" It makes me groan to think that probably I shall never again have the exquisite pleasure of making out some new district, of evolving geological light

* John Lindley (b. 1799, d. 1865) was the son of a nurseryman near Norwich, through whose failure in business he was thrown at the age of twenty on his own resources. He was befriended by Sir W. Hooker, and employed as assistant librarian by Sir J. Banks. He seems to have had enormous capacity for work, and is said to have translated Richard's *Analyse du Fruit* at one sitting of two days and three nights. He became Assistant-Secretary to the Horticultural Society, and in 1829 was appointed Professor of Botany at University College, a post which he held for upwards of thirty years. His writings are numerous; the best known being perhaps his *Vegetable Kingdom*, published in 1846.

† Shortly afterwards he received a fresh mark of esteem from his warm-hearted friend: " Hooker's book (*Himalayan Journal*) is out, and *most beautifully* got up. He has honoured me beyond measure by dedicating it to me ! "

‡ In 1860 he wrote to Lyell, " Is not Krohn a good fellow ? I have long meant to write to him. He has been working at Cirripedes, and has detected two or three gigantic blunders, about which, I thank Heaven, I spoke rather doubtfully. Such difficult dissection that even Huxley failed. It is chiefly the interpretation which I put on parts that is so wrong, and not the parts which I describe. But they were gigantic blunders, and why I say all this is because Krohn, instead of crowing at all, pointed out my errors with the utmost gentleness and pleasantness."

There are two papers by Aug. Krohn, one on the Cement Glands, and the other on the development of Cirripedes, *Weigmann's Archiv*, xxv. and xxvi. See *Autobiography*, p. 39, where my father remarks, " I blundered dreadfully about the cement glands."

out of some troubled dark region. So I must make the best of my Cirripedia. . . ."

October 1849.—" I have of late been at work at mere species describing, which is much more difficult than I expected, and has much the same sort of interest as a puzzle has; but I confess I often feel wearied with the work, and cannot help sometimes asking myself what is the good of spending a week or fortnight in ascertaining that certain just perceptible differences blend together and constitute varieties and not species. As long as I am on anatomy I never feel myself in that disgusting, horrid, *cui bono*, inquiring humour. What miserable work, again, it is searching for priority of names. I have just finished two species, which possess seven generic, and twenty-four specific names! My chief comfort is, that the work must be sometimes done, and I may as well do it, as any one else."

October 1852.—" I am at work at the second volume of the Cirripedia, of which creatures I am wonderfully tired. I hate a Barnacle as no man ever did before, not even a sailor in a slow-sailing ship. My first volume is out; the only part worth looking at is on the sexes of Ibla and Scalpellum. I hope by next summer to have done with my tedious work."

July 1853.—" I am *extremely* glad to hear that you approved of my cirripedial volume. I have spent an almost ridiculous amount of labour on the subject, and certainly would never have undertaken it had I foreseen what a job it was."

In September, 1854, his Cirripede work was practically finished, and he wrote to Sir J. Hooker:

" I have been frittering away my time for the last several weeks in a wearisome manner, partly idleness, and odds and ends, and sending ten thousand Barnacles * out of the house all over the world. But I shall now in a day or two begin to look over my old notes on species. What a deal I shall have to discuss with you; I shall have to look sharp that I do not 'progress' into one of the greatest bores in life, to the few like you with lots of knowledge."

---

* The duplicate type-specimens of my father's Cirripedes are in the Liverpool Free Public Museum, as I learn from the Rev. H. H. Higgins.

CHAPTER IX.

To give an account of the development of the chief work of my father's life—the *Origin of Species*—it will be necessary to return to an earlier date, and to weave into the story letters and other material, purposely omitted from the chapters dealing with the voyage and with his life at Down.

To be able to estimate the greatness of the work, we must know something of the state of knowledge on the species question at the time when the germs of the Darwinian theory were forming in my father's mind.

For the brief sketch which I can here insert, I am largely indebted to vol. ii., chapter v., of the *Life and Letters*—a discussion on the *Reception of the Origin of Species* which Mr. Huxley was good enough to write for me, also to the masterly obituary essay on my father, which the same writer contributed to the Proceedings of the Royal Society.*

Mr. Huxley has well said : †

"To any one who studies the signs of the times, the emergence of the philosophy of Evolution, in the attitude of claimant to the throne of the world of thought, from the limbo of hated and, as many hoped, forgotten things, is the most portentous event of the nineteenth century."

In the autobiographical chapter, my father has given an account of his share in this great work : the present chapter does little more than expand that story.

Two questions naturally occur to one : (1)—When and how did Darwin become convinced that species are mutable? How (that is to say) did he begin to believe in evolution? And (2)—When and how did he conceive the manner in which species are modified; when did he begin to believe in Natural Selection?

The first question is the more difficult of the two to

---

* Vol. xliv. No. 269.   † *Life and Letters*, vol. ii. p. 180.

answer. He has said in the *Autobiography* (p. 39) that
certain facts observed by him in South America seemed to be
explicable only on the " supposition that species gradually
become modified." He goes on to say that the subject
" haunted him " ; and I think it is especially worthy of note
that this " haunting "—this unsatisfied dwelling on the subject was connected with the desire to explain *how* species
can be modified. It was characteristic of him to feel, as he
did, that it was " almost useless " to endeavour to prove the
general truth of evolution, unless the cause of change could
be discovered. I think that throughout his life the questions 1 and 2 were intimately—perhaps unduly so—connected in his mind. It will be shown, however, that after
the publication of the *Origin*, when his views were being
weighed in the balance of scientific opinion, it was to the
acceptance of Evolution, not of Natural Selection, that he
attached importance.

An interesting letter (Feb. 24, 1877) to Dr. Otto Zacha-
rias,* gives the same impression as the *Autobiography* :—

" When I was on board the *Beagle* I believed in the permanence of species, but as far as I can remember, vague
doubts occasionally flitted across my mind. On my return
home in the autumn of 1836, I immediately began to prepare my Journal for publication, and then saw how many
facts indicated the common descent of species, so that in
July, 1837, I opened a note-book to record any facts which
might bear on the question. But I did not become convinced that species were mutable until, I think, two or three
years had elapsed."

Two years bring us to 1839, at which date the idea of
natural selection had already occurred to him—a fact which
agrees with what has been said above. How far the idea
that evolution is conceivable came to him from earlier writers
it is not possible to say. He has recorded in the *Autobiography* (p. 38) the " silent astonishment with which, about
the year 1825, he heard Grant expound the Lamarckian
philosophy." He goes on :—

" I had previously read the *Zoonomia* of my grandfather,
in which similar views are maintained, but without producing any effect on me. Nevertheless, it is probable that the
hearing rather early in life such views maintained and

---

* This letter was unaccountably overlooked in preparing the *Life and
Letters* for publication.

praised, may have favoured my upholding them under a different form in my *Origin of Species*. At this time I admired greatly the *Zoonomia ;* but on reading it a second time after an interval of ten or fifteen years, I was much disappointed ; the proportion of speculation being so large to the facts given."

Mr. Huxley has well said (*Obituary Notice*, p. ii.) : " Erasmus Darwin was in fact an anticipator of Lamarck, and not of Charles Darwin ; there is no trace in his works of the conception by the addition of which his grandson metamorphosed the theory of evolution as applied to living things, and gave it a new foundation."

On the whole it seems to me that the effect on his mind of the earlier evolutionists was inappreciable, and as far as concerns the history of the *Origin of Species*, it is of no particular importance, because, as before said, evolution made no progress in his mind until the cause of modification was conceivable.

I think Mr. Huxley is right in saying * that " it is hardly too much to say that Darwin's greatest work is the outcome of the unflinching application to biology of the leading idea, and the method applied in the *Principles* to Geology." Mr. Huxley has elsewhere † admirably expressed the bearing of Lyell's work in this connection :—

" I cannot but believe that Lyell, for others, as for myself, was the chief agent in smoothing the road for Darwin. For consistent uniformitarianism postulates evolution as much in the organic as in the inorganic world. The origin of a new species by other than ordinary agencies would be a vastly greater ' catastrophe ' than any of those which Lyell successfully eliminated from sober geological speculation. . . .

" Lyell,‡ with perfect right, claims this position for himself. He speaks of having ' advocated a law of continuity even in the organic world, so far as possible without adopting Lamarck's theory of transmutation. . . .

" ' But while I taught,' Lyell goes on, ' that as often as certain forms of animals and plants disappeared, for reasons quite intelligible to us, others took their place by virtue of

---

* *Obituary Notice*, p. viii.

† *Life and Letters*, vol. ii. p. 190. In Mr. Huxley's chapter the passage beginning " Lyell with perfect right . . . ." is given as a footnote : it will be seen that I have incorporated it with Mr. Huxley's text.

‡ Lyell's *Life and Letters*, Letter to Haeckel, vol. ii. p. 436. Nov. 23, 1868.

a causation which was beyond our comprehension; it remained for Darwin to accumulate proof that there is no break between the incoming and the outgoing species, that they are the work of evolution, and not of special creation. . . . I had certainly prepared the way in this country, in six editions of my work before the *Vestiges of Creation* appeared in 1842 [1844], for the reception of Darwin's gradual and insensible evolution of species.'"

Mr. Huxley continues :—

"If one reads any of the earlier editions of the *Principles* carefully (especially by the light of the interesting series of letters recently published by Sir Charles Lyell's biographer), it is easy to see that, with all his energetic opposition to Lamarck, on the one hand, and to the ideal quasi-progressionism of·Agassiz, on the other, Lyell, in his own mind, was strongly disposed to account for the origination of all past and present species of living things by natural causes. But he would have liked, at the same time, to keep the name of creation for a natural process which he imagined to be incomprehensible."

The passage above given refers to the influence of Lyell in preparing men's minds for belief in the *Origin*, but I cannot doubt that it "smoothed the way" for the author of that work in his early searchings, as well as for his followers. My father spoke prophetically when he wrote the dedication to Lyell of the second edition of the *Journal of Researches* (1845).

"To Charles Lyell, Esq., F.R.S., this second edition is dedicated with grateful pleasure—as an acknowledgment that the chief part of whatever scientific merit this journal and the other work of the author may possess, has been derived from studying the well-known and admirable *Principles of Geology*."

Professor Judd, in some reminiscences of my father which he was so good as to give me, quotes him as saying that, "It was the reading of the *Principles of Geology* which did most towards moulding his mind and causing him to take up the line of investigation to which his life was devoted."

The *rôle* that Lyell played as a pioneer makes his own point of view as to evolution all the more remarkable. As the late H. C. Watson wrote to my father (December 21, 1859) :—

"Now these novel views are brought fairly before the

scientific public, it seems truly remarkable how so many of them could have failed to see their right road sooner. How could Sir C. Lyell, for instance, for thirty years read, write, and think, on the subject of species *and their succession*, and yet constantly look down the wrong road !

"A quarter of a century ago, you and I must have been in something like the same state of mind on the main question. But you were able to see and work out the *quo modo* of the succession, the all-important thing, while I failed to grasp it."

In his earlier attitude towards evolution, my father was on a par with his contemporaries. He wrote in the *Autobiography* :—

"I occasionally sounded not a few naturalists, and never happened to come across a single one who seemed to doubt about the permanence of species : " and it will be made abundantly clear by his letters that in supporting the opposite view he felt himself a terrible heretic.

Mr. Huxley * writes in the same sense :

"Within the ranks of biologists, at that time [1851–58], I met with nobody, except Dr. Grant, of University College, who had a word to say for Evolution—and his advocacy was not calculated to advance the cause. Outside these ranks, the only person known to me whose knowledge and capacity compelled respect, and who was, at the same time, a thoroughgoing evolutionist, was Mr. Herbert Spencer, whose acquaintance I made, I think, in 1852, and then entered into the bonds of a friendship which, I am happy to think, has known no interruption. Many and prolonged were the battles we fought on this topic. But even my friend's rare dialectic skill and copiousness of apt illustration could not drive me from my agnostic position. I took my stand upon two grounds : firstly, that up to that time, the evidence in favour of transmutation was wholly insufficient ; and, secondly, that no suggestion respecting the causes of the transmutation assumed, which had been made, was in any way adequate to explain the phenomena. Looking back at the state of knowledge at that time, I really do not see that any other conclusion was justifiable."

These two last citations refer of course to a period much later than the time, 1836–37, at which the Darwinian theory

---

* *Life and Letters*, vol. ii. p. 188.

was growing in my father's mind.   The same thing is however true of earlier days.

So much for the general problem : the further question
as to the growth of Darwin's theory of natural selection is a
less complex one, and I need add but little to the history
given in the *Autobiography* of how he came by that great
conception by the help of which he was able to revivify " the
oldest of all philosophies—that of evolution."

The first point in the slow journey towards the *Origin
of Species* was the opening of that note-book of 1837 of
which mention has been already made.   The reader who is
curious on the subject will find a series of citations from
this most interesting note-book, in the *Life and Letters*,
vol. ii. p. 5, *et seq.*

The two following extracts show that he applied the
theory of evolution to the " whole organic kingdom " from
plants to man.

" If we choose to let conjecture run wild, then animals,
our fellow brethren in pain, disease, death, suffering and
famine—our slaves in the most laborious works, our companions in our amusements—they may partake [of] our
origin in one common ancestor—we may be all melted
together."

" The different intellects of man and animals not so great
as between living things without thought (plants), and living things with thought (animals).

Speaking of intermediate forms, he remarks :

" Opponents will say—*show them me.*   I will answer yes,
if you will show me every step between bulldog and grey-
hound."

Here we see that the argument from domestic animals
was already present in his mind as bearing on the production of natural species, an argument which he afterwards
used with such signal force in the *Origin*.

A comparison of the two editions of the *Naturalists'
Voyage* is instructive, as giving some idea of the development of his views on evolution.   It does not give us a true
index of the mass of conjecture which was taking shape in
his mind, but it shows us that he felt sure enough of the
truth of his belief to allow a stronger tinge of evolution to
appear in the second edition.   He has mentioned in the
Autobiography (p. 40), that it was not until he read Malthus
that he got a clear view of the potency of natural selection.
This was in 1838—a year after he finished the first edition

(it was not published until 1839), and seven years before the second edition was issued (1845). Thus the turning-point in the formation of his theory took place between the writing of the two editions. Yet the difference between the two editions is not very marked; it is another proof of the author's caution and self-restraint in the treatment of his ideas. After reading the second edition of the *Voyage* we remember with a strong feeling of surprise how far advanced were his views when he wrote it.

These views are given in the manuscript volume of 1844, mentioned in the *Autobiography.* I give from my father's Pocket-book the entries referring to the preliminary sketch of this historic essay.

"1842, *May* 18,—Went to Maer. *June* 15—to Shrewsbury, and 18th to Capel Curig. During my stay at Maer and Shrewsbury. . . . wrote pencil sketch of species theory." *

In 1844, the pencil-sketch was enlarged to one of 230 folio pages, which is a wonderfully complete presentation of the arguments familiar to us in the *Origin.*

The following letter shows in a striking manner the value my father put on this piece of work.

C. D. to Mrs. Darwin. Down [July 5, 1844].

. . . I have just finished my sketch of my species theory. If, as I believe, my theory in time be accepted even by one competent judge, it will be a considerable step in science.

I therefore write this in case of my sudden death, as my most solemn and last request, which I am sure you will consider the same as if legally entered in my will, that you will devote £400 to its publication, and further, will yourself, or through Hensleigh,† take trouble in promoting it. I wish that my sketch be given to some competent person, with this sum to induce him to take trouble in its improvement and enlargement. I give to him all my books on Natural History, which are either scored or have references at the end to the pages, begging him carefully to look over and consider such passages as actually bearing, or by possibility bearing on this subject. I wish you to make a list of all such

---

* I have discussed in the *Life and Letters* the statement often made that the first sketch of his theory was written in 1839.

† The late Mr. H. Wedgwood.

books as some temptation to an editor. I also request that you will hand over [to] him all those scraps roughly divided in eight or ten brown paper portfolios. The scraps, with copied quotations from various works, are those which may aid my editor. I also request that you, or some amanuensis will aid in deciphering any of the scraps which the editor may think possibly of use. I leave to the editor's judgment whether to interpolate these facts in the text, or as notes, or under appendices. As the looking over the references and scraps will be a long labour, and as the *correcting* and enlarging and altering my sketch will also take considerable time, I leave this sum of £400 as some remuneration, and any profits from the work. I consider that for this the editor is bound to get the sketch published either at a publisher's or his own risk. Many of the scraps in the portfolios contain mere rude suggestions and early views, now useless, and many of the facts will probably turn out as having no bearing on my theory.

With respect to editors, Mr. Lyell would be the best if he would undertake it; I believe he would find the work pleasant, and he would learn some facts new to him. As the editor must be a geologist as well as a naturalist, the next best editor would be Professor Forbes of London. The next best (and quite best in many respects) would be Professor Henslow. Dr. Hooker would be *very* good. The next Mr. Strickland.* If none of these would undertake it, I request you to consult with Mr. Lyell, or some other capable man for some editor, a geologist and naturalist. Should one other hundred pounds make the difference of procuring a good editor, I request earnestly that you will raise £500.

My remaining collections in Natural History may be given to any one or any museum where [they] would be accepted. . . .

The following note seems to have formed part of the original letter, but may have been of later date :

" Lyell, especially with the aid of Hooker (and of any good zoological aid), would be best of all. Without an editor will pledge himself to give up time to it, it would be of no use paying such a sum."

" If there should be any difficulty in getting an editor

---

* After Mr. Strickland's name comes the following sentence, which has been erased, but remains legible : " Professor Owen would be very good ; but I presume he would not undertake such a work."

who would go thoroughly into the subject, and think of the bearing of the passages marked in the books and copied out [on?] scraps of paper, then let my sketch be published as it is, stating that it was done several years ago * and from memory without consulting any works, and with no intention of publication in its present form."

The idea that the Sketch of 1844 might remain, in the event of his death, as the only record of his work, seems to have been long in his mind, for in August 1854, when he had finished with the Cirripedes, and was thinking of beginning his "species work," he added on the back of the above letter, " Hooker by far best man to edit my species volume. August 1854."

---

* The words " several years ago and," seem to have been added at a later date.

CHAPTER X.

THE history of the years 1843–1858 is here related in an extremely abbreviated fashion. It was a period of minute labour on a variety of subjects, and the letters accordingly abound in detail. They are in many ways extremely interesting, more especially so to professed naturalists, and the picture of patient research which they convey is of great value from a biographical point of view. But such a picture must either be given in a complete series of unabridged letters, or omitted altogether. The limits of space compel me to the latter choice. The reader must imagine my father corresponding on problems in geology, geographical distribution, and classification; at the same time collecting facts on such varied points as the stripes on horses' legs, the floating of seeds, the breeding of pigeons, the form of bees' cells and the innumerable other questions to which his gigantic task demanded answers.

The concluding letter of the last chapter has shown how strong was his conviction of the value of his work. It is impressive evidence of the condition of the scientific atmosphere, to discover, as in the following letters to Sir Joseph Hooker, how small was the amount of encouragement that he dared to hope for from his brother-naturalists.

[January 11th, 1844.]

. . . I have been now ever since my return engaged in a very presumptuous work, and I know no one individual who would not say a very foolish one. I was so struck with the distribution of the Galapagos organisms, &c. &c., and with the character of the American fossil mammifers, &c. &c., that I determined to collect blindly every sort of fact, which could bear any way on what are species. I have read heaps of agricultural and horticultural books, and have never ceased collecting facts. At last gleams of light have come, and I am almost convinced (quite contrary to the opinion I started with) that species are not (it is like confessing a murder) immutable. Heaven forfend me from Lamarck nonsense of a "tendency to progression," "adaptations from the slow willing of animals," &c.! But the conclusions I am led to are not widely different from his; though the means of change are wholly so. I think I have found out (here's presumption!) the simple way by which species become exquisitely adapted to various ends. You will now groan, and think to yourself, "on what a man have I been wasting my time and writing to." I should, five years ago, have thought so. . . .

And again (1844) :—

"In my most sanguine moments, all I expect, is that I shall be able to show even to sound Naturalists, that there are two sides to the question of the immutability of species —that facts can be viewed and grouped under the notion of allied species having descended from common stocks. With respect to books on this subject, I do not know of any systematical ones, except Lamarck's which is veritable rubbish: but there are plenty, as Lyell, Pritchard, &c., on the view of the immutability. Agassiz lately has brought the strongest argument in favour of immutability. Isidor G. St. Hilaire has written some good Essays, tending towards the mutability-side, in the *Suites à Buffon*, entitled *Zoolog. Générale.* Is it not strange that the author of such a book as the *Animaux sans Vertèbres* should have written that insects, which never see their eggs, should *will* (and plants, their seeds) to be of particular forms, so as to become attached to particular objects. The other common (specially Germanic) notion is hardly less absurd, viz. that climate, food, &c., should make a Pediculus formed to climb hair, or a wood-pecker to climb trees. I believe all these absurd views arise from no one having, as far as I know, approached the subject on the side of variation under domestication, and having studied all that is known about domestication."

"I hate arguments from results, but on my views of descent, really Natural History becomes a sublimely grand result-giving subject (now you may quiz me for so foolish an escape of mouth). . . ."

C. D. to L. Jenyns.* Down, Oct. 12th [1845].

MY DEAR JENYNS—Thanks for your note. I am sorry
to say I have not even the tail-end of a fact in English Zoology to communicate. I have found that even trifling
observations require, in my case, some leisure and energy,
[of] both of which ingredients I have had none to spare, as
writing my Geology thoroughly expends both. I had always
thought that I would keep a journal and record everything,
but in the way I now live I find I observe nothing to record.
Looking after my garden and trees, and occasionally a very
little walk in an idle frame of my mind, fills up every afternoon in the same manner. I am surprised that with all
your parish affairs, you have had time to do all that which
you have done. I shall be very glad to see your little work †
(and proud should I have been if I could have added a single
fact to it). My work on the species question has impressed
me very forcibly with the importance of all such works as
your intended one, containing what people are pleased generally to call trifling facts. These are the facts which make
one understand the working or economy of nature. There
is one subject, on which I am very curious, and which perhaps you may throw some light on, if you have ever thought
on it ; namely, what are the checks and what the periods of
life—by which the increase of any given species is limited.
Just calculate the increase of any bird, if you assume that
only half the young are reared, and these breed : within the
natural (*i.e.* if free from accidents) life of the parents the
number of individuals will become enormous, and I have
been much surprised to think how great destruction *must*
annually or occasionally be falling on every species, yet the
means and period of such destruction are scarcely perceived
by us.

I have continued steadily reading and collecting facts on
variation of domestic animals and plants, and on the question of what are species. I have a grand body of facts, and
I think I can draw some sound conclusions. The general
conclusions at which I have slowly been driven from a di-

---

* Rev. L. Blomefield.
† Mr. Jenyns' *Observations in Natural History*. It is prefaced by an In-
troduction on " Habits of observing as connected with the study of Natural
History," and followed by a " Calendar of Periodic Phenomena in Natural
History," with " Remarks on the importance of such Registers."

rectly opposite conviction, is that species are mutable, and that allied species are co-descendants from common stocks. I know how much I open myself to reproach for such a conclusion, but I have at least honestly and deliberately come to it. I shall not publish on this subject for several years.

C. Darwin to L. Jenyns.* Down [1845?]

With respect to my far distant work on species I must have expressed myself with singular inaccuracy if I led you to suppose that I meant to say that my conclusions were inevitable. They have become so, after years of weighing puzzles, to myself *alone ;* but in my wildest day-dream, I never expect more than to be able to show that there are two sides to the question of the immutability of species, *i.e.* whether species are *directly* created or by intermediate laws (as with the life and death of individuals). I did not approach the subject on the side of the difficulty in determining what are species and what are varieties, but (though why I should give you such a history of my doings it would be hard to say) from such facts as the relationship between the living and extinct mammifers in South America, and between those living on the Continent and on adjoining islands, such as the Galapagos. It occurred to me that a collection of all such analogous facts would throw light either for or against the view of related species being co-descendants from a common stock. A long searching amongst agricultural and horticultural books and people makes me believe (I well know how absurdly presumptuous this must appear) that I see the way in which new varieties become exquisitely adapted to the external conditions of life and to other surrounding beings. I am a bold man to lay myself open to being thought a complete fool, and a most deliberate one. From the nature of the grounds which make me believe that species are mutable in form, these grounds cannot be restricted to the closest-allied species ; but how far they extend I cannot tell, as my reasons fall away by degrees, when applied to species more and more remote from each other. Pray do not think that I am so blind as not to see that there are numerous immense difficulties in my notions, but they appear to me less than on the common view. I have drawn up a sketch and had it

---

* Rev. L. Blomefield.

copied (in 200 pages) of my conclusions; and if I thought at some future time that you would think it worth reading, I should, of course, be most thankful to have the criticism of so competent a critic. Excuse this very long and egotistical and ill-written letter, which by your remarks you have led me into.

C. D. to J. D. Hooker. Down [1849–50?]

. . . . How painfully (to me) true is your remark, that no one has hardly a right to examine the question of species who has not minutely described many. I was, however, pleased to hear from Owen (who is vehemently opposed to any mutability in species), that he thought it was a very fair subject, and that there was a mass of facts to be brought to bear on the question, not hitherto collected. My only comfort is (as I mean to attempt the subject), that I have dabbled in several branches of Natural History, and seen good specific men work out my species, and know something of geology (an indispensable union); and though I shall get more kicks than half-pennies, I will, life serving, attempt my work. Lamarck is the only exception, that I can think of, of an accurate describer of species at least in the Invertebrate Kingdom, who has disbelieved in permanent species, but he in his absurd though clever work has done the subject harm, as has Mr. Vestiges, and, as (some future loose naturalist attempting the same speculations will perhaps say) has Mr. D. . . .

C. D. to J. D. Hooker. September 25th [1853].

In my own Cirripedial work (by the way, thank you for the dose of soft solder; it does one—or at least me—a great deal of good)—in my own work I have not felt conscious that disbelieving in the mere *permanence* of species has made much difference one way or the other; in some few cases (if publishing avowedly on the doctrine of non-permanence), I should *not* have affixed names, and in some few cases should have affixed names to remarkable varieties. Certainly I have felt it humiliating, discussing and doubting, and examining over and over again, when in my own mind the only doubt has been whether the form varied *to-day or yesterday* (not to put too fine a point on it, as Snagsby *

---

* In *Bleak House.*

would say). After describing a set of forms as distinct species, tearing up my MS., and making them one species, tearing that up and making them separate, and then making them one again (which has happened to me), I have gnashed my teeth, cursed species, and asked what sin I had committed to be so punished. But I must confess that perhaps nearly the same thing would have happened to me on any scheme of work.

C. D. to J. D. Hooker.   Down, March 26th [1854].

MY DEAR HOOKER.—I had hoped that you would have had a little breathing-time after your Journal,* but this seems to be very far from the case; and I am the more obliged (and somewhat contrite) for the long letter received this morning, *most* juicy with news and *most* interesting to me in many ways. I am very glad indeed to hear of the reforms, &c., in the Royal Society. With respect to the Club,† I am deeply interested; only two or three days ago, I was regretting to my wife, how I was letting drop and being dropped by nearly all my acquaintances, and that I would endeavour to go oftener to London; I was not then thinking of the Club, which, as far as one thing goes, would answer my exact object in keeping up old and making some new acquaintances. I will therefore come up to London for every (with rare exceptions) Club-day, and then my head, I think, will allow me on an average to go to every other meeting. But it is grievous how often any change knocks me up. I will further pledge myself, as I told Lyell, to resign after a year, if I did not attend pretty often, so that I should *at worst* encumber the Club temporarily. If you can get me elected, I certainly shall be very much

---

* Sir Joseph Hooker's *Himalayan Journal.*
† The Philosophical Club, to which my father was elected (as Professor Bonney is good enough to inform me) on April 24, 1854. He resigned his membership in 1864. The club was founded in 1847. The number of members being limited to 47, it was proposed to christen it " the Club of 47," but the name was never adopted. The nature of the Club may be gathered from its first rule : " The purpose of the Club is to promote as much as possible the scientific objects of the Royal Society ; to facilitate intercourse between those Fellows who are actively engaged in cultivating the various branches of Natural Science, and who have contributed to its progress ; to increase the attendance at the evening meetings, and to encourage the contribution and discussion of papers." The Club met for dinner at 6, and the chair was to be quitted at 8.15, it being expected that members would go to the Royal Society. Of late years the dinner has been at 6.30, the Society meeting in the afternoon.

pleased. . . . I am particularly obliged to you for sending me Asa Gray's letter; how very pleasantly he writes. To see his and your caution on the species-question ought to overwhelm me in confusion and shame; it does make me feel deuced uncomfortable. . . . I was pleased and surprised to see A. Gray's remarks on crossing obliterating varieties, on which, as you know, I have been collecting facts for these dozen years. How awfully flat I shall feel, if, when I get my notes together on species, &c. &c., the whole thing explodes like an empty puff-ball. Do not work yourself to death.   Ever yours most truly.

To work out the problem of the Geographical Distribution of animals and plants on evolutionary principles, Darwin had to study the means by which seeds, eggs, &c., can be transported across wide spaces of ocean. It was this need which gave an interest to the class of experiment to which the following letters refer.

C. D. to J. D. Hooker.   April 13th [1855].

. . . I have had one experiment some little time in progress which will, I think, be interesting, namely, seeds in salt water, immersed in water of 32°–33°, which I have and shall long have, as I filled a great tank with snow. When I wrote last I was going to triumph over you, for my experiment had in a slight degree succeeded; but this, with infinite baseness, I did not tell, in hopes that you would say that you would eat all the plants which I could raise after immersion. It is very aggravating that I cannot in the least remember what you did formerly say that made me think you scoffed at the experiments vastly; for you now seem to view the experiment like a good Christian. I have in small bottles out of doors, exposed to variation of temperature, cress, radish, cabbages, lettuces, carrots, and celery, and onion seed. These, after immersion for exactly one week, have all germinated, which I did not in the least expect (and thought how you would sneer at me); for the water of nearly all, and of the cress especially, smelt very badly, and the cress seed emitted a wonderful quantity of mucus (the *Vestiges* * would have expected them to turn into tadpoles), so as to adhere in a mass; but these seeds

---

* *The Vestiges of Creation*, by R. Chambers.

germinated and grew splendidly. The germination of all (especially cress and lettuces) has been accelerated, except the cabbages, which have come up very irregularly, and a good many, I think, dead. One would have thought, from their native habitat, that the cabbage would have stood well. The Umbelliferæ and onions seem to stand the salt well. I wash the seed before planting them. I have written to the *Gardeners' Chronicle*,* though I doubt whether it was worth while. If my success seems to make it worth while, I will send a seed list, to get you to mark some different classes of seeds. To-day I replant the same seeds as above after fourteen days' immersion. As many sea-currents go a mile an hour, even in a week they might be transported 168 miles; the Gulf Stream is said to go fifty and sixty miles a day. So much and too much on this head; but my geese are always swans. . . .

C. D. to J. D. Hooker. [April 14th, 1855].

. . . You are a good man to confess that you expected the cress would be killed in a week, for this gives me a nice little triumph. The children at first were tremendously eager, and asked me often, "whether I should beat Dr. Hooker!" The cress and lettuce have just vegetated well after twenty-one days' immersion. But I will write no more, which is a great virtue in me; for it is to me a very great pleasure telling you everything I do.

. . . If you knew some of the experiments (if they may be so called) which I am trying, you would have a good right to sneer, for they are so *absurd* even in *my* opinion that I dare not tell you.

Have not some men a nice notion of experimentising? I have had a letter telling me that seeds *must* have *great* power of resisting salt water, for otherwise how could they get to islands? This is the true way to solve a problem?

Experiments on the transportal of seeds through the agency of animals, also gave him much labour. He wrote to Fox (1855):—

" All nature is perverse and will not do as I wish it; and

---

* A few words asking for information. The results were published in the *Gardeners' Chronicle*, May 26, Nov. 24, 1855. In the same year (p. 789) he sent a postscript to his former paper, correcting a misprint and adding a few words on the seeds of the Leguminosæ. A fuller paper on the germination of seeds after treatment in salt water, appeared in the *Linnean Soc. Journal*, 1857, p. 130.

just at present I wish I had my old barnacles to work at, and nothing new."

And to Hooker :—

" Everything has been going wrong with me lately : the fish at the Zoolog. Soc. ate up lots of soaked seeds, and in imagination they had in my mind been swallowed, fish and all, by a heron, had been carried a hundred miles, been voided on the banks of some other lake and germinated splendidly, when lo and behold, the fish ejected vehemently, and with disgust equal to my own, *all* the seeds from their mouths."

THE UNFINISHED BOOK.

In his Autobiographical sketch (p. 41) my father wrote : —" Early in 1856 Lyell advised me to write out my views pretty fully, and I began at once to do so on a scale three or four times as extensive as that which was afterwards followed in my *Origin of Species ;* yet it was only an abstract of the materials which I had collected."   The remainder of the present chapter is chiefly concerned with the preparation of this unfinished book.

The work was begun on May 14th, and steadily continued up to June 1858, when it was interrupted by the arrival of Mr. Wallace's MS.   During the two years which we are now considering, he wrote ten chapters (that is about one-half) of the projected book.

C. D. to J. D. Hooker.   May 9th [1856].

. . . I very much want advice and *truthful* consolation if you can give it.   I had a good talk with Lyell about my species work, and he urges me strongly to publish something.   I am fixed against any periodical or Journal, as I positively will *not* expose myself to an Editor or a Council allowing a publication for which they might be abused.   If I publish anything it must be a *very thin* and little volume, giving a sketch of my views and difficulties ; but it is really dreadfully unphilosophical to give a *résumé*, without exact references, of an unpublished work.   But Lyell seemed to think I might do this, at the suggestion of friends, and on the ground, which I might state, that I had been at work for eighteen* years, and yet could not publish for several

---

* The interval of eighteen years, from 1837 when he began to collect facts, would bring the date of this letter to 1855, not 1856, nevertheless the latter seems the more probable date.

years, and especially as I could point out difficulties which seemed to me to require especial investigation. Now what think you? I should be really grateful for advice. I thought of giving up a couple of months and writing such a sketch, and trying to keep my judgment open whether or no to publish it when completed. It will be simply impossible for me to give exact references; anything important I should state on the authority of the author generally; and instead of giving all the facts on which I ground my opinion, I could give by memory only one or two. In the Preface I would state that the work could not be considered strictly scientific, but a mere sketch or outline of a future work in which full references, &c., should be given. Eheu, eheu, I believe I should sneer at any one else doing this, and my only comfort is, that I *truly* never dreamed of it, till Lyell suggested it, and seems deliberately to think it advisable.

I am in a peck of troubles, and do pray forgive me for troubling you.

Yours affectionately.

He made an attempt at a sketch of his views, but as he wrote to Fox in October 1856 :—

" I found it such unsatisfactory work that I have desisted, and am now drawing up my work as perfect as my materials of nineteen years' collecting suffice, but do not intend to stop to perfect any line of investigation beyond current work."

And in November he wrote to Sir Charles Lyell :—

" I am working very steadily at my big book; I have found it quite impossible to publish any preliminary essay or sketch; but am doing my work as completely as my present materials allow without waiting to perfect them. And this much acceleration I owe to you."

Again to Mr. Fox, in February, 1857 :—

" I am got most deeply interested in my subject; though I wish I could set less value on the bauble fame, either present or posthumous, than I do, but not I think, to any extreme degree; yet, if I know myself, I would work just as hard, though with less gusto, if I knew that my book would be published for ever anonymously."

C. D. to A. R. Wallace. Moor Park, May 1st, 1857.

MY DEAR SIR—I am much obliged for your letter of October 10th, from Celebes, received a few days ago; in a laborious undertaking, sympathy is a valuable and real encouragement. By your letter and even still more by your paper* in the Annals, a year or more ago, I can plainly see that we have thought much alike and to a certain extent have come to similar conclusions. In regard to the Paper in the Annals, I agree to the truth of almost every word of your paper; and I dare say that you will agree with me that it is very rare to find oneself agreeing pretty closely with any theoretical paper; for it is lamentable how each man draws his own different conclusions from the very same facts. This summer will make the 20th year (!) since I opened my first note-book, on the question how and in what way do species and varieties differ from each other. I am now preparing my work for publication, but I find the subject so very large, that though I have written many chapters, I do not suppose I shall go to press for two years. I have never heard how long you intend staying in the Malay Archipelago; I wish I might profit by the publication of your Travels there before my work appears, for no doubt you will reap a large harvest of facts. I have acted already in accordance with your advice of keeping domestic varieties, and those appearing in a state of nature, distinct; but I have sometimes doubted of the wisdom of this, and therefore I am glad to be backed by your opinion. I must confess, however, I rather doubt the truth of the now very prevalent doctrine of all our domestic animals having descended from several wild stocks; though I do not doubt that it is so in some cases. I think there is rather better evidence on the sterility of hybrid animals than you seem to admit; and in regard to plants the collection of carefully recorded facts by Kölreuter and Gaertner (and Herbert) is *enormous.* I most entirely agree with you on the little effects of " climatal conditions," which one sees referred to *ad nauseam* in all books: I suppose some very little effect must be attributed to such influences, but I fully believe that they are very slight. It is really *impossible* to explain my views (in

---

* " On the Law that has regulated the Introduction of New Species."— *Ann. Nat. Hist.*, 1855.

the compass of a letter), on the causes and means of variation in a state of nature; but I have slowly adopted a distinct and tangible idea,—whether true or false others must judge; for the firmest conviction of the truth of a doctrine by its author, seems, alas, not to be the slightest guarantee of truth! . . .

In December 1857 he wrote to the same correspondent:—

"You ask whether I shall discuss 'man.' I think I shall avoid the whole subject, as so surrounded with prejudices; though I fully admit that it is the highest and most interesting problem for the naturalist. My work, on which I have now been at work more or less for twenty years, will not fix or settle anything; but I hope it will aid by giving a large collection of fact, with one definite end. I get on very slowly, partly from ill-health, partly from being a very slow worker. I have got about half written; but I do not suppose I shall publish under a couple of years. I have now been three whole months on one chapter on Hybridism!

I am astonished to see that you expect to remain out three or four years more. What a wonderful deal you will have seen, and what interesting areas—the grand Malay Archipelago and the richest parts of South America! I infinitely admire and honour your zeal and courage in the good cause of Natural Science; and you have my very sincere and cordial good wishes for success of all kinds, and may all your theories succeed, except that on Oceanic Islands, on which subject I will do battle to the death."

And to Fox in February 1858:—

"I am working very hard at my book, perhaps too hard. It will be very big, and I am become most deeply interested in the way facts fall into groups. I am like Crœsus overwhelmed with my riches in facts, and I mean to make my book as perfect as ever I can. I shall not go to press at soonest for a couple of years."

The letter which follows, written from his favourite resting place, the Water-Cure Establishment at Moor Park, comes in like a lull before the storm,—the upset of all his plans by the arrival of Mr. Wallace's manuscript, a phase in the history of his life to which the next chapter is devoted.

C. D. to Mrs. Darwin. Moor Park, April [1858].

The weather is quite delicious. Yesterday, after writing to you, I strolled a little beyond the glade for an hour and

a half, and enjoyed myself—the fresh yet dark green of the
grand Scotch firs, the brown of the catkins of the old
birches, with their white stems, and a fringe of distant
green from the larches, made an excessively pretty view.
At last I fell fast asleep on the grass, and awoke with a
chorus of birds singing around me, and squirrels running
up the trees, and some woodpeckers laughing, and it was as
pleasant and rural a scene as ever I saw, and I did not care
one penny how any of the beasts or birds had been formed.
I sat in the drawing-room till after eight, and then went
and read the Chief-Justice's summing up, and thought Bernard * guilty, and then read a bit of my novel, which is
feminine, virtuous, clerical, philanthropical, and all that
sort of thing, but very decidedly flat.   I say feminine, for
the author is ignorant about money matters, and not much
of a lady—for she makes her men say, "My Lady."   I like
Miss Craik very much, though we have some battles, and
differ on every subject.   I like also the Hungarian; a thorough gentleman, formerly attaché at Paris, and then in the
Austrian cavalry, and now a pardoned exile, with broken
health.   He does not seem to like Kossuth, but says, he is
certain [he is] a sincere patriot, most clever and eloquent,
but weak, with no determination of character. . . .

---

* Simon Bernard was tried in April 1858 as an accessory to Orsini's
attempt on the life of the Emperor of the French.   The verdict was " not
guilty."

CHAPTER XI.

THE WRITING OF THE 'ORIGIN OF SPECIES.'

" I have done my best. If you had all my material I am sure you would have made a splendid book."—*From a letter to Lyell, June* 21, 1859.

JUNE 18, 1858, TO NOVEMBER 1859.

C. D. to C. Lyell.   Down, 18th [June 1858].

MY DEAR LYELL—Some year or so ago you recommended me to read a paper by Wallace in the *Annals*,* which had interested you, and as I was writing to him, I knew this would please him much, so I told him. He has to-day sent me the enclosed, and asked me to forward it to you. It seems to me well worth reading. Your words have come true with a vengeance—that I should be forestalled. You said this, when I explained to you here very briefly my views of 'Natural Selection' depending on the struggle for existence. I never saw a more striking coincidence; if Wallace had my MS. sketch written out in 1842, he could not have made a better short abstract! Even his terms now stand as heads of my chapters. Please return me the MS., which he does not say he wishes me to publish, but I shall, of course, at once write and offer to send to any journal. So all my originality, whatever it may amount to, will be smashed, though my book, if it will ever have any value, will not be deteriorated; as all the labour consists in the application of the theory.

I hope you will approve of Wallace's sketch, that I may tell him what you say.

My dear Lyell, yours most truly.

---

* *Annals and Mag. of Nat. Hist.*, 1855.

C. D. to C. Lyell. Down [June 25, 1858].

MY DEAR LYELL—I am very sorry to trouble you, busy as you are, in so merely personal an affair; but if you will give me your deliberate opinion, you will do me as great a service as ever man did, for I have entire confidence in your judgment and honour. . . .

There is nothing in Wallace's sketch which is not written out much fuller in my sketch, copied out in 1844, and read by Hooker some dozen years ago.   About a year ago I sent a short sketch, of which I have a copy, of my views (owing to correspondence on several points) to Asa Gray, so that I could most truly say and prove that I take nothing from Wallace.   I should be extremely glad now to publish a sketch of my general views in about a dozen pages or so; but I cannot persuade myself that I can do so honourably. Wallace says nothing about publication, and I enclose his letter.   But as I had not intended to publish any sketch, can I do so honourably, because Wallace has sent me an outline of his doctrine?   I would far rather burn my whole book, than that he or any other man should think that I had behaved in a paltry spirit.   Do you not think his having sent me this sketch ties my hands? . . . .   If I could honourably publish I would state that I was induced now to publish a sketch (and I should be very glad to be permitted to say, to follow your advice long ago given) from Wallace having sent me an outline of my general conclusions.   We differ only [in] that I was led to my views from what artificial selection has done for domestic animals.   I would send Wallace a copy of my letter to Asa Gray, to show him that I had not stolen his doctrine.   But I cannot tell whether to publish now would not be base and paltry.   This was my first impression, and I should have certainly acted on it had it not been for your letter.

This is a trumpery affair to trouble you with, but you cannot tell how much obliged I should be for your advice.

By the way, would you object to send this and your answer to Hooker to be forwarded to me? for then I shall have the opinion of my two best and kindest friends.   This letter is miserably written, and I write it now that I may for a time banish the whole subject; and I am worn out with musing. . . .

My good dear friend, forgive me. This is a trumpery letter, influenced by trumpery feelings.

Yours most truly.

I will never trouble you or Hooker on the subject again.

C. D. to C. Lyell.   Down, 26th [June 1858].

MY DEAR LYELL—Forgive me for adding a P.S. to make the case as strong as possible against myself.

Wallace might say, "You did not intend publishing an abstract of your views till you received my communication. Is it fair to take advantage of my having freely, though unasked, communicated to you my ideas, and thus prevent me forestalling you?" The advantage which I should take being that I am induced to publish from privately knowing that Wallace is in the field. It seems hard on me that I should be thus compelled to lose my priority of many years' standing, but I cannot feel at all sure that this alters the justice of the case. First impressions are generally right, and I at first thought it would be dishonourable in me now to publish.

Yours most truly.

P.S.—I have always thought you would make a first-rate Lord Chancellor; and I now appeal to you as a Lord Chancellor.

C. D. to J. D. Hooker.   Tuesday night [June 29, 1858].

MY DEAR HOOKER—I have just read your letter, and see you want the papers at once. I am quite prostrated,* and can do nothing, but I send Wallace, and the abstract † of my letter to Asa Gray, which gives most imperfectly only the means of change, and does not touch on reasons for believing that species do change. I dare say all is too late. I hardly care about it. But you are too generous to sacrifice so much time and kindness. It is most generous, most kind. I send my sketch of 1844 solely that you may see by your own handwriting that you did read it. I really cannot bear to look at it. Do not waste much time. It is miserable in me to care at all about priority.

---

* After the death, from scarlet fever, of his infant child.

† "Abstract" is here used in the sense of "extract;" in this sense also it occurs in the *Linnean Journal*, where the sources of my father's paper are described.

The table of contents will show what it is.

I would make a similar, but shorter and more accurate sketch for the *Linnean Journal*.

I will do anything. God bless you, my dear kind friend. I can write no more. I send this by my servant to Kew.

The joint paper * of Mr. Wallace and my father was read at the Linnean Society on the evening of July 1st. Mr. Wallace's Essay bore the title, " On the Tendency of Varieties to depart indefinitely from the Original Type."

My father's contribution to the paper consisted of (1) Extracts from the sketch of 1844; (2) part of a letter addressed to Dr. Asa Gray, dated September 5, 1857. The paper was " communicated " to the Society by Sir Charles Lyell and Sir Joseph Hooker, in whose prefatory letter a clear account of the circumstances of the case is given.

Referring to Mr. Wallace's Essay, they wrote :—

" So highly did Mr. Darwin appreciate the value of the views therein set forth, that he proposed, in a letter to Sir Charles Lyell, to obtain Mr. Wallace's consent to allow the Essay to be published as soon as possible. Of this step we highly approved, provided Mr. Darwin did not withhold from the public, as he was strongly inclined to do (in favour of Mr. Wallace), the memoir which he had himself written on the same subject, and which, as before stated, one of us had perused in 1844, and the contents of which we had both of us been privy to for many years. On representing this to Mr. Darwin, he gave us permission to make what use we thought proper of his memoir, &c. ; and in adopting our present course, of presenting it to the Linnean Society, we have explained to him that we are not solely considering the relative claims to priority of himself and his friend, but the interests of science generally."

Sir Charles Lyell and Sir J. D. Hooker were present at the reading of the paper, and both, I believe, made a few remarks, chiefly with a view of impressing on those present the necessity of giving the most careful consideration to what they had heard. There was, however, no semblance of a discussion. Sir Joseph Hooker writes to me : " The interest excited was intense, but the subject was too novel

---

* " On the tendency of Species to form Varieties and on the Perpetuation of Varieties and Species by Natural Means of Selection."—*Linnean Society's Journal*, iii. p. 53.

and too ominous for the old school to enter the lists, before armouring. After the meeting it was talked over with bated breath : Lyell's approval and perhaps in a small way mine, as his lieutenant in the affair, rather overawed the Fellows, who would otherwise have flown out against the doctrine. We had, too, the vantage ground of being familiar with the authors and their theme."

Mr. Wallace has, at my request, been so good as to allow me to publish the following letter. Professor Newton, to whom the letter is addressed, had submitted to Mr. Wallace his recollections of what the latter had related to him many years before, and had asked Mr. Wallace for a fuller version of the story. Hence the few corrections in Mr. Wallace's letter, for instance *bed* for *hammock*.

A. R. Wallace to A. Newton. Frith Hill, Godalming, Dec. 3rd, 1887.

MY DEAR NEWTON—I had hardly heard of Darwin before going to the East, except as connected with the voyage of the *Beagle*, which I *think* I had read. I saw him *once* for a few minutes in the British Museum before I sailed. Through Stevens, my agent, I heard that he wanted curious *varieties* which he was studying. I *think* I wrote to him about some varieties of ducks I had sent, and he must have written once to me. I find on looking at his " Life " that his *first* letter to me is given in vol. ii. p. 95, and another at p. 109, both after the publication of my first paper. I must have heard from some notices in the *Athenæum*, I think (which I had sent me), that he was studying varieties and species, and as I was continually thinking of the subject, I wrote to him giving some of my notions, and making some suggestions. But at that time I had not the remotest notion that he had already arrived at a definite theory—still less that it was the same as occurred to me, suddenly, in Ternate in 1858. The most interesting coincidence in the matter, I think, is, that I, *as well as Darwin*, was led to the theory itself through Malthus—in my case it was his elaborate account of the action of " preventive checks " in keeping down the population of savage races to a tolerably fixed but scanty number. This had strongly impressed me, and it suddenly flashed upon me that all animals are necessarily thus kept down—" the struggle for existence "—while *variations*, on which I was always thinking, must necessarily often be *beneficial*, and would then cause those varieties to increase while the injurious variations diminished.* You are quite at liberty to mention the circumstances, but I think you have coloured them a little highly, and introduced some slight errors. I was lying on my bed (no hammocks in the East) in the hot fit of intermittent fever, when the idea suddenly came to me. I thought it almost all out before the fit was over, and the moment I got up began to write it down, and I believe finished the first draft the next day.

I had no idea whatever of " dying,"—as it was not a serious illness,—but I *had* the idea of working it out, so far as I was able, when I returned home, not at all expecting that Darwin had so long anticipated me. I can truly say *now*, as I said many years ago, that I am glad it was so; for I have not the love of *work, experiment* and *detail* that was so pre-eminent in Darwin, and without which anything I could have written would never have convinced the world. If you do refer to me at any length, can you send me a proof and I will return it to you at once ?

Yours faithfully
ALFRED R. WALLACE.

C. D. to J. D. Hooker. Miss Wedgwood's, Hartfield, Tunbridge Wells [July 13th, 1858].

MY DEAR HOOKER—Your letter to Wallace seems to me perfect, quite clear and most courteous. I do not think it could possibly be improved, and I have to-day forwarded it with a letter of my own. I always thought it very possible that I might be forestalled, but I fancied that I had a grand enough soul not to care; but I found myself mistaken and punished; I had, however, quite resigned myself, and had written half a letter to Wallace to give up all priority to him, and should certainly not have changed had it not been for Lyell's and your quite extraordinary kindness.

---

* This passage was published as a footnote in a review of the *Life and Letters of Charles Darwin* which appeared in the *Quarterly Review*, Jan. 1888. In the new edition (1891) of *Natural Selection and Tropical Nature* (p. 20), Mr. Wallace has given the facts above narrated. There is a slight and quite unimportant discrepancy between the two accounts, viz. that in the narrative of 1891 Mr. Wallace speaks of the " cold fit " instead of the " hot fit " of his ague attack.

I assure you I feel it, and shall not forget it.   I am *more* than satisfied at what took place at the Linnean Society.   I had thought that your letter and mine to Asa Gray were to be only an appendix to Wallace's paper.

We go from here in a few days to the sea-side, probably to the Isle of Wight, and on my return (after a battle with pigeon skeletons) I will set to work at the abstract, though how on earth I shall make anything of an abstract in thirty pages of the Journal, I know not, but will try my best. . . .

I must try and see you before your journey; but do not think I am fishing to ask you to come to Down, for you will have no time for that.

You cannot imagine how pleased I am that the notion of Natural Selection has acted as a purgative on your bowels of immutability.   Whenever naturalists can look at species changing as certain, what a magnificent field will be open,—on all the laws of variation,—on the genealogy of all living beings,—on their lines of migration, &c. &c. Pray thank Mrs. Hooker for her very kind little note, and pray say how truly obliged I am, and in truth ashamed to think that she should have had the trouble of copying my ugly MS.   It was extraordinarily kind in her.   Farewell, my dear kind friend.

Yours affectionately.

P.S.—I have had some fun here in watching a slave-making ant; for I could not help rather doubting the wonderful stories, but I have now seen a defeated marauding party, and I have seen a migration from one nest to another of the slave-makers, carrying their slaves (who are *house*, and not field niggers) in their mouths!

C. D. to C. Lyell.   King's Head Hotel, Sandown, Isle of Wight.   July 18th [1858].

. . . We are established here for ten days, and then go on to Shanklin, which seems more amusing to one, like myself, who cannot walk.   We hope much that the sea may do H. and L. good.   And if it does, our expedition will answer, but not otherwise.

I have never half thanked you for all the extraordinary trouble and kindness you showed me about Wallace's affair. Hooker told me what was done at the Linnean Society, and I am far more than satisfied, and I do not think that Wallace can think my conduct unfair in allowing you and Hooker to do whatever you thought fair. I certainly was a little annoyed to lose all priority, but had resigned myself to my fate. I am going to prepare a longer abstract; but it is really impossible to do justice to the subject, except by giving the facts on which each conclusion is grounded, and that will, of course, be absolutely impossible. Your name and Hooker's name appearing as in any way the least interested in my work will, I am certain, have the most important bearing in leading people to consider the subject without prejudice. I look at this as so very important, that I am almost glad of Wallace's paper for having led to this.

My dear Lyell, yours most gratefully.

The following letter refers to the proof-sheets of the Linnean paper. The 'introduction' means the prefatory letter signed by Sir C. Lyell and Sir J. D. Hooker.

C. D. to J. D. Hooker.   King's Head Hotel, Sandown, Isle of Wight.   July 21st [1858].

MY DEAR HOOKER.—I received only yesterday the proof-sheets, which I now return. I think your introduction cannot be improved.

I am disgusted with my bad writing. I could not improve it, without rewriting all, which would not be fair or worth while, as I have begun on a better abstract for the Linnean Society. My excuse is that it *never* was intended for publication. I have made only a few corrections in the style; but I cannot make it decent, but I hope moderately intelligible. I suppose some one will correct the revise. (Shall I?)

Could I have a clean proof to send to Wallace?

I have not yet fully considered your remarks on big genera (but your general concurrence is of the *highest possible* interest to me); nor shall I be able till I re-read my MS.; but you may rely on it that you never make a remark to me which is lost from *inattention*. I am particularly glad you do not object to my stating your objections in a modified form, for they always struck me as very important, and as having much inherent value, whether or no they were fatal to my notions. I will consider and reconsider all your remarks. . . .

I am very glad at what you say about my Abstract, but

you may rely on it that I will condense to the utmost. I would aid in money if it is too long.* In how many ways you have aided me!

Yours affectionately.

The "Abstract" mentioned in the last sentence of the preceding letter was in fact the *Origin of Species,* on which he now set to work. In his *Autobiography* (p. 41) he speaks of beginning to write in September, but in his Diary he wrote, "July 20 to Aug. 12, at Sandown, began Abstract of Species book." "Sep. 16, Recommenced Abstract." The book was begun with the idea that it would be published as a paper, or series of papers, by the Linnean Society, and it was only in the late autumn that it became clear that it must take the form of an independent volume.

C. D. to J. D. Hooker. Norfolk House, Shanklin, Isle of Wight. [August 1858.]

MY DEAR HOOKER,—I write merely to say that the MS. came safely two or three days ago. I am much obliged for the correction of style : I find it unutterably difficult to write clearly. When we meet I must talk over a few points on the subject.

You speak of going to the sea-side somewhere ; we think this the nicest sea-side place which we have ever seen, and we like Shanklin better than other spots on the south coast of the island, though many are charming and prettier, so that I would suggest your thinking of this place. We are on the actual coast ; but tastes differ so much about places.

If you go to Broadstairs, when there is a strong wind from the coast of France and in fine, dry, warm weather, look out and you will *probably* (!) see thistle-seeds blown across the Channel. The other day I saw one blown right inland, and then in a few minutes a second one and then a third ; and I said to myself, God bless me, how many thistles there must be in France : and I wrote a letter in imagination to you. But I then looked at the *low* clouds, and noticed that they were not coming inland, so I feared a screw was loose. I then walked beyond the headland and found the wind parallel to the coast, and on this very headland a noble

---

* That is to say, he would help to pay for the printing, if it should prove too long for the Linnean Society.

bed of thistles, which by every wide eddy were blown far out
to sea, and then came right in at right angles to the shore !
One day such a number of insects were washed up by the
tide, and I brought to life thirteen species of Coleoptera;
not that I suppose these came from France. But do you
watch for thistle-seed as you saunter along the coast. . . .

C. D. to J. D. Hooker. [Down] Oct. 6th, 1858.

" . . . If you have or can make leisure, I should very
much like to hear news of Mrs. Hooker, yourself, and the
children. Where did you go, and what did you do and are
doing? There is a comprehensive text.

" You cannot tell how I enjoyed your little visit here. It
did me much good. If Harvey * is still with you, pray remember me very kindly to him.

" . . . I am working most steadily at my Abstract [*Origin
of Species*], but it grows to an inordinate length; yet fully
to make my view clear (and never giving briefly more than
a fact or two, and slurring over difficulties), I cannot make
it shorter. It will yet take me three or four months; so
slow do I work, though never idle. You cannot imagine
what a service you have done me in making me make this
Abstract; for though I thought I had got all clear, it has
clarified my brains very much, by making me weigh the
relative importance of the several elements."

He was not so fully occupied but that he could find time
to help his boys in their collecting. He sent a short notice
to the *Entomologists' Weekly Intelligencer*, June 25th, 1859,
recording the capture of *Licinus silphoides, Clytus mysti-
cus, Panagæus* 4-*pustulatus.* The notice begins with the
words, " We three very young collectors having lately taken
in the parish of Down," &c., and is signed by three of his
boys, but was clearly not written by them. I have a vivid
recollection of the pleasure of turning out my bottle of dead
beetles for my father to name, and the excitement, in which
he fully shared, when any of them proved to be uncommon
ones. The following letter to Mr. Fox (Nov. 13th, 1858),
illustrates this point :—

" I am reminded of old days by my third boy having just
begun collecting beetles, and he caught the other day
Brachinus crepitans, of immortal Whittlesea Mere memory.

---

* W. H. Harvey, born 1811, died 1866 : a well-known botanist.

My blood boiled with old ardour when he caught a Licinus
—a prize unknown to me."

And again to Sir John Lubbock :—

" I feel like an old war-horse at the sound of the trum-
pet when I read about the capturing of rare beetles—is not
this a magnanimous simile for a decayed entomologist?—It
really almost makes me long to begin collecting again.
Adios.

" ' Floreat Entomologia ' !—to which toast at Cambridge
I have drunk many a glass of wine. So again, ' Floreat
Entomologia.'—N.B. I have *not* now been drinking any
glasses full of wine."

C. D. to J. D. Hooker. Down, Jan. 23rd, 1859.

. . . I enclose letters to you and me from Wallace. I
admire extremely the spirit in which they are written. I
never felt very sure what he would say. He must be an
amiable man. Please return that to me, and Lyell ought
to be told how well satisfied he is. These letters have vividly
brought before me how much I owe to your and Lyell's
most kind and generous conduct in all this affair.

. . . How glad I shall be when the Abstract is finished,
and I can rest ! . . .

C. D. to A. R. Wallace. Down, Jan. 25th [1859].

MY DEAR SIR,—I was extremely much pleased at receiving three days ago your letter to me and that to Dr.
Hooker. Permit me to say how heartily I admire the spirit
in which they are written. Though I had absolutely nothing whatever to do in leading Lyell and Hooker to what
they thought a fair course of action, yet I naturally could
not but feel anxious to hear what your impression would be.
I owe indirectly much to you and them ; for I almost think
that Lyell would have proved right, and I should never have
completed my larger work, for I have found my Abstract
[*Origin of Species*] hard enough with my poor health, but
now, thank God, I am in my last chapter but one. My Abstract will make a small volume of 400 or 500 pages. Whenever published, I will, of course, send you a copy, and then
you will see what I mean about the part which I believe selection has played with domestic productions. It is a very
different part, as you suppose, from that played by " Natural
Selection." I sent off, by the same address as this note, a

copy of the *Journal of the Linnean Society*, and subsequently I have sent some half-dozen copies of the paper. I have many other copies at your disposal. . . .

I am glad to hear that you have been attending to birds' nests. I have done so, though almost exclusively under one point of view, viz. to show that instincts vary, so that selection could work on and improve them. Few other instincts, so to speak, can be preserved in a Museum.

Many thanks for your offer to look after horses' stripes; if there are any donkeys, pray add them. I am delighted to hear that you have collected bees' combs. . . . . This is an especial hobby of mine, and I think I can throw a light on the subject. If you can collect duplicates at no very great expense, I should be glad of some specimens for myself with some bees of each kind. Young, growing, and irregular combs, and those which have not had pupæ, are most valuable for measurements and examination. Their edges should be well protected against abrasion.

Every one whom I have seen has thought your paper very well written and interesting. It puts my extracts (written in 1839,* now just twenty years ago!), which I must say in apology were never for an instant intended for publication, into the shade.

You ask about Lyell's frame of mind. I think he is somewhat staggered, but does not give in, and speaks with horror, often to me, of what a thing it would be, and what a job it would be for the next edition of *The Principles*, if he " were perverted." But he is most candid and honest, and I think will end by being perverted. Dr. Hooker has become almost as heterodox as you or I, and I look at Hooker as *by far* the most capable judge in Europe.

Most cordially do I wish you health and entire success in all your pursuits, and, God knows, if admirable zeal and energy deserve success, most amply do you deserve it. I look at my own career as nearly run out. If I can publish my Abstract and perhaps my greater work on the same subject, I shall look at my course as done.

Believe me, my dear Sir, yours very sincerely.

In March 1859 the work was telling heavily on him. He wrote to Fox :—

---

* See a discussion on the date of the earliest sketch of the *Origin* in the *Life and Letters*, ii. p. 10.

" I can see daylight through my work, and am now finally correcting my chapters for the press; and I hope in a month or six weeks to have proof-sheets. I am weary of my work. It is a very odd thing that I have no sensation that I overwork my brain; but facts compel me to conclude that my brain was never formed for much thinking. We are resolved to go for two or three months, when I have finished, to Ilkley, or some such place, to see if I can anyhow give my health a good start, for it certainly has been wretched of late, and has incapacitated me for everything. You do me injustice when you think that I work for fame; I value it to a certain extent; but if I know myself, I work from a sort of instinct to try to make out truth."

C. D. to C. Lyell. Down, March 28th [1859].

MY DEAR LYELL,—If I keep decently well, I hope to be able to go to press with my volume early in May. This being so, I want much to beg a little advice from you. From an expression in Lady Lyell's note, I fancy that you have spoken to Murray. Is it so? And is he willing to publish my Abstract? * If you will tell me whether anything, and what has passed, I will then write to him. Does he know at all of the subject of the book? Secondly, can you advise me whether I had better state what terms of publication I should prefer, or first ask him to propose terms? And what do you think would be fair terms for an edition? Share profits, or what?

Lastly, will you be so very kind as to look at the enclosed title and give me your opinion and any criticisms; you must remember that, if I have health, and it appears worth doing, I have a much larger and full book on the same subject nearly ready.

My Abstract will be about five hundred pages of the size of your first edition of the *Elements of Geology.*

Pray forgive me troubling you with the above queries; and you shall have no more trouble on the subject. I hope the world goes well with you, and that you are getting on with your various works.

I am working very hard for me, and long to finish and be free and try to recover some health.

My dear Lyell, ever yours.

---

* *The Origin of Species.*

P.S.—Would you advise me to tell Murray that my book is not more *un*-orthodox than the subject makes inevitable. That I do not discuss the origin of man. That I do not bring in any discussion about Genesis, &c. &c., and only give facts, and such conclusions from them as seem to me fair.

Or had I better say *nothing* to Murray, and assume that he cannot object to this much unorthodoxy, which in fact is not more than any Geological Treatise which runs slap counter to Genesis.

Enclosure.

AN ABSTRACT OF AN ESSAY
ON THE
ORIGIN
OF
SPECIES AND VARIETIES
THROUGH NATURAL SELECTION

BY

CHARLES DARWIN, M.A.

FELLOW OF THE ROYAL, GEOLOGICAL, AND LINNEAN SOCIETIES.

LONDON:

&c.   &c.   &c.   &c.

1859.

C. D. to C. Lyell.   Down, March 30th [1859].

MY DEAR LYELL,—You have been uncommonly kind in all you have done. You not only have saved me much trouble and some anxiety, but have done all incomparably better than I could have done it. I am much pleased at all you say about Murray. I will write either to-day or to-morrow to him, and will send shortly a large bundle of MS., but unfortunately I cannot for a week, as the first three chapters are in the copyists' hands.

I am sorry about Murray objecting to the term Abstract, as I look at it as the only possible apology for *not* giving references and facts in full, but I will defer to him and you. I am also sorry about the term "natural selection." I hope to retain it with explanation somewhat as thus :—

"Through natural selection, or the preservation of favoured races."

Why I like the term is that it is constantly used in all works on breeding, and I am surprised that it is not familiar to Murray; but I have so long studied such works that I have ceased to be a competent judge.

I again most truly and cordially thank you for your really valuable assistance.

Yours most truly.

C. D. to J. D. Hooker.   Down, April 2nd [1859].

. . . I wrote to him [Mr. Murray] and gave him the headings of the chapters, and told him he could not have the MS. for ten days or so; and this morning I received a letter, offering me handsome terms, and agreeing to publish without seeing the MS.! So he is eager enough; I think I should have been cautious, anyhow, but owing to your letter I told him most *explicitly* that I accept his offer solely on condition that, after he has seen part or all the MS. he has full power of retracting. You will think me presumptuous, but I think my book will be popular to a certain extent (enough to ensure [against] heavy loss) amongst scientific and semi-scientific men; why I think so is because I have found in conversation so great and surprising an interest amongst such men, and some 0-scientific [non-scientific] men on this subject, and all my chapters are not *nearly* so dry and dull as that which you have read on geographical distribution. Anyhow, Murray ought to be the best judge, and if he chooses to publish it, I think I may wash my hands of all responsibility. I am sure my friends, *i.e.* Lyell and you, have been *extraordinarily* kind in troubling yourselves on the matter.

I shall be delighted to see you the day before Good Friday; there would be one advantage for you in any other day—as I believe both my boys come home on that day—and it would be almost impossible that I could send the carriage for you. There will, I believe, be some relations in the house—but I hope you will not care for that, as we shall easily get as much talking as my *imbecile state* allows. I shall deeply enjoy seeing you.

. . . I am tired, so no more.

P.S.—Please to send, well *tied up* with strong string, my Geographical MS. towards the latter half of next week

—*i.e.* 7th or 8th—that I may send it with more to Murray; and God help him if he tries to read it.

. . . I cannot help a little doubting whether Lyell would take much pains to induce Murray to publish my book; this was not done at my request, and it rather grates against my pride.

I know that Lyell has been *infinitely* kind about my affair, but your dashed [*i.e.* underlined] "*induce*" gives the idea that Lyell had unfairly urged Murray.

C. D. to J. Murray.   Down, April 5th [1859].

MY DEAR SIR,—I send by this post, the Title (with some remarks on a separate page), and the first three chapters. If you have patience to read all Chapter I., I honestly think you will have a fair notion of the interest of the whole book. It may be conceit, but I believe the subject will interest the public, and I am sure that the views are original. If you think otherwise, I must repeat my request that you will freely reject my work; and though I shall be a little disappointed, I shall be in no way injured.

If you choose to read Chapters II. and III., you will have a dull and rather abstruse chapter, and a plain and interesting one, in my opinion.

As soon as you have done with the MS., please to send it by *careful messenger, and plainly directed*, to Miss G. Tollett,* 14, Queen Anne Street, Cavendish Square.

This lady, being an excellent judge of style, is going to look out for errors for me.

You must take your own time, but the sooner you finish, the sooner she will, and the sooner I shall get to press, which I so earnestly wish.

I presume you will wish to see Chapter IV.,† the keystone of my arch, and Chapters X. and XI., but please to inform me on this head.

My dear Sir, yours sincerely.

On April 11th he wrote to Hooker :—

" I write one line to say that I heard from Murray yesterday, and he says he has read the first three chapters of [my] MS. (and this includes a very dull one), and he abides

---

* Miss Tollett was an old friend of the family.
† In the first edition Chapter iv. was on Natural Selection.

by his offer.   Hence he does not want more MS., and you can send my Geographical chapter when it pleases you."

Part of the MS. seems to have been lost on its way back to my father.   He wrote (April 14) to Sir J. D. Hooker:—

"I have the old MS., otherwise the loss would have killed me!   The worst is now that it will cause delay in getting to press, and *far worst* of all, I lose all advantage of your having looked over my chapter,* except the third part returned.   I am very sorry Mrs. Hooker took the trouble of copying the two pages."

C. D. to J. D. Hooker.   [April or May, 1859.]

. . . Please do not say to any one that I thought my book on species would be fairly popular, and have a fairly remunerative sale (which was the height of my ambition), for if it prove a dead failure, it would make me the more ridiculous.

I enclose a criticism, a taste of the future—

Rev. S. Haughton's Address to the Geological Society, Dublin.†

" This speculation of Messrs. Darwin and Wallace would not be worthy of notice were it not for the weight of authority of the names (*i.e.* Lyell's and yours), under whose auspices it has been brought forward.   If it means what it says, it is a truism; if it means anything more, it is contrary to fact."   Q. E. D.

C. D. to J. D. Hooker.   Down, May 11th [1859].

MY DEAR HOOKER,—Thank you for telling me about obscurity of style.   But on my life no nigger with lash over him could have worked harder at clearness than I have done.   But the very difficulty to me, of itself leads to the probability that I fail.   Yet one lady who has read all my MS. has found only two or three obscure sentences; but Mrs. Hooker having so found it, makes me tremble.   I will

---

* The following characteristic acknowledgment of the help he received occurs in a letter to Hooker, of about this time: "I never did pick any one's pocket, but whilst writing my present chapter I keep on feeling (even when differing most from you) just as if I were stealing from you, so much do I owe to your writings and conversation, so much more than mere acknowledgments show."

† Feb. 9th, 1858.

do my best in proofs. You are a good man to take the trouble to write about it.

With respect to our mutual muddle,* I never for a moment thought we could not make our ideas clear to each other by talk, or if either of us had time to write *in extenso*.

I imagine from some expressions (but if you ask me what, I could not answer) that you look at variability as some necessary contingency with organisms, and further that there is some necessary tendency in the variability to go on diverging in character or degree. *If you do*, I do not agree. " Reversion " again (a form of inheritance), I look at as in no way directly connected with Variation, though of course inheritance is of fundamental importance to us, for if a variation be not inherited, it is of no signification to us. It was on such points as these *I fancied* that we perhaps started differently.

I fear that my book will not deserve at all the pleasant things you say about it, and Good Lord, how I do long to have done with it!

Since the above was written, I have received and have been *much interested* by A. Gray. I am delighted at his note about my and Wallace's paper. He will go round, for it is futile to give up very many species, and stop at an arbitrary line at others. It is what my father called Unitarianism, " a feather-bed to catch a falling Christian. . . .''

C. D. to J. Murray. Down, June 14th [1859].

MY DEAR SIR,—The diagram will do very well, and I will send it shortly to Mr. West to have a few trifling corrections made.

I get on very slowly with proofs. I remember writing to you that I thought there would be not much correction. I honestly wrote what I thought, but was most grievously mistaken. I find the style incredibly bad, and most difficult to make clear and smooth. I am extremely sorry to say, on account of expense, and loss of time for me, that the corrections are very heavy, as heavy as possible. But from casual glances, I still hope that later chapters are not so badly written. How I could have written so badly is

---

* " When I go over the chapter I will see what I can do, but I hardly know how I am obscure, and I think we are somehow in a mutual muddle with respect to each other, from starting from some fundamentally different notions."—Letter of May 6th, 1859.

quite inconceivable, but I suppose it was owing to my whole attention being fixed on the general line of argument, and not on details. All I can say is, that I am very sorry.

Yours very sincerely.

C. D. to J. D. Hooker. Down [Sept.] 11th [1859].

MY DEAR HOOKER,—I corrected the last proof yesterday, and I have now my revises, index, &c., which will take me near to the end of the month. So that the neck of my work, thank God, is broken.

I write now to say that I am uneasy in my conscience about hesitating to look over your proofs,* but I was feeling miserably unwell and shattered when I wrote. I do not suppose I could be of hardly any use, but if I could, pray send me any proofs. I should be (and fear I was) the most ungrateful man to hesitate to do anything for you after some fifteen or more years' help from you.

As soon as ever I have fairly finished I shall be off to Ilkley, or some other Hydropathic establishment. But I shall be some time yet, as my proofs have been so utterly obscured with corrections, that I have to correct heavily on revises.

Murray proposes to publish the first week in November. Oh, good heavens, the relief to my head and body to banish the whole subject from my mind !

I hope you do not think me a brute about your proof-sheets.

Farewell, yours affectionately.

The following letter is interesting as showing with what a very moderate amount of recognition he was satisfied,—and more than satisfied.

Sir Charles Lyell was President of the Geological section at the meeting of the British Association at Aberdeen in 1859. In his address he said :—" On this difficult and mysterious subject [Evolution] a work will very shortly appear by Mr. Charles Darwin, the result of twenty years of observations and experiments in Zoology, Botany, and Geology, by which he has been led to the conclusion that those powers of nature which give rise to races and permanent varieties in animals and plants, are the same as those which in

---

* Of Hooker's *Flora of Australia.*

much longer periods produce species, and in a still longer series of ages give rise to differences of generic rank. He appears to me to have succeeded by his investigations and reasonings in throwing a flood of light on many classes of phenomena connected with the affinities, geographical distribution, and geological succession of organic beings, for which no other hypothesis has been able, or has even attempted to account."

My father wrote :—

" You once gave me intense pleasure, or rather delight, by the way you were interested, in a manner I never expected, in my Coral Reef notions, and now you have again given me similar pleasure by the manner you have noticed my species work. Nothing could be more satisfactory to me, and I thank you for myself, and even more for the subject's sake, as I know well that the sentence will make many fairly consider the subject, instead of ridiculing it."

And again, a few days later :—

" I do thank you for your eulogy at Aberdeen. I have been so wearied and exhausted of late that I have for months doubted whether I have not been throwing away time and labour for nothing. But now I care not what the universal world says ; I have always found you right, and certainly on this occasion I am not going to doubt for the first time. Whether you go far, or but a very short way with me and others who believe as I do, I am contented, for my work cannot be in vain. You would laugh if you knew how often I have read your paragraph, and it has acted like a little dram."

C. D. to C. Lyell.   Down, Sept. 30th [1859].

MY DEAR LYELL,—I sent off this morning the last sheets, but without index, which is not in type. I look at you as my Lord High Chancellor in Natural Science, and therefore I request you, after you have finished, just to *re-run* over the heads in the recapitulation-part of the last chapter. I shall be deeply anxious to hear what you decide (if you are able to decide) on the balance of the pros and contras given in my volume, and of such other pros and contras as may occur to you. I hope that you will think that I have given the difficulties fairly. I feel an entire conviction that if you are now staggered to any moderate extent, you will come more and more round, the longer you keep the subject at all before your mind. I remember well how many long years it was before I could look into the face of some of the difficulties and not feel quite abashed. I fairly struck my colours before the case of neuter insects.*

I suppose that I am a very slow thinker, for you would be surprised at the number of years it took me to see clearly what some of the problems were which had to be solved, such as the necessity of the principle of divergence of character, the extinction of intermediate varieties, on a continuous area, with graduated conditions; the double problem of sterile first crosses and sterile hybrids, &c. &c.

Looking back, I think it was more difficult to see what the problems were than to solve them, so far as I have succeeded in doing, and this seems to me rather curious. Well, good or bad, my work, thank God, is over; and hard work, I can assure you, I have had, and much work which has never borne fruit. You can see, by the way I am scribbling, that I have an idle and rainy afternoon. I was not able to start for Ilkley yesterday as I was too unwell; but I hope to get there on Tuesday or Wednesday. Do, I beg you, when you have finished my book and thought a little over it, let me hear from you. Never mind and pitch into me, if you think it requisite; some future day, in London possibly, you may give me a few criticisms in detail, that is, if you have scribbled any remarks on the margin, for the chance of a second edition.

Murray has printed 1250 copies, which seems to me rather too large an edition, but I hope he will not lose.

I make as much fuss about my book as if it were my first. Forgive me, and believe me, my dear Lyell,

Yours most sincerely.

The book was at last finished and printed, and he wrote to Mr. Murray :—

Ilkley, Yorkshire [1859].

MY DEAR SIR,—I have received your kind note and the copy; I am infinitely pleased and proud at the appearance of my child.

I quite agree to all you propose about price. But you

---

* *Origin of Species*, 6th Edition, vol. ii. p. 357. " But with the working ant we have an insect differing greatly from its parents, yet absolutely sterile, so that it could never have transmitted successively acquired modifications of structure or instinct to its progeny. It may well be asked how is it possible to reconcile this case with the theory of natural selection ? "

are really too generous about the, to me, scandalously heavy corrections. Are you not acting unfairly towards yourself? Would it not be better at least to share the £72 8*s.*? I shall be fully satisfied, for I had no business to send, though quite unintentionally and unexpectedly, such badly composed MS. to the printers.

Thank you for your kind offer to distribute the copies to my friends and assisters as soon as possible. Do not trouble yourself much about the foreigners, as Messrs. Williams and Norgate have most kindly offered to do their best, and they are accustomed to send to all parts of the world.

I will pay for my copies whenever you like. I am so glad that you were so good as to undertake the publication of my book.

My dear Sir, yours very sincerely,
CHARLES DARWIN.

The further history of the book is given in the next chapter.

CHAPTER XII.

" Remember that your verdict will probably have more influence than my book in deciding whether such views as I hold will be admitted or rejected at present; in the future I cannot doubt about their admittance, and our posterity will marvel as much about the current belief as we do about fossil shells having been thought to have been created as we now see them."— From a letter to Lyell, Sept. 1859.

OCTOBER 3RD, 1859, TO DECEMBER 31ST, 1859.

UNDER the date of October 1st, 1859, in my father's Diary occurs the entry :—" Finished proofs (thirteen months and ten days) of Abstract on *Origin of Species ;* 1250 copies printed. The first edition was published on November 24th, and all copies sold first day."

In October he was, as we have seen in the last chapter, at Ilkley, near Leeds : there he remained with his family until December, and on the 9th of that month he was again at Down. The only other entry in the Diary for this year is as follows :—" During end of November and beginning of December, employed in correcting for second edition of 3000 copies ; multitude of letters."

The first and a few of the subsequent letters refer to proof-sheets, and to early copies of the *Origin* which were sent to friends before the book was published.

C. Lyell to C. Darwin. October 3rd, 1859.

MY DEAR DARWIN,—I have just finished your volume, and right glad I am that I did my best with Hooker to persuade you to publish it without waiting for a time which probably could never have arrived, though you lived to the age of a hundred, when you had prepared all your facts on which you ground so many grand generalizations.

It is a splendid case of close reasoning, and long substantial argument throughout so many pages ; the condensation immense, too great perhaps for the uninitiated, but an effective and important preliminary statement, which will admit, even before your detailed proofs appear, of some occasional useful exemplification, such as your pigeons and cirripedes, of which you make such excellent use.

I mean that, when, as I fully expect, a new edition is soon called for, you may here and there insert an actual case to relieve the vast number of abstract propositions. So far as I am concerned, I am so well prepared to take your statements of facts for granted, that I do not think the " pièces justificatives " when published will make much difference, and I have long seen most clearly that if any concession is made, all that you claim in your concluding pages will follow. It is this which has made me so long hesitate, always feeling that the case of Man and his races, and of other animals, and that of plants is one and the same, and that if a " vera causa " be admitted for one, instead of a purely unknown and imaginary one, such as the word " Creation," all the consequences must follow.

I fear I have not time to-day, as I am just leaving this place to indulge in a variety of comments, and to say how much I was delighted with Oceanic Islands—Rudimentary Organs—Embryology—the genealogical key to the Natural System, Geographical Distribution, and if I went on I should be copying the heads of all your chapters. But I will say a word of the Recapitulation, in case some slight alteration, or, at least, omission of a word or two be still possible in that.

In the first place, at p. 480, it cannot surely be said that the most eminent naturalists have rejected the view of the mutability of species? You do not mean to ignore G. St. Hilaire and Lamarck. As to the latter, you may say, that in regard to animals you substitute natural selection for volition to a certain considerable extent, but in his theory of the changes of plants he could not introduce volition; he may, no doubt, have laid an undue comparative stress on changes in physical conditions, and too little on those of contending organisms. He at least was for the universal mutability of species and for a genealogical link between the first and the present. The men of his school also appealed to domesticated varieties. (Do you mean *living* naturalists?) *

---

* In his next letter to Lyell my father writes : " The omission of 'living'

The first page of this most important summary gives the adversary an advantage, by putting forth so abruptly and crudely such a startling objection as the formation of " the eye," * not by means analogous to man's reason, or rather by some power immeasurably superior to human reason, but by superinduced variation like those of which a cattle-breeder avails himself. Pages would be required thus to state an objection and remove it. It would be better, as you wish to persuade, to say nothing. Leave out several sentences, and in a future edition bring it out more fully.

. . . But these are small matters, mere spots on the sun. Your comparison of the letters retained in words, when no longer wanted for the sound, to rudimentary organs is excellent, as both are truly genealogical. . . .

You enclose your sheets in old MS., so the Post Office very properly charge them, as letters, 2*d*. extra. I wish all their fines on MS. were worth as much. I paid 4*s*. 6*d*. for such wash the other day from Paris, from a man who can prove 300 deluges in the valley of Seine.

With my hearty congratulations to you on your grand work, believe me,

Ever very affectionately yours.

C. D. to L. Agassiz.† Down, November 11th [1859].

MY DEAR SIR,—I have ventured to send you a copy of my book (as yet only an abstract) on the *Origin of Species.* As the conclusions at which I have arrived on several points differ so widely from yours, I have thought (should you at

---

before ' eminent ' naturalists was a dreadful blunder." In the first edition, as published, the blunder is corrected by the addition of the word " living."

* Darwin wrote to Asa Gray in 1860 :—" The eye to this day gives me a cold shudder, but when I think of the fine known gradations, my reason tells me I ought to conquer the cold shudder."

† Jean Louis Rodolphe Agassiz, born at Mortier, on the lake of Morat in Switzerland, on May 28th, 1807. He emigrated to America in 1846, where he spent the rest of his life, and died Dec. 14th, 1873. His *Life*, written by his widow, was published in 1885. The following extract from a letter to Agassiz (1850) is worth giving, as showing how my father regarded him, and it may be added that his cordial feeling towards the great American naturalist remained strong to the end of his life :—

"I have seldom been more deeply gratified than by receiving your most kind present of *Lake Superior.* I had heard of it, and had much wished to read it, but I confess that it was the very great honour of having in my possession a work with your autograph as a presentation copy, that has given me such lively and sincere pleasure. I cordially thank you for it. I have begun to read it with uncommon interest, which I see will increase as I go on."

any time read my volume) that you might think that I had
sent it to you out of a spirit of defiance or bravado; but I
assure you that I act under a wholly different frame of
mind. I hope that you will at least give me credit, however
erroneous you may think my conclusions, for having earnestly endeavoured to arrive at the truth. With sincere respect, I beg leave to remain,

Yours very faithfully.

He sent copies of the *Origin*, accompanied by letters
similar to the last, to M. De Candolle, Dr. Asa Gray, Falconer, and Mr. Jenyns (Blomefield).

To Henslow he wrote (Nov. 11th, 1859):—

" I have told Murray to send a copy of my book on Species to you, my dear old master in Natural History; I fear,
however, that you will not approve of your pupil in this
case. The book in its present state does not show the
amount of labour which I have bestowed on the subject.

" If you have time to read it carefully, and would take
the trouble to point out what parts seem weakest to you and
what best, it would be a most material aid to me in writing
my bigger book, which I hope to commence in a few
months. You know also how highly I value your judgment. But I am not so unreasonable as to wish or expect
you to write detailed and lengthy criticisms, but merely a
few general remarks, pointing out the weakest parts.

" If you are *in ever so slight a degree* staggered (which
I hardly expect) on the immutability of species, then I am
convinced with further reflection you will become more and
more staggered, for this has been the process through which
my mind has gone."

C. D. to A. R. Wallace. Ilkley, November 13th, 1859.

MY DEAR SIR,—I have told Murray to send you by post
(if possible) a copy of my book, and I hope that you will
receive it at nearly the same time with this note. (N.B. I
have got a bad finger, which makes me write extra badly.)
If you are so inclined, I should very much like to hear your
general impression of the book, as you have thought so profoundly on the subject, and in so nearly the same channel
with myself. I hope there will be some little new to you,
but I fear not much. Remember it is only an abstract, and
very much condensed. God knows what the public will

think. No one has read it, except Lyell, with whom I have had much correspondence. Hooker thinks him a complete convert, but he does not seem so in his letters to me; but is evidently deeply interested in the subject. I do not think your share in the theory will be overlooked by the real judges, as Hooker, Lyell, Asa Gray, &c. I have heard from Mr. Sclater that your paper on the Malay Archipelago has been read at the Linnean Society, and that he was *extremely* much interested by it.

I have not seen one naturalist for six or nine months, owing to the state of my health, and therefore I really have no news to tell you. I am writing this at Ilkley Wells, where I have been with my family for the last six weeks, and shall stay for some few weeks longer. As yet I have profited very little. God knows when I shall have strength for my bigger book.

I sincerely hope that you keep your health; I suppose that you will be thinking of returning * soon with your magnificent collections, and still grander mental materials. You will be puzzled how to publish. The Royal Society fund will be worth your consideration. With every good wish, pray believe me,

Yours very sincerely.

P.S.—I think that I told you before that Hooker is a complete convert. If I can convert Huxley I shall be content.

C. Darwin to W. B. Carpenter. November 19th [1859].

. . . If, after reading my book, you are able to come to a conclusion in any degree definite, will you think me very unreasonable in asking you to let me hear from you? I do not ask for a long discussion, but merely for a brief idea of your general impression. From your widely extended knowledge, habit of investigating the truth, and abilities, I should value your opinion in the very highest rank. Though I, of course, believe in the truth of my own doctrine, I suspect that no belief is vivid until shared by others. As yet I know only one believer, but I look at him as of the greatest authority, viz. Hooker. When I think of the many cases of men who have studied one subject for years, and

---

* Mr. Wallace was in the Malay Archipelago.

have persuaded themselves of the truth of the foolishest doctrines, I feel sometimes a little frightened, whether I may not be one of these monomaniacs.

Again pray excuse this, I fear, unreasonable request. A short note would suffice, and I could bear a hostile verdict, and shall have to bear many a one.

Yours very sincerely.

C. D. to J. D. Hooker.   Ilkley, Yorkshire.   [November, 1859.]

MY DEAR HOOKER,—I have just read a review on my book in the *Athenæum*,* and it excites my curiosity much who is the author. If you should hear who writes in the *Athenæum* I wish you would tell me. It seems to me well done, but the reviewer gives no new objections, and, being hostile, passes over every single argument in favour of the doctrine. . . . I fear, from the tone of the review, that I have written in a conceited and cocksure style,† which shames me a little. There is another review of which I should like to know the author, viz. of H. C. Watson in the *Gardeners' Chronicle.*‡ Some of the remarks are like yours, and he does deserve punishment; but surely the review is too severe. Don't you think so? . . .

I have heard from Carpenter, who, I think, is likely to be a convert. Also from Quatrefages, who is inclined to go a long way with us. He says that he exhibited in his lecture a diagram closely like mine!

J. D. Hooker to C. Darwin.   Monday [Nov. 21, 1859].

MY DEAR DARWIN,—I am a sinner not to have written you ere this, if only to thank you for your glorious book— what a mass of close reasoning on curious facts and fresh phenomena—it is capitally written, and will be very successful. I say this on the strength of two or three plunges into as many chapters, for I have not yet attempted to read it. Lyell, with whom we are staying, is perfectly enchanted, and is absolutely gloating over it. I must accept your compliment

---

* Nov. 19, 1859.

† The Reviewer speaks of the author's " evident self-satisfaction," and of his disposing of all difficulties " more or less confidently."

‡ A review of the fourth volume of Watson's *Cybele Britannica*, *Gard. Chron.*, 1859, p. 911.

to me, and acknowledgment of supposed assistance * from me, as the warm tribute of affection from an honest (though deluded) man, and furthermore accept it as very pleasing to my vanity; but, my dear fellow, neither my name nor my judgment nor my assistance deserved any such compliments, and if I am dishonest enough to be pleased with what I don't deserve, it must just pass. How different the *book* reads from the MS. I see I shall have much to talk over with you. Those lazy printers have not finished my luckless Essay: which, beside your book, will look like a ragged handkerchief beside a Royal Standard. . . .

C. D. to J. D. Hooker. [November, 1859.]

My dear Hooker,—I cannot help it, I must thank you for your affectionate and most kind note. My head will be turned. By Jove, I must try and get a bit modest. I was a little chagrined by the review.† I hope it was *not* ——. As advocate, he might think himself justified in giving the argument only on one side. But the manner in which he drags in immortality, and sets the priests at me, and leaves me to their mercies, is base. He would, on no account, burn me, but he will get the wood ready, and tell the black beasts how to catch me. . . . It would be unspeakably grand if Huxley were to lecture on the subject, but I can see this is a mere chance; Faraday might think it too unorthodox.

. . . I had a letter from [Huxley] with such tremendous praise of my book, that modesty (as I am trying to cultivate that difficult herb) prevents me sending it to you, which I should have liked to have done, as he is very modest about himself.

You have cockered me up to that extent, that I now feel I can face a score of savage reviewers. I suppose you are still with the Lyells. Give my kindest remembrance to them. I triumph to hear that he continues to approve.

Believe me, your would-be modest friend.

The following passage from a letter to Lyell shows how strongly he felt on the subject of Lyell's adherence:—" I

---

* See the *Origin*, first edition, p. 3, where Sir J. D. Hooker's help is conspicuously acknowledged.

† This refers to the review in the *Athenæum*, Nov. 19th, 1859, where the reviewer, after touching on the theological aspects of the book, leaves the author to " the mercies of the Divinity Hall, the College, the Lecture Room, and the Museum."

rejoice profoundly that you intend admitting the doctrine of modification in your new edition; * nothing, I am convinced, could be more important for its success.   I honour you most sincerely.   To have maintained in the position of a master, one side of a question for thirty years, and then deliberately give it up, is a fact to which I much doubt whether the records of science offer a parallel.   For myself, also I rejoice profoundly; for, thinking of so many cases of men pursuing an illusion for years, often and often a cold shudder has run through me, and I have asked myself whether I may not have devoted my life to a phantasy.   Now I look at it as morally impossible that investigators of truth, like you and Hooker, can be wholly wrong, and therefore I rest in peace."

T. H. Huxley† to *C. Darwin.*   Jermyn Street, W.   November 23rd, 1859.

MY DEAR DARWIN,—I finished your book yesterday, a lucky examination having furnished me with a few hours of continuous leisure.

Since I read Von Bär's ‡ essays, nine years ago, no work on Natural History Science I have met with has made so great an impression upon me, and I do most heartily thank you for the great store of new views you have given me. Nothing, I think, can be better than the tone of the book, it impresses those who know nothing about the subject.   As for your doctrine, I am prepared to go to the stake, if requisite, in support of Chapter IX.,‡ and most parts of Chap-

---

* It appears from Sir Charles Lyell's published letters that he intended to admit the doctrine of evolution in a new edition of the *Manual*, but this was not published till 1865. He was, however, at work on the *Antiquity of Man* in 1860, and had already determined to discuss the *Origin* at the end of the book.

† In a letter written in October, my father had said, " I am intensely curious to hear Huxley's opinion of my book. I fear my long discussion on classification will disgust him, for it is much opposed to what he once said to me." He may have remembered the following incident told by Mr. Huxley in his chapter of the *Life and Letters*, ii. p. 196 :—" I remember, in the course of my first interview with Mr. Darwin, expressing my belief in the sharpness of the lines of demarcation between natural groups and in the absence of transitional forms, with all the confidence of youth and imperfect knowledge. . I was not aware, at that time, that he had then been many years brooding over the species question ; and the humorous smile which accompanied his gentle answer, that such was not altogether his view, long haunted and puzzled me."

‡ Karl Ernst von Baer, b. 1792, d. at Dorpat 1876—one of the most distinguished biologists of the century. He practically founded the modern science of embryology.

In the first edition of the *Origin*, Chap. IX. is on the ' Imperfection of

ters X., XI., XII.; and Chapter XIII. contains much that is most admirable, but on one or two points I enter a *caveat* until I can see further into all sides of the question.

As to the first four chapters, I agree thoroughly and fully with all the principles laid down in them. I think you have demonstrated a true cause for the production of species, and have thrown the *onus probandi*, that species did not arise in the way you suppose, on your adversaries.

But I feel that I have not yet by any means fully realized the bearings of those most remarkable and original Chapters III., IV. and V., and I will write no more about them just now.

The only objections that have occurred to me are, 1st that you have loaded yourself with an unnecessary difficulty in adopting *Natura non facit saltum* so unreservedly. . . . And 2nd, it is not clear to me why, if continual physical conditions are of so little moment as you suppose, variation should occur at all.

However, I must read the book two or three times more before I presume to begin picking holes.

I trust you will not allow yourself to be in any way disgusted or annoyed by the considerable abuse and misrepresentation which, unless I greatly mistake, is in store for you. Depend upon it you have earned the lasting gratitude of all thoughtful men. And as to the curs which will bark and yelp, you must recollect that some of your friends, at any rate, are endowed with an amount of combativeness which (though you have often and justly rebuked it) may stand you in good stead.

I am sharpening up my claws and beak in readiness.

Looking back over my letter, it really expresses so feebly all I think about you and your noble book that I am half ashamed of it; but you will understand that, like the parrot in the story, "I think the more."

Ever yours faithfully.

C. D. to T. H. Huxley. Ilkley, Nov. 25 [1859].

MY DEAR HUXLEY,—Your letter has been forwarded to me from Down. Like a good Catholic who has received

---

the Geological Record;' Chap. X., on the ' Geological Succession of Organic Beings;' Chaps. XI. and XII., on ' Geographical Distribution;' Chap. XIII., on ' Mutual Affinities of Organic Beings; Morphology; Embryology; Rudimentary Organs.'

extreme unction, I can now sing " nunc dimittis." I should have been more than contented with one quarter of what you have said. Exactly fifteen months ago, when I put pen to paper for this volume, I had awful misgivings; and thought perhaps I had deluded myself, like so many have done, and I then fixed in my mind three judges, on whose decision I determined mentally to abide. The judges were Lyell, Hooker, and yourself. It was this which made me so excessively anxious for your verdict. I am now contented, and can sing my " nunc dimittis." What a joke it would be if I pat you on the back when you attack some immovable creationists! You have most cleverly hit on one point, which has greatly troubled me; if, as I must think, external conditions produce little *direct* effect, what the devil determines each particular variation? What makes a tuft of feathers come on a cock's head, or moss on a moss-rose? I shall much like to talk over this with you. . . .

My dear Huxley, I thank you cordially for your letter.

Yours very sincerely.

Erasmus Darwin* to C. Darwin.   November 23rd [1859].

DEAR CHARLES,—I am so much weaker in the head, that I hardly know if I can write, but at all events I will jot down a few things that the Dr.† has said. He has not read much above half, so, as he says, he can give no definite conclusion, and keeps stating that he is not tied down to either view, and that he has always left an escape by the way he has spoken of varieties. I happened to speak of the eye before he had read that part, and it took away his breath —utterly impossible—structure—function, &c., &c., &c., but when he had read it he hummed and hawed, and perhaps it was partly conceivable, and then he fell back on the bones of the ear, which were beyond all probability or conceivability. He mentioned a slight blot, which I also observed, that in speaking of the slave-ants carrying one another, you change the species without giving notice first, and it makes one turn back. . . .

. . . For myself I really think it is the most interesting book I ever read, and can only compare it to the first knowledge of chemistry, getting into a new world or rather behind the scenes. To me the geographical distribution, I mean

---

* His brother.   † Dr., afterwards Sir Henry, Holland.

the relation of islands to continents is the most convincing of the proofs, and the relation of the oldest forms to the existing species. I dare say I don't feel enough the absence of varieties, but then I don't in the least know if everything now living were fossilized whether the palæontologists could distinguish them. In fact the *à priori* reasoning is so entirely satisfactory to me that if the facts won't fit in, why so much the worse for the facts is my feeling. My ague has left me in such a state of torpidity that I wish I had gone through the process of natural selection.

Yours affectionately.

A. Sedgwick * *to C. Darwin.* [November?], 1859.

MY DEAR DARWIN,—I write to thank you for your work on the *Origin of Species.* It came, I think, in the latter part of last week; but it *may* have come a few days sooner, and been overlooked among my book-parcels, which often remain unopened when I am lazy or busy with any work before me. So soon as I opened it I began to read it, and I finished it, after many interruptions, on Tuesday. Yesterday I was employed—1st, in preparing for my lecture; 2ndly, in attending a meeting of my brother Fellows to discuss the final propositions of the Parliamentary Commissioners; 3rdly, in lecturing; 4thly, in hearing the conclusion of the discussion and the College reply, whereby, in conformity with my own wishes, we accepted the scheme of the Commissioners; 5thly, in dining with an old friend at Clare College; 6thly, in adjourning to the weekly meeting of the Ray Club, from which I returned at 10 P.M., dog-tired, and hardly able to climb my staircase. Lastly, in looking through the *Times* to see what was going on in the busy world.

I do not state this to fill space (though I believe that Nature does abhor a vacuum), but to prove that my reply and my thanks are sent to you by the earliest leisure I have, though that is but a very contracted opportunity. If I did not think you a good-tempered and truth-loving man, I should not tell you that (spite of the great knowledge, store of facts, capital views of the correlation of the various parts of organic nature, admirable hints about the diffusion,

---

* Rev. Adam Sedgwick, Woodwardian Professor of Geology in the University of Cambridge. Born 1785, died 1873.

through wide regions, of many related organic beings, &c.
&c.) I have read your book with more pain than pleasure.
Parts of it I admired greatly, parts I laughed at till my sides
were almost sore; other parts I read with absolute sorrow,
because I think them utterly false and grievously mischievous. You have *deserted*—after a start in that tram-road
of all solid physical truth—the true method of induction,
and started us in machinery as wild, I think, as Bishop
Wilkins's locomotive that was to sail with us to the moon.
Many of your wide conclusions are based upon assumptions
which can neither be proved nor disproved, why then express them in the language and arrangement of philosophical
induction? As to your grand principle—*natural selection*—
what is it but a secondary consequence of supposed, or known,
primary facts? Development is a better word, because
more close to the cause of the fact? For you do not deny
causation. I call (in the abstract) causation the will of God;
and I can prove that He acts for the good of His creatures.
He also acts by laws which we can study and comprehend.
Acting by law, and under what is called final causes, compre-
hends, I think, your whole principle. You write of "natural
selection" as if it were done consciously by the selecting
agent. 'Tis but a consequence of the pre-supposed development, and the subsequent battle for life. This view of
nature you have stated admirably, though admitted by all
naturalists and denied by no one of common-sense. We all
admit development as a fact of history: but how came it
about? Here, in language, and still more in logic, we are
point-blank at issue. There is a moral or metaphysical
part of nature as well as a physical. A man who denies
this is deep in the mire of folly. 'Tis the crown and glory
of organic science that it *does* through *final cause*, link
material and moral; and yet *does not* allow us to mingle
them in our first conception of laws, and our classification
of such laws, whether we consider one side of nature or the
other. You have ignored this link; and, if I do not mistake your meaning, you have done your best in one or two
pregnant cases to break it. Were it possible (which, thank
God, it is not) to break it, humanity, in my mind, would
suffer a damage that might brutalize it, and sink the human
race into a lower grade of degradation than any into which
it has fallen since its written records tell us of its history.
Take the case of the bee-cells. If your development produced the successive modification of the bee and its cells

(which no mortal can prove), final cause would stand good as the directing cause under which the successive generations acted and gradually improved. Passages in your book, like that to which I have alluded (and there are others almost as bad), greatly shocked my moral taste. I think, in speculating on organic descent, you *over*-state the evidence of geology; and that you *under*-state it while you are talking of the broken links of your natural pedigree : but my paper is nearly done, and I must go to my lecture-room. Lastly, then, I greatly disliked the concluding chapter—not as a summary, for in that light it appears good—but I dislike it from the tone of triumph and confidence in which you appeal to the rising generation (in a tone I condemned in the author of the *Vestiges*) and prophesy of things not yet in the womb of time, nor (if we are to trust the accumulated experience of human sense and the inferences of its logic) ever likely to be found anywhere but in the fertile womb of man's imagination. And now to say a word about a son of a monkey and an old friend of yours : I am better, far better, than I was last year. I have been lecturing three days a week (formerly I gave six a week) without much fatigue, but I find by the loss of activity and memory, and of all productive powers, that my bodily frame is sinking slowly towards the earth. But I have visions of the future. They are as much a part of myself as my stomach and my heart, and these visions are to have their antitype in solid fruition of what is best and greatest. But on one condition only—that I humbly accept God's revelation of Himself both in His works and in His word, and do my best to act in conformity with that knowledge which He only can give me, and He only can sustain me in doing. If you and I do all this, we shall meet in heaven.

I have written in a hurry, and in a spirit of brotherly love, therefore forgive any sentence you happen to dislike; and believe me, spite of any disagreement in some points of the deepest moral interest, your true-hearted old friend,

A. SEDGWICK.

The following extract from a note to Lyell (Nov. 24) gives an idea of the conditions under which the second edition was prepared : " This morning I heard from Murray that he sold the whole edition * the first day to the trade.

---

* First edition, 1250 copies.

He wants a new edition instantly, and this utterly confounds me. Now, under water-cure, with all nervous power directed to the skin, I cannot possibly do head-work, and I must make only actually necessary corrections. But I will, as far as I can without my manuscript, take advantage of your suggestions: I must not attempt much. Will you send me one line to say whether I must strike out about the secondary whale,* it goes to my heart. About the rattle-snake, look to my Journal, under Trigonocephalus, and you will see the probable origin of the rattle, and generally in transitions it is the *premier pas qui coûte.*"

Here follows a hint of the coming storm (from a letter to Lyell, Dec. 2) :—

"Do what I could, I fear I shall be greatly abused. In answer to Sedgwick's remark that my book would be 'mischievous,' I asked him whether truth can be known except by being victorious over all attacks. But it is no use. H. C. Watson tells me that one zoologist says he will read my book, 'but I will never believe it.' What a spirit to read any book in ! ·Crawford † writes to me that his notice will be hostile, but that ' he will not calumniate the author.' He says he has read my book, 'at least such parts as he could understand.' ‡ He sent me some notes and suggestions (quite unimportant), and they show me that I have unavoidably done harm to the subject, by publishing an abstract. . . . I have had several notes from ——, very civil and less decided. Says he shall not pronounce against me without much reflection, *perhaps will say nothing* on the subject. X. says he will go to that part of hell, which Dante tells us is appointed for those who are neither on God's side nor on that of the devil."

---

* The passage was omitted in the second edition.

† John Crawford, orientalist, ethnologist, &c., b. 1783, d. 1868. The review appeared in the *Examiner*, and, though hostile, is free from bigotry, as the following citation will show: " We cannot help saying that piety must be fastidious indeed that objects to a theory the tendency of which is to show that all organic beings, man included, are in a perpetual progress of amelioration and that is expounded in the reverential language which we have quoted."

‡ A letter of Dec. 14, gives a good example of the manner in which some naturalists received and understood it. " Old J. E. Gray of the British Museum attacked me in fine style : ' You have just reproduced Lamarck's doctrine, and nothing else, and here Lyell and others have been attacking him for twenty years, and because *you* (with a sneer and laugh) say the very same thing, they are all coming round ; it is the most ridiculous inconsistency, &c. &c.' "

But his friends were preparing to fight for him. Huxley gave, in *Macmillan's Magazine* for December, an analysis of the *Origin*, together with the substance of his Royal Institution lecture, delivered before the publication of the book.

Carpenter was preparing an essay for the *National Review*, and negotiating for a notice in the *Edinburgh* free from any taint of *odium theologicum.*

C. D. to C. Lyell. Down [December 12th, 1859].

. . I had very long inverviews with——, which perhaps you would like to hear about. . . . I infer from several expressions that, at bottom, he goes an immense way with us. . . .

He said to the effect that my explanation was the best ever published of the manner of formation of species. I said I was very glad to hear it. He took me up short: "You must not at all suppose that I agree with you in all respects." I said I thought it no more likely that I should be right in nearly all points, than that I should toss up a penny and get heads twenty times running. I asked him what he thought the weakest part. He said he had no particular objection to any part. He added :—

"If I must criticise, I should say, we do not want to know what Darwin believes and is convinced of, but what he can prove." I agreed most fully and truly that I have probably greatly sinned in this line, and defended my general line of argument of inventing a theory and seeing how many classes of facts the theory would explain. I added that I would endeavour to modify the "believes" and "convinceds." He took me up short: "You will then spoil your book, the charm of it is that it is Darwin himself." He added another objection, that the book was too *teres atque rotundus*—that it explained everything, and that it was improbable in the highest degree that I should succeed in this. I quite agree with this rather queer objection, and it comes to this that my book must be very bad or very good. . . .

I have heard, by a roundabout channel, that Herschel says my book "is the law of higgledy-piggledy." What this exactly means I do not know, but it is evidently very contemptuous. If true this is a great blow and discouragement.

J. D. Hooker to C. Darwin.   Kew [1859].

DEAR DARWIN,—You have, I know, been drenched with letters since the publication of your book, and I have hence forborne to add my mite.* I hope now that you are well through Edition II., and I have heard that you were flourishing in London. I have not yet got half-through the book, not from want of will, but of time—for it is the very hardest book, to read, to full profits, that I ever tried—it is so cram-full of matter and reasoning.† I am all the more glad that you have published in this form, for the three volumes, unprefaced by this, would have choked any Naturalist of the nineteenth century, and certainly have softened my brain in the operation of assimilating their contents. I am perfectly tired of marvelling at the wonderful amount of facts you have brought to bear, and your skill in marshalling them and throwing them on the enemy; it is also extremely clear as far as I have gone, but very hard to fully appreciate. Somehow it reads very different from the MS., and I often fancy that I must have been very stupid not to have more fully followed it in MS. Lyell told me of his criticisms. I did not appreciate them all, and there are many little matters I hope one day to talk over with you. I saw a highly flattering notice in the *English Churchman*, short and not at all entering into discussion, but praising you and your book, and talking patronizingly of the doctrine! . . . Bentham and Henslow will still shake their heads, I fancy. . . .

Ever yours affectionately.

C. D. to T. H. Huxley.   Down, Dec. 28th [1859.]

MY DEAR HUXLEY,—Yesterday evening, when I read the *Times* of a previous day, I was amazed to find a splendid essay and review of me. Who can the author be? I am intensely curious. It included an eulogium of me which quite touched me, though I am not vain enough to think it all deserved. The author is a literary man and German scholar. He has read my book very attentively; but what

---

* See, however, p. 211.
† Mr. Huxley has made a similar remark : " Long occupation with the work has led the present writer to believe that the *Origin of Species* is one of the hardest of books to master."—*Obituary Notice, Proc. R. Soc.* No. 269, p. xvii.

is very remarkable, it seems that he is a profound naturalist. He knows my Barnacle book, and appreciates it too highly. Lastly he writes and thinks with quite uncommon force and clearness; and what is even still rarer, his writing is seasoned with most pleasant wit. We all laughed heartily over some of the sentences. . . . Who can it be? Certainly I should have said that there was only one man in England who could have written this essay, and that *you* were the man. But I suppose I am wrong, and that there is some hidden genius of great calibre. For how could you influence Jupiter Olympus and make him give three and a half columns to pure science? The old fogies will think the world will come to an end. Well, whoever the man is, he has done great service to the cause, far more than by a dozen reviews in common periodicals. The grand way he soars above common religious prejudices, and the admission of such views into the *Times*, I look at as of the highest importance, quite independently of the mere question of species. If you should happen to be *acquainted* with the author, for Heaven-sake tell me who he is?

My dear Huxley, yours most sincerely.

There can be no doubt that this powerful essay, appearing in the leading daily Journal, must have had a strong influence on the reading public. Mr. Huxley allows me to quote from a letter an account of the happy chance that threw into his hands the opportunity of writing it:—

" The *Origin* was sent to Mr. Lucas, one of the staff of the *Times* writers at that day, in what I suppose was the ordinary course of business. Mr. Lucas, though an excellent journalist, and, at a later period, editor of *Once a Week*, was as innocent of any knowledge of science as a babe, and bewailed himself to an acquaintance on having to deal with such a book. Whereupon he was recommended to ask me to get him out of his difficulty, and he applied to me accordingly, explaining, however, that it would be necessary for him formally to adopt anything I might be disposed to write, by prefacing it with two or three paragraphs of his own.

" I was too anxious to seize upon the opportunity thus offered of giving the book a fair chance with the multitudinous readers of the *Times* to make any difficulty about conditions; and being then very full of the subject, I wrote the article faster, I think, than I ever wrote anything in my life, and sent it to Mr. Lucas, who duly prefixed his opening sentences.

" When the article appeared, there was much speculation as to its authorship. The secret leaked out in time, as all secrets will, but not by my aid ; and then I used to derive a good deal of innocent amusement from the vehement assertions of some of my more acute friends, that they knew it was mine from the first paragraph !

" As the *Times* some years since referred to my connection with the review, I suppose there will be no breach of confidence in the publication of this little history, if you think it worth the space it will occupy."

CHAPTER XIII.

THE 'ORIGIN OF SPECIES'—REVIEWS AND CRITICISMS—ADHESIONS AND ATTACKS.

"You are the greatest revolutionist in natural history of this century, if not of all centuries."—H. C. Watson to C. Darwin, Nov. 21, 1859.

1860.

THE second edition, 3000 copies, of the *Origin* was published on January 7th; on the 10th, he wrote with regard to it, to Lyell:—

C. D. to C. Lyell. Down, January 10th [1860].

. . . It is perfectly true that I owe nearly all the corrections to you, and several verbal ones to you and others; I am heartily glad you approve of them, as yet only two things have annoyed me; those confounded millions * of years (not that I think it is probably wrong), and my not having (by inadvertence) mentioned Wallace towards the close of the book in the summary, not that any one has noticed this to me. I have now put in Wallace's name at p. 484 in a conspicuous place. I shall be truly glad to read carefully any MS. on man, and give my opinion. You used to caution me to be cautious about man. I suspect I shall have to return the caution a hundred fold! Yours will, no doubt, be a grand discussion; but it will horrify the world at first more than my whole volume; although by the sentence (p. 489, new edition †) I show that I believe man is in the same predicament with other animals. It is in fact impossible to doubt it. I have thought (only

---

* This refers to the passage in the *Origin of Species* (2nd edit. p. 285) in which the lapse of time implied by the denudation of the Weald is discussed. The discussion closes with the sentence: "So that it is not improbable that a longer period than 300 million years has elapsed since the latter part of the Secondary period." This passage is omitted in the later editions of the *Origin*, against the advice of some of his friends, as appears from the pencil notes in my father's copy of the 2nd edition.

† In the first edition, the passages occur on p. 488.

vaguely) on man. With respect to the races, one of my best chances of truth has broken down from the impossibility of getting facts. I have one good speculative line, but a man must have entire credence in Natural Selection before he will even listen to it. Psychologically, I have done scarcely anything. Unless, indeed, expression of countenance can be included, and on that subject I have collected a good many facts, and speculated, but I do not suppose I shall ever publish, but it is an uncommonly curious subject.

A few days later he wrote again to the same correspondent :

"What a grand immense benefit you conferred on me by getting Murray to publish my book. I never till to-day realised that it was getting widely distributed ; for in a letter from a lady to-day to E., she says she heard a man enquiring for it at the *Railway Station ! ! !* at Waterloo Bridge ; and the bookseller said that he had none till the new edition was out. The bookseller said he had not read it, but had heard it was a very remarkable book ! ! ! "

C. D. to J. D. Hooker.   Down, 14th [January, 1860].

. . . . I heard from Lyell this morning, and he tells me a piece of news. You are a good-for-nothing man ; here you are slaving yourself to death with hardly a minute to spare, and you must write a review on my book ! I thought it * a very good one, and was so much struck with it, that I sent it to Lyell. But I assumed, as a matter of course, that it was Lindley's. Now that I know it is yours, I have re-read it, and my kind and good friend, it has warmed my heart with all the honourable and noble things you say of me and it. I was a good deal surprised at Lindley hitting on some of the remarks, but I never dreamed of you. I admired it chiefly as so well adapted to tell on the readers of the *Gardeners' Chronicle ;* but now I admire it in another spirit. Farewell, with hearty thanks. . . .

Asa Gray to J. D. Hooker.   Cambridge, Mass.,
January 5th, 1860.

MY DEAR HOOKER,—Your last letter, which reached me just before Christmas, has got mislaid during the upturn-

---

* *Gardeners' Chronicle,* 1860. Sir J. D. Hooker took the line of complete impartiality, so as not to commit the editor, Lindley.

ings in my study which take place at that season, and has not yet been discovered. I should be very sorry to lose it, for there were in it some botanical mems. which I had not secured. . .

The principal part of your letter was high laudation of Darwin's book.

Well, the book has reached me, and I finished its careful perusal four days ago ; and I freely say that your laudation is not out of place.

It is done in a *masterly manner.* It might well have taken twenty years to produce it. It is crammed full of most interesting matter — thoroughly digested—well expressed—close, cogent, and taken as a system it makes out a better case than I had supposed possible. . . .

Agassiz, when I saw him last, had read but a part of it. He says it is *poor—very poor ! !* (entre nous). The fact [is] he is very much annoyed by it, . . . and I do not wonder at it. To bring all *ideal* systems within the domain of science, and give good physical or natural explanations of all his capital points, is as bad as to have Forbes take the glacier materials . . . and give scientific explanation of all the phenomena.

Tell Darwin all this. I will write to him when I get a chance. As I have promised, he and you shall have fair-play here. . . . I must myself write a review * of Darwin's book for *Silliman's Journal* (the more so that I suspect Agassiz means to come out upon it) for the next (March) number, and I am now setting about it (when I ought to be every moment working the Expl[oring] Expedition Compositæ, which I know far more about). And really it is no easy job as you may well imagine.

I doubt if I shall please you altogether. I know I shall not please Agassiz at all. I hear another reprint is in the Press, and the book will excite much attention here, and some controversy. . . .

---

* On Jan. 23 Gray wrote to Darwin : " It naturally happens that my review of your book does not exhibit anything like the full force of the impression the book has made upon me. Under the circumstances I suppose I do your theory more good here, by bespeaking for it a fair and favourable consideration, and by standing non-committed as to its full conclusions, than I should if I announced myself a convert; nor could I say the latter, with truth. . . .

" What seems to me the weakest point in the book is the attempt to account for the formation of organs, the making of eyes, &c., by natural selection. Some of this reads quite Lamarckian."

C. D. to Asa Gray.   Down, January 28th [1860].

MY DEAR GRAY,—Hooker has forwarded to me your letter to him ; and I cannot express how deeply it has gratified me. To receive the approval of a man whom one has long sincerely respected, and whose judgment and knowledge are most universally admitted, is the highest reward an author can possibly wish for ; and I thank you heartily for your most kind expressions.

I have been absent from home for a few days, and so could not earlier answer your letter to me of the 10th of January. You have been extremely kind to take so much trouble and interest about the edition. It has been a mistake of my publisher not thinking of sending over the sheets. I had entirely and utterly forgotten your offer of receiving the sheets as printed off. But I must not blame my publisher, for had I remembered your most kind offer I feel pretty sure I should not have taken advantage of it ; for I never dreamed of my book being so successful with general readers : I believe I should have laughed at the idea of sending the sheets to America.*

After much consideration, and on the strong advice of Lyell and others, I have resolved to leave the present book as it is (excepting correcting errors, or here and there inserting short sentences), and to use all my strenth, *which is but little*, to bring out the first part (forming a separate volume, with index, &c.) of the three volumes which will make my bigger work ; so that I am very unwilling to take up time in making corrections for an American edition. I enclose a list of a few corrections in the second reprint, which you will have received by this time complete, and I could send four or five corrections or additions of equally small importance, or rather of equal brevity. I also intend to write a *short* preface with a brief history of the subject. These I will set about, as they must some day be done, and I will send them to you in a short time—the few corrections first, and the preface afterwards, unless I hear that you have given up all idea of a separate edition. You will then be able to judge whether it is worth having the new edition with *your*

---

* In a letter to Mr. Murray, 1860, my father wrote :—" I am amused by Asa Gray's account of the excitement my book has made amongst naturalists in the U. States. Agassiz has denounced it in a newspaper, but yet in such terms that it is in fact a fine advertisement ! " This seems to refer to a lecture given before the Mercantile Library Association.

review prefixed. Whatever be the nature of your review, I assure you I should feel it a *great* honour to have my book thus preceded. . . .

C. D. to C. Lyell. Down [February 15th, 1860].

. . . I am perfectly convinced (having read it this morning) that the review in the *Annals* * is by Wollaston; no one else in the world would have used so many parentheses. I have written to him, and told him that the " pestilent " fellow thanks him for his kind manner of speaking about him. I have also told him that he would be pleased to hear that the Bishop of Oxford says it is the most unphilosophical † work he ever read. The review seems to me clever, and only misinterprets me in a few places. Like all hostile men, he passes over the explanation given of Classification, Morphology, Embryology, and Rudimentary Organs, &c. I read Wallace's paper in MS., ‡ and thought it admirably good ; he does not know that he has been anticipated about the depth of intervening sea determining distribution. . . . The most curious point in the paper seems to me that about the African character of the Celebes productions, but I should require further confirmation. . . .

Henslow is staying here; I have had some talk with him; he is in much the same state as Bunbury,# and will go a very little way with us, but brings up no real argument against going further. He also shudders at the eye! It is really curious (and perhaps is an argument in our favour) how differently different opposers view the subject. Henslow used to rest his opposition on the imperfection of the Geological Record, but he now thinks nothing of this, and says I have got well out of it; I wish I could quite agree

---

* *Annals and Mag. of Nat. Hist.* third series, vol. v., p. 132. My father has obviously taken the expression " pestilent " from the following passage (p. 138) . " But who is this Nature, we have a right to ask, who has such tremendous power, and to whose efficiency such marvellous performances are ascribed ? What are her image and attributes, when dragged from her wordy lurking-place ? Is she ought but a pestilent abstraction, like dust cast in our eyes to obscure the workings of an Intelligent First Cause of all ? " The reviewer pays a tribute to my father's candour " so manly and outspoken as almost to ' cover a multitude of sins.' " The parentheses (to which allusion is made above) are so frequent as to give a characteristic appearance to Mr. Wollaston's pages.

† Another version of the words is given by Lyell, to whom they were spoken, viz. " the most illogical book ever written."—*Life and Letters of Sir C. Lyell,* vol. ii. p. 358.

‡ " On the Zoological Geography of the Malay Archipelago."—*Linn. Soc. Journ.* 1860.

The late Sir Charles Bunbury, well known as a Palæo-botanist.

with him. Baden Powell says he never read anything so conclusive as my statement about the eye !! A stranger writes to me about sexual selection, and regrets that I boggle about such a trifle as the brush of hair on the male turkey, and so on. As L. Jenyns has a really philosophical mind, and as you say you like to see everything, I send an old letter of his. In a later letter to Henslow, which I have seen, he is more candid than any opposer I have heard of, for he says, though he *cannot* go so far as I do, yet he can give no good reason why he should not. It is funny how each man draws his own imaginary line at which to halt. It reminds me so vividly [of] what I was told * about you when I first commenced geology—to believe a *little*, but on no account to believe all.

Ever yours affectionately.

With regard to the attitude of the more liberal representatives of the Church, the following letter from Charles Kingsley is of interest :

C. Kingsley to C. Darwin. Eversley Rectory, Winchfield, November 18th, 1859.

DEAR SIR,—I have to thank you for the unexpected honour of your book. That the Naturalist whom, of all naturalists living, I most wish to know and to learn from, should have sent a scientist like me his book, encourages me at least to observe more carefully, and think more slowly.

I am so poorly (in brain), that I fear I cannot read your book just now as I ought. All I have seen of it *awes* me ; both with the heap of facts and the prestige of your name, and also with the clear intuition, that if you be right, I must give up much that I have believed and written.

In that I care little. Let God be true, and every man a liar ! Let us know what *is*, and, as old Socrates has it, ἕπεσθαι τῷ λόγῳ—follow up the villainous shifty fox of an argument, into whatsoever unexpected bogs and brakes he may lead us, if we do but run into him at last.

From two common superstitions, at least, I shall be free while judging of your book :—

(1.) I have long since, from watching the crossing of domesticated animals and plants, learnt to disbelieve the dogma of the permanence of species.

---

* By Professor Henslow.

(2.) I have gradually learnt to see that it is just as noble a conception of Deity, to believe that He created primal forms capable of self-development into all forms needful *pro tempore* and *pro loco*, as to believe that He required a fresh act of intervention to supply the *lacunas* which He himself had made. I question whether the former be not the loftier thought.

Be it as it may, I shall prize your book, both for itself, and as a proof that you are aware of the existence of such a person as

Your faithful servant,

C. KINGSLEY.

My father's old friend, the Rev. J. Brodie Innes, of Milton Brodie, who was for many years Vicar of Down, in some reminiscences of my father which he was so good as to give me, writes in the same spirit :

"We never attacked each other. Before I knew Mr. Darwin I had adopted, and publicly expressed, the principle that the study of natural history, geology, and science in general, should be pursued without reference to the Bible. That the Book of Nature and Scripture came from the same Divine source, ran in paralled lines, and when properly understood would never cross. . . .

"In [a] letter, after I had left Down, he [Darwin] writes, ' We often differed, but you are one of those rare mortals from whom one can differ and yet feel no shade of animosity, and that is a thing [of] which I should feel very proud if any one could say [it] of me.'

"On my last visit to Down, Mr. Darwin said, at his dinner-table, ' Innes and I have been fast friends for thirty years, and we never thoroughly agreed on any subject but once, and then we stared hard at each other, and thought one of us must be very ill.' "

The following extract from a letter to Lyell, Feb. 23, 1860, has a certain bearing on the points just touched on :

"With respect to Bronn's * objection that it cannot be shown how life arises, and likewise to a certain extent Asa Gray's remark that natural selection is not a *vera causa*, I was much interested by finding accidentally in Brewster's *Life of Newton*, that Leibnitz objected to the law of gravity

---

* The translator of the first German edition of the *Origin*.

because Newton could not show what gravity itself is. As it has chanced, I have used in letters this very same argument, little knowing that any one had really thus objected to the law of gravity. Newton answers by saying that it is philosophy to make out the movements of a clock, though you do not know why the weight descends to the ground. Leibnitz further objected .that the law of gravity was opposed to Natural Religion ! Is this not curious ? I really think I shall use the facts for some introductory remarks for my bigger book."

C. D. to J. D. Hooker. Down, March 3rd [1860].

. . . I think you expect too much in regard to change of opinion on the subject of Species. One large class of men, more especially I suspect of naturalists, never will care about *any* general question, of which old Gray, of the British Museum, may be taken as a type ; and secondly, nearly all men past a moderate age, either in actual years or in mind are, I am fully convinced, incapable of looking at facts under a new point of view. Seriously, I am astonished and rejoiced at the progress which the subject has made ; look at the enclosed memorandum. —— says my book will be forgotten in ten years, perhaps so ; but, with such a list, I feel convinced the subject will not.

[Here follows the memorandum referred to :]

Geologists.	Zoologists and Palæontologists.	Physiologists.	Botanists.
Lyell. Ramsay.* Jukes.† H. D. Rogers.‡	Huxley. J. Lubbock. L. Jenyns (to large extent). Searles Wood.#	Carpenter. Sir H. Holland (to large extent).	Hooker. H. C. Watson. Asa Gray (to some extent). Dr. Boott (to large extent). Thwaites.‖

* Andrew Ramsay, late Director-General of the Geological Survey.
† Joseph Beete Jukes, M.A., F.R.S., born 1811, died 1869. He was educated at Cambridge, and from 1842 to 1846 he acted as naturalist to H.M.S. *Fly*, on an exploring expedition in Australia and New Guinea. He was afterwards appointed Director of the Geological Survey of Ireland. He was the author of many papers, and of more than one good handbook of geology.
‡ Professor of Geology in the University of Glasgow. Born in the United States 1809, died 1866.
# Searles Valentine Wood, died 1880. Chiefly known for his work on the Mollusca of the *Crag*.
‖ Dr. G. H. K. Thwaites, F.R.S., was born in 1811, or about that date, and

C. D. to Asa Gray. Down, April 3 [1860].

. . . I remember well the time when the thought of the eye made me cold all over, but I have got over this stage of the complaint, and now small trifling particulars of structure often make me very uncomfortable. The sight of a feather in a peacock's tail, whenever I gaze at it, makes me sick ! . . .

You may like to hear about reviews on my book. Sedgwick (as I and Lyell feel *certain* from internal evidence) has reviewed me savagely and unfairly in the *Spectator.** The notice includes much abuse, and is hardly fair in several respects. He would actually lead any one, who was ignorant of geology, to suppose that I had invented the great gaps between successive geological formations, instead of its being an almost universally admitted dogma. But my dear old friend Sedgwick, with his noble heart, is old, and is rabid with indignation. . . . There has been one prodigy of a review, namely, an *opposed* one (by Pictet,† the palæontologist, in the *Bib. Universelle* of Geneva) which is *perfectly* fair and just, and I agree to every word he says ; our only difference being that he attaches less weight to arguments in favour, and more to arguments opposed, than I do. Of all the opposed reviews, I think this the only quite fair one, and I never expected to see one. Please observe that I do not class your review by any means as opposed, though you think so yourself ! It has done me *much* too good service ever to appear in that rank in my eyes. But I fear I shall weary you with so much about my book. I should rather think there was a good chance of my becoming the most egotistical man in all Europe ! What a proud

---

died in Ceylon, September 11, 1882. He began life as a Notary, but his passion for Botany and Entomology ultimately led to his taking to Science as a profession. He became lecturer on Botany at the Bristol School of Medicine, and in 1849 he was appointed Director of the Botanic Gardens at Peradeniya, which he made "the most beautiful tropical garden in the world." He is best known through his important discovery of conjugation in the Diatomaceæ (1847). His *Enumeratio Plantarum Zeylaniæ* (1858–64] was "the first complete account, on modern lines, of any definitely circumscribed tropical area." (From a notice in *Nature*, October 26, 1882.)

* *Spectator*, March 24, 1860. There were favourable notices of the *Origin* by Huxley in the *Westminster Review*, and Carpenter in the *Medico-Chir. Review*, both in the April numbers.

† François Jules Pictet, in the *Archives des Sciences de la Bibliothèque Universelle*, Mars 1860.

pre-eminence! Well, you have helped to make me so, and therefore you must forgive me if you can.

My dear Gray, ever yours most gratefully.

C. D. to C. Lyell.   Down, April 10th [1860].

I have just read the *Edinburgh*,* which without doubt is by ——. It is extremely malignant, clever, and I fear will be very damaging. He is atrociously severe on Huxley's lecture, and very bitter against Hooker. So we three *enjoyed* it together. Not that I really enjoyed it, for it made me uncomfortable for one night; but I have got quite over it to-day. It requires much study to appreciate all the bitter spite of many of the remarks against me; indeed I did not discover all myself. It scandalously misrepresents many parts. He misquotes some passages, altering words within inverted commas. . . .

It is painful to be hated in the intense degree with which —— hates me.

Now for a curious thing about my book, and then I have done. In last Saturday's *Gardeners' Chronicle*,† a Mr. Patrick Matthew publishes a long extract from his work on *Naval Timber and Arboriculture* published in 1831, in which he briefly but completely anticipates the theory of Natural Selection. I have ordered the book, as some few passages are rather obscure, but it is certainly, I think, a complete but not developed anticipation! Erasmus always said that surely this would be shown to be the case some day. Anyhow, one may be excused in not having discovered the fact in a work on Naval Timber.

C. D. to J. D. Hooker.   Down [April 13th, 1860].

My dear Hooker,—Questions of priority so often lead to odious quarrels, that I should esteem it a great favour if you would read the enclosed.‡ If you think it proper that

---

* *Edinburgh Review*, April, 1860.
† April 7, 1860.
‡ My father wrote (*Gardeners' Chronicle*, April 21, 1860, p. 362): "I have been much interested by Mr. Patrick Matthew's communication in the number of your paper dated April 7th. I freely acknowledge that Mr. Matthew has anticipated by many years the explanation which I have offered of the origin of species, under the name of natural selection. I think that no one will feel surprised that neither I, nor apparently any other naturalist, had heard of Mr. Matthew's views, considering how briefly they are given, and that they appeared in the appendix to a work on Naval Timber and Arbori-

I should send it (and of this there can hardly be any question), and if you think it full and ample enough, please alter the date to the day on which you post it, and let that be soon. The case in the *Gardeners' Chronicle* seems a *little* stronger than in Mr. Matthew's book, for the passages are therein scattered in three places; but it would be mere hair-splitting to notice that. If you object to my letter, please return it; but I do not expect that you will, but I thought that you would not object to run your eye over it. My dear Hooker, it is a great thing for me to have so good, true, and old a friend as you. I owe much for science to my friends.

. . . I have gone over [the *Edinburgh*] review again, and compared passages, and I am astonished at the misrepresentations. But I am glad I resolved not to answer. Perhaps it is selfish, but to answer and think more on the subject is too unpleasant. I am so sorry that Huxley by my means has been thus atrociously attacked. I do not suppose you much care about the gratuitous attack on you.

Lyell in his letter remarked that you seemed to him as if you were overworked. Do, pray, be cautious, and remember how many and many a man has done this—who thought it absurd till too late. I have often thought the same. You know that you were bad enough before your Indian journey.

C. D. to C. Lyell. Down, April [1860].

. . . I was particularly glad to hear what you thought about not noticing [the *Edinburgh*] review. Hooker and Huxley thought it a sort of duty to point out the alteration of quoted citations, and there is truth in this remark; but I so hated the thought that I resolved not to do so. I shall

---

culture. I can do no more than offer my apologies to Mr. Matthew for my entire ignorance of his publication. If another edition of my work is called for, I will insert to the foregoing effect." In spite of my father's recognition of his claims, Mr. Matthew remained unsatisfied, and complained that an article in the *Saturday Analyst and Leader*, Nov. 24, 1860, was "scarcely fair in alluding to Mr. Darwin as the parent of the origin of species, seeing that I published the whole that Mr. Darwin attempts to prove, more than twenty-nine years ago." It was not until later that he learned that Matthew had also been forestalled. In October 1865, he wrote Sir J. D. Hooker:—"Talking of the *Origin*, a Yankee has called my attention to a paper attached to Dr. Wells' famous *Essay on Dew*, which was read in 1813 to the Royal Soc., but not [then] printed, in which he applies most distinctly the principle of Natural Selection to the races of Man. So poor old Patrick Matthew is not the first, and he cannot, or ought not, any longer to put on his title-pages, 'Discoverer of the principle of Natural Selection'!"

come up to London on Saturday the 14th, for Sir B. Brodie's party, as I have an accumulation of things to do in London, and will (if I do not hear to the contrary) call about a quarter before ten on Sunday morning, and sit with you at breakfast, but will not sit long, and so take up much of your time. I must say one more word about our quasi-theological controversy about natural selection, and let me have your opinion when we meet in London. Do you consider that the successive variations in the size of the crop of the Pouter Pigeon, which man has accumulated to please his caprice, have been due to "the creative and sustaining powers of Brahma?" In the sense that an omnipotent and omniscient Deity must order and know everything, this must be admitted; yet, in honest truth, I can hardly admit it. It seems preposterous that a maker of a universe should care about the crop of a pigeon solely to please man's silly fancies. But if you agree with me in thinking such an interposition of the Deity uncalled for, I can see no reason whatever for believing in such interpositions in the case of natural beings, in which strange and admirable peculiarities have been naturally selected for the creature's own benefit. Imagine a Pouter in a state of nature wading into the water and then, being buoyed up by its inflated crop, sailing about in search of food. What admiration this would have excited—adaptation to the laws of hydrostatic pressure, &c. &c. For the life of me, I cannot see any difficulty in natural selection producing the most exquisite structure, *if such structure can be arrived at by gradation,* and I know from experience how hard it is to name any structure towards which at least some gradations are not known.

Ever yours.

P. S.—The conclusion at which I have come, as I have told Asa Gray, is that such a question, as is touched on in this note, is beyond the human intellect, like "predestination and free will," or the "origin of evil."

C. D. to J. D. Hooker. Down [May 15th, 1860].

. . . How paltry it is in such men as X., Y. and Co. not reading your essay. It is incredibly paltry. They may all attack me to their hearts' content. I am got case-hardened. As for the old fogies in Cambridge,* it really signifies noth-

---

* This refers to a "savage onslaught" on the *Origin* by Sedgwick at the

ing. I look at their attacks as a proof that our work is worth the doing. It makes me resolve to buckle on my armour. I see plainly that it will be a long uphill fight. But think of Lyell's progress with Geology. One thing I see most plainly, that without Lyell's, yours, Huxley's and Carpenter's aid, my book would have been a mere flash in the pan. But if we all stick to it, we shall surely gain the day. And I now see that the battle is worth fighting. I deeply hope that you think so.

C. D. to Asa Gray. Down, May 22nd [1860].

MY DEAR GRAY,—Again I have to thank you for one of your very pleasant letters of May 7th, enclosing a very pleasant remittance of £22. I am in simple truth astonished at all the kind trouble you have taken for me. I return Appletons' account. For the chance of your wishing for a formal acknowledgment I send one. If you have any further communication to the Appletons, pray express my acknowledgment for [their] generosity; for it is generosity in my opinion. I am not at all surprised at the sale diminishing; my extreme surprise is at the greatness of the sale. No doubt the public has been *shamefully* imposed on! for they bought the book thinking it would be nice easy reading. I expect the sale to stop soon in England, yet Lyell wrote to me the other day that calling at Murray's he heard that fifty copies had gone in the previous forty-eight hours. I am extremely glad that you will notice in *Silliman* the additions in the *Origin.** Judging from letters (and I have just seen one from Thwaites to Hooker), and from remarks, the most serious omission in my book was not explaining how it is, as I believe, that all forms do not necessarily advance, how there can now be *simple* organisms still existing. . . . I hear there is a *very* severe review on me in the *North British* by a Rev. Mr. Dunns,† a Free Kirk minister, and dabbler in Natural History. In the *Saturday Review* (one

---

Cambridge Philosophical Society. Henslow defended his old pupil, and maintained that "the subject was a legitimate one for investigation."

* "The battle rages furiously in the United States. Gray says he was preparing a speech which would take 1¼ hours to deliver, and which he fondly hoped would be a stunner.' He is fighting splendidly, and there seem to have been many discussions with Agassiz and others, at the meetings. Agassiz pities me much at being so deluded."—From a letter to Hooker, May 30th, 1860.

† The statement as to authorship was made on the authority of Robert Chambers.

of our cleverest periodicals) of May 5th, p. 573, there is a nice article on [the *Edinburgh*] review, defending Huxley, but not Hooker; and the latter, I think, [the *Edinburgh* reviewer] treats most ungenerously.* But surely you will get sick unto death of me and my reviewers.

With respect to the theological view of the question. This is always painful to me. I am bewildered. I had no intention to write atheistically. But I own that I cannot see as plainly as others do, and as I should wish to do, evidence of design and beneficence on all sides of us. There seems to me too much misery in the world. I cannot persuade myself that a beneficent and omnipotent God would have designedly created the Ichneumonidæ with the express intention of their feeding within the living bodies of caterpillars, or that a cat should play with mice. Not believing this, I see no necessity in the belief that the eye was expressly designed. On the other hand, I cannot anyhow be contented to view this wonderful universe, and especially the nature of man, and to conclude that everything is the result of brute force. I am inclined to look at everything as resulting from designed laws, with the details, whether good or bad, left to the working out of what we may call chance. Not that this notion *at all* satisfies me. I feel most deeply that the whole subject is too profound for the human intellect. A dog might as well speculate on the mind of Newton. Let each man hope and believe what he can. Certainly I agree with you that my views are not at all necessarily atheistical. The lightning kills a man, whether a good one or bad one, owing to the excessively complex action of natural laws. A child (who may turn out an idiot) is born by the action of even more complex laws, and I can see no reason why a man, or other animal, may not have been aboriginally produced by other laws, and that all these laws may have been expressly designed by an omniscient Creator, who foresaw every future event and consequence. But the more I think the more bewildered I become; as indeed I have probably shown by this letter.

Most deeply do I feel your generous kindness and interest.
Yours sincerely and cordially.

---

* In a letter to Mr. Huxley my father wrote :—" Have you seen the last *Saturday Review ?* I am very glad of the defence of you and of myself. I wish the reviewer had noticed Hooker. The reviewer, whoever he is, is a jolly good fellow, as this review and the last on me showed. He writes

The meeting of the British Association at Oxford in 1860 is famous for two pitched battles over the *Origin of Species*. Both of them originated in unimportant papers. On Thursday, June 28th, Dr. Daubeny of Oxford made a communication to Section D : " On the final causes of the sexuality of plants, with particular reference to Mr. Darwin's work on the *Origin of Species*." Mr. Huxley was called on by the President, but tried (according to the *Athenæum* report) to avoid a discussion, on the ground " that a general audience, in which sentiment would unduly interfere with intellect, was not the public before which such a discussion should be carried on." However, the subject was not allowed to drop. Sir R. Owen (I quote from the *Athenæum*, July 7th, 1860), who " wished to approach this subject in the spirit of the philosopher," expressed his " conviction that there were facts by which the public could come to some conclusion with regard to the probabilities of the truth of Mr. Darwin's theory." He went on to say that the brain of the gorilla " presented more differences, as compared with the brain of man, than it did when compared with the brains of the very lowest and most problematical of the Quadrumana." Mr. Huxley replied, and gave these assertions a " direct and unqualified contradiction," pledging himself to " justify that unusual procedure elsewhere," * a pledge which he amply fulfilled.† On Friday there was peace, but on Saturday 30th, the battle arose with redoubled fury, at a conjoint meeting of three Sections, over a paper by Dr. Draper of New York, on the " Intellectual development of Europe considered with reference to the views of Mr. Darwin."

The following account is from an eye-witness of the scene.

" The excitement was tremendous. The Lecture-room, in which it had been arranged that the discussion should be held, proved far too small for the audience, and the meeting adjourned to the Library of the Museum, which was crammed to suffocation long before the champions entered the lists. The numbers were estimated at from 700 to 1000. Had it been term-time, or had the general public been admitted, it would have been impossible to have ac-

---

capitally, and understands well his subject. I wish he had slapped [the *Edinburgh* reviewer] a little bit harder."

* *Man's Place in Nature*, by T. H. Huxley, 1863, p. 114.

† See the *Nat. Hist. Review*, 1861.

commodated the rush to hear the oratory of the bold
Bishop. * Prof. Henslow, the President of Section D,
occupied the chair, and wisely announced *in limine* that
none who had not valid arguments to bring forward
on one side or the other, would be allowed to address
the meeting : a caution that proved necessary, for no
fewer than four combatants had their utterances burked
by him, because of their indulgence in vague declama-
tion.

"The Bishop was up to time, and spoke for full half-an-
hour with inimitable spirit, emptiness and unfairness. It
was evident from his handling of the subject that he had
been 'crammed' up to the throat, and that he knew nothing at first hand; in fact, he used no argument not to be
found in his *Quarterly* article.† He ridiculed Darwin
badly, and Huxley savagely, but all in such dulcet tones, so
persuasive a manner, and in such well-turned periods, that I
who had been inclined to blame the President for allowing
a discussion that could serve no scientific purpose, now for-
gave him from the bottom of my heart."

What follows is from notes most kindly supplied by the
Hon. and Rev. W. H. Fremantle, who was an eye-witness of
the scene.

"The Bishop of Oxford attacked Darwin, at first play-
fully but at last in grim earnest. It was known that the
Bishop had written an article against Darwin in the last
Quarterly Review : it was also rumoured that Prof. Owen
had been staying at Cuddesden and had primed the Bishop,
who was to act as mouthpiece to the great Palæontologist,
who did not himself dare to enter the lists. The Bishop,
however, did not show himself master of the facts, and
made one serious blunder. A fact which had been much
dwelt on as confirmatory of Darwin's idea of variation, was
that a sheep had been born shortly before in a flock in the
North of England, having an addition of one to the verte-
bræ of the spine. The Bishop was declaring with rhetorical
exaggeration that there was hardly any actual evidence on
Darwin's side. 'What have they to bring forward?' he exclaimed. 'Some rumoured statement about a long-legged
sheep.' But he passed on to banter : 'I should like to ask
Professor Huxley, who is sitting by me, and is about to tear

---

* It was well known that Bishop Wilberforce was going to speak.
† *Quarterly Review,* July 1860.

me to pieces when I have sat down, as to his belief in being descended from an ape. Is it on his grandfather's or his grandmother's side that the ape ancestry comes in?' And then taking a graver tone, he asserted in a solemn peroration that Darwin's views were contrary to the revelations of God in the Scriptures. Professor Huxley was unwilling to respond : but he was called for and spoke with his usual incisiveness and with some scorn. 'I am here only in the interest of science,' he said, 'and I have not heard anything which can prejudice the case of my august client.' Then after showing how little competent the Bishop was to enter upon the discussion, he touched on the question of Creation. 'You say that development drives out the Creator. But you assert that God made you; and yet you know that you yourself were originally a little piece of matter no bigger than the end of this gold pencil-case.' Lastly as to the descent from a monkey, he said: 'I should feel it no shame to have risen from such an origin. But I should feel it a shame to have sprung from one who prostituted the gifts of culture and of eloquence to the service of prejudice and of falsehood.'

" Many others spoke. Mr. Gresley, an old Oxford don, pointed out that in human nature at least orderly development was not the necessary rule ; Homer was the greatest of poets, but he lived 3000 years ago, and has not produced his like.

" Admiral Fitz-Roy was present, and said that he had often expostulated with his old comrade of the *Beagle* for entertaining views which were contradictory to the First Chapter of Genesis.

" Sir John Lubbock declared that many of the arguments by which the permanence of species was supported came to nothing, and instanced some wheat which was said to have come off an Egyptian mummy and was sent to him to prove that wheat had not changed since the time of the Pharaohs; but which proved to be made of French chocolate.* Sir Joseph (then Dr.) Hooker spoke shortly, saying that he had found the hypothesis of Natural Selection so helpful in explaining the phenomena of his own subject of Botany, that he had been constrained to accept it. After a few words from Darwin's old friend Professor Henslow who occupied

---

* Sir John Lubbock also insisted on the embryological evidence for evolution.—F. D.

the chair, the meeting broke up, leaving the impression that
those most capable of estimating the arguments of Darwin
in detail saw their way to accept his conclusions."

Many versions of Mr. Huxley's speech were current: the
following report of his conclusion is from a letter addressed
by the late John Richard Green, then an undergraduate, to
a fellow-student, now Professor Boyd Dawkins :—" I asserted, and I repeat, that a man has no reason to be ashamed of
having an ape for his grandfather. If there were an ancestor
whom I should feel shame in recalling, it would be a *man*, a
man of restless and versatile intellect, who, not content with
an equivocal success in his own sphere of activity, plunges
into scientific questions with which he has no real acquaintance, only to obscure them by an aimless rhetoric, and dis-
tract the attention of his hearers from the real point at issue
by eloquent digressions, and skilled appeals to religious prejudice." *

The following letter shows that Mr. Huxley's presence at
this remarkable scene depended on so slight a chance as that
of meeting a friend in the street; that this friend should
have been Robert Chambers, so that the author of the
Vestiges should have sounded the war-note for the battle of
the *Origin*, adds interest to the incident. I have to thank
Mr. Huxley for allowing the story to be told in words of his
not written for publication.

T. H. Huxley to Francis Darwin.

June 27, 1861.

. . . I should say that Fremantle's account is substantially correct; but that Green has the passage of my speech
more accurately. However, I am certain that I did not use
the word " equivocal." †

The odd part of the business is that I should not have
been present except for Robert Chambers. I had heard of
the Bishop's intention to utilise the occasion. I knew he
had the reputation of being a first-rate controversialist, and
I was quite aware that if he played his cards properly, we
should have little chance, with such an audience, of making
an efficient defence. Moreover, I was very tired, and wanted

---

* Mr. Fawcett wrote (*Macmillan's Magazine*, 1860) :—
" The retort was so justly deserved and so inimitable in its manner, that
no one who was present can ever forget the impression that it made."
† This agrees with Professor Victor Carus's recollection.

to join my wife at her brother-in-law's country house near Reading, on the Saturday. On the Friday I met Chambers in the street, and in reply to some remark of his about the meeting, I said that I did not mean to attend it; did not see the good of giving up peace and quietness to be episcopally pounded. Chambers broke out into vehement remonstrances and talked about my deserting them. So I said, " Oh ! if you take it that way, I'll come and have my share of what is going on."

So I came, and chanced to sit near old Sir Benjamin Brodie. The Bishop began his speech, and, to my astonishment, very soon showed that he was so ignorant that he did not know how to manage his own case. My spirits rose proportionally, and when he turned to me with his insolent question, I said to Sir Benjamin, in an undertone, " The Lord hath delivered him into mine hands."

That sagacious old gentleman stared at me as if I had lost my senses. But, in fact, the Bishop had justified the severest retort I could devise, and I made up my mind to let him have it. I was careful, however, not to rise to reply, until the meeting called for me—then I let myself go.

In justice to the Bishop, I am bound to say he bore no malice, but was always courtesy itself when we occasionally met in after years. Hooker and I walked away from the meeting together, and I remember saying to him that this experience had changed my opinion as to the practical value of the art of public speaking, and that, from that time forth, I should carefully cultivate it, and try to leave off hating it. I did the former, but never quite succeeded in the latter effort.

I did not mean to trouble you with such a long scrawl when I began about this piece of ancient history.

Ever yours very faithfully

T. H. HUXLEY.

The eye-witness above quoted (p. 250) continues:—

" There was a crowded conversazione in the evening at the rooms of the hospitable and genial Professor of Botany, Dr. Daubeny, where the almost sole topic was the battle of the *Origin*, and I was much struck with the fair and unprejudiced way in which the black coats and white cravats of Oxford discussed the question, and the frankness with which they offered their congratulations to the winners in the combat." *

C. D. to J. D. Hooker.   Monday night [July 2nd, 1860].

MY DEAR HOOKER,—I have just received your letter. I have been very poorly, with almost continuous bad headache for forty-eight hours, and I was low enough, and thinking what a useless burthen I was to myself and all others, when your letter came, and it has so cheered me; your kindness and affection brought tears into my eyes. Talk of fame, honour, pleasure, wealth, all are dirt compared with affection; and this is a doctrine with which, I know, from your letter, that you will agree with from the bottom of your heart. . . . How I should have liked to have wandered about Oxford with you, if I had been well enough; and how still more I should have liked to have heard you triumphing over the Bishop. I am astonished at your success and audacity. It is something unintelligible to me how any one can argue in public like orators do. I had no idea you had this power. I have read lately so many hostile views, that I was beginning to think that perhaps I was wholly in the wrong, and that —— was right when he said the whole subject would be forgotten in ten years; but now that I hear that you and Huxley will fight publicly (which I am sure I never could do), I fully believe that our cause will, in the long-run, prevail. I am glad I was not in Oxford, for I should have been overwhelmed, with my [health] in its present state.

C. D. to J. D. Hooker.   [July 1860.]

. . . I have just read the *Quarterly.*† It is uncommonly clever; it picks out with skill all the most conjectural parts,

---

* See Professor Newton's interesting *Early Days of Darwinism* in *Macmillan's Magazine*, Feb. 1888, where the battle at Oxford is briefly described.

† *Quarterly Review*, July 1860. The article in question was by Wilberforce, Bishop of Oxford, and was afterwards published in his *Essays Contributed to the Quarterly Review*, 1874. In the *Life and Letters*, ii. p. 182, Mr. Huxley has given some account of this article. I quote a few lines :—" Since Lord Brougham assailed Dr. Young, the world has seen no such specimen of the insolence of a shallow pretender to a Master in Science as this remarkable production, in which one of the most exact of observers, most cautious of reasoners, and most candid of expositors, of this or any other age, is held up to scorn as a 'flighty' person, who endeavours 'to prop up his utterly rotten fabric of guess and speculation,' and whose 'mode of dealing with nature' is

and brings forward well all the difficulties. It quizzes me quite splendidly by quoting the *Anti-Jacobin* versus my Grandfather. You are not alluded to, nor, strange to say, Huxley ; and I can plainly see, here and there, ——'s hand. The concluding pages will make Lyell shake in his shoes. By Jove, if he sticks to us, he will be a real hero. Goodnight. Your well-quizzed, but not sorrowful, and affectionate friend.　　　　　　　　　　　　　　　　　　**C. D.**

I can see there has been some queer tampering with the review, for a page has been cut out and reprinted.

The following extract from a letter of Sept. 1st, 1860, is of interest, not only as showing that Lyell was still conscientiously working out his conversion, but also and especially as illustrating the remarkable fact that hardly any of my father's critics gave him any new objections—so fruitful had been his ponderings of twenty years :—

"I have been much interested by your letter of the 28th, received this morning. It has *delighted* me, because

---

reprobated as 'utterly dishonourable to Natural Science.'" The passage from the *Anti-Jacobin*, referred to in the letter, gives the history of the evolution of space from the "primæval point or *punctum saliens* of the universe," which is conceived to have moved "forward in a right line, *ad infinitum*, till it grew tired ; after which the right line, which it had generated, would begin to put itself in motion in a lateral direction, describing an area of infinite extent. This area, as soon as it became conscious of its own existence, would begin to ascend or descend according as its specific gravity would determine it, forming an immense solid space filled with vacuum, and capable of containing the present universe."

The following (p. 263) may serve as an example of the passages in which the reviewer refers to Sir Charles Lyell :—"That Mr. Darwin should have wandered from this broad highway of nature's works into the jungle of fanciful assumption is no small evil. We trust that he is mistaken in believing that he may count Sir C. Lyell as one of his converts. We know, indeed, the strength of the temptations which he can bring to bear upon his geological brother. . . . Yet no man has been more distinct and more logical in the denial of the transmutation of species than Sir C. Lyell, and that not in the infancy of his scientific life, but in its full vigour and maturity." The Bishop goes on to appeal to Lyell, in order that with his help " this flimsy speculation may be as completely put down as was what in spite of all denials we must venture to call its twin though less instructed brother, the *Vestiges of Creation*."

With reference to this article, Mr. Brodie Innes, my father's old friend and neighbour, writes :—" Most men would have been annoyed by an article written with the Bishop's accustomed vigour, a mixture of argument and ridicule. Mr. Darwin was writing on some parish matter, and put a postscript— 'If you have not seen the last *Quarterly*, do get it ; the Bishop of Oxford has made such capital fun of me and my grandfather.' By a curious coincidence, when I received the letter, I was staying in the same house with the Bishop, and showed it to him. He said, ' I am very glad he takes it in that way, he is such a capital fellow.' "

it demonstrates that you have thought a good deal lately on Natural Selection. Few things have surprised me more than the entire paucity of objections and difficulties new to me in the published reviews. Your remarks are of a different stamp and new to me."

C. D. to Asa Gray. [Hartfield, Sussex] July 22nd [1860].

MY DEAR GRAY,—Owing to absence from home at water-cure and then having to move my sick girl to whence I am now writing, I have only lately read the discussion in *Proc. American Acad.*,* and now I cannot resist expressing my sincere admiration of your most clear powers of reasoning. As Hooker lately said in a note to me, you are more than *any one* else the thorough master of the subject. I declare that you know my book as well as I do myself; and bring to the question new lines of illustration and argument in a manner which excites my astonishment and almost my envy! † I admire these discussions, I think, almost more than your article in *Silliman's Journal.* Every single word seems weighed carefully, and tells like a 32-pound shot. It makes me much wish (but I know that you have not time) that you could write more in detail, and give, for instance, the facts on the variability of the American wild fruits. The *Athenæum* has the largest circulation, and I have sent my copy to the editor with a request that he would republish the first discussion; I much fear he will not, as he reviewed the subject in so hostile a spirit. . . . I shall be curious [to see], and will order the August number, as soon as I know that it contains your review of reviews. My conclusion is that you have made a mistake in being a botanist, you ought to have been a lawyer.

The following passages from a letter to Huxley (Dec. 2nd, 1860) may serve to show what was my father's view of the position of the subject, after a year's experience of reviewer's, critics and converts :—

---

* April 10th, 1860. Dr Gray criticised in detail "several of the positions taken at the preceding meeting by Mr. [J. A.] Lowell, Prof. Bowen and Prof. Agassiz." It was reprinted in the *Athenæum*, Aug. 4th, 1860.

† On Sept. 26th, 1860, he wrote in the same sense to Gray :—"You never touch the subject without making it clearer. I look at it as even more extraordinary that you never say a word or use an epithet which does not express fully my meaning. Now Lyell, Hooker, and others, who perfectly understand my book, yet sometimes use expressions to which I demur."

" I have got fairly sick of hostile reviews. Nevertheless, they have been of use in showing me when to expatiate a little and to introduce a few new discussions.

" I entirely agree with you, that the difficulties on my notions are terrific, yet having seen what all the Reviews have said against me, I have far more confidence in the *general* truth of the doctrine than I formerly had. Another thing gives me confidence, viz. that some who went half an inch with me now go further, and some who were bitterly opposed are now less bitterly opposed. . . . I can pretty plainly see that, if my view is ever to be generally adopted, it will be by young men growing up and replacing the old workers, and then young ones finding that they can group facts and search out new lines of investigation better on the notion of descent than on that of creation."

CHAPTER XIV.

1861—1871.

THE beginning of the year 1861 saw my father engaged on the third edition (2000 copies) of the *Origin*, which was largely corrected and added to, and was published in April, 1861.

On July 1, he started, with his family, for Torquay, where he remained until August 27—a holiday which he characteristically enters in his diary as "eight weeks and a day." The house he occupied was in Hesketh Crescent, a pleasantly placed row of houses close above the sea, somewhat removed from what was then the main body of the town, and not far from the beautiful cliffed coast-line in the neighbourhood of Anstey's Cove.

During the Torquay holiday, and for the remainder of the year, he worked at the fertilisation of orchids. This part of the year 1861 is not dealt with in the present chapter, because (as explained in the preface) the record of his life, seems to become clearer when the whole of his botanical work is placed together and treated separately. The present chapter will, therefore, include only the progress of his work in the direction of a general amplification of the *Origin of Species—e. g.,* the publication of *Animals and Plants* and the *Descent of Man.* It will also give some idea of the growth of belief in evolutionary doctrines.

With regard to the third edition, he wrote to Mr. Murray in December, 1860 :—

"I shall be glad to hear when you have decided how many copies you will print off—the more the better for me in all ways, as far as compatible with safety; for I hope never again to make so many corrections, or rather additions, which I have made in hopes of making my many rather stupid reviewers at least understand what is meant. I hope and think I shall improve the book considerably."

An interesting feature of the new edition was the " Historical Sketch of the Recent Progress of Opinion on the Origin of Species," * which now appeared for the first time, and was continued in the later editions of the work. It bears a strong impress of the author's personal character in the obvious wish to do full justice to all his predecessors— though even in this respect it has not escaped some adverse criticism.

A passage in a letter to Hooker (March 27, 1861) gives the history of one of his corrections.

" Here is a good joke : H. C. Watson (who, I fancy and hope, is going to review the new edition of the *Origin*) says that in the first four paragraphs of the introduction, the words ' I,' ' me,' ' my,' occur forty-three times ! I was dimly conscious of the accursed fact. He says it can be explained phrenologically, which I suppose civilly means, that I am the most egotistically self-sufficient man alive; perhaps so. I wonder whether he will print this pleasing fact; it beats hollow the parentheses in Wollaston's writing.

"*I* am, *my* dear Hooker, ever yours,

"C. DARWIN.

" P.S.—Do not spread this pleasing joke; it is rather too biting."

He wrote a couple of years later, 1863, to Asa Gray, in a manner which illustrates his use of the personal pronoun in the earlier editions of the *Origin* :—

" You speak of Lyell as a judge; now what I complain of is that he declines to be a judge. . . . I have sometimes almost wished that Lyell had pronounced against me. When I say ' me,' I only mean *change of species by descent*. That seems to me the turning-point. Personally, of course, I care much about Natural Selection ; but that seems to me utterly unimportant, compared to the question of Creation *or* Modification."

He was, at first, alone, and felt himself to be so in main-

---

* The Historical Sketch had already appeared in the first German edition (1860) and the American edition. Bronn states in the German edition (footnote, p. 1) that it was his critique in the *N. Jahrbuch für Mineralogie* that suggested to my father the idea of such a sketch.

taining rational workable theory of Evolution. It was therefore perfectly natural that he should speak of "my" theory.

Towards the end of the present year (1861) the final arrangements for the first French edition of the *Origin* were completed, and in September a copy of the third English edition was despatched to Mdlle. Clémence Royer, who undertook the work of translation. The book was now spreading on the Continent, a Dutch edition had appeared, and, as we have seen, a German translation had been published in 1860. In a letter to Mr. Murray (September 10, 1861), he wrote, "My book seems exciting much attention in Germany, judging from the number of discussions sent me." The silence had been broken, and in a few years the voice of German science was to become one of the strongest of the advocates of Evolution.

A letter, June 23, 1861, gave a pleasant echo from the Continent of the growth of his views :—

Hugh Falconer * *to C. Darwin.*   31 Sackville St., W., June 23, 1861.

MY DEAR DARWIN,—I have been to Adelsberg cave and brought back with me a live *Proteus anguinus*, designed for you from the moment I got it; *i.e.* if you have got an aquarium and would care to have it. I only returned last night from the Continent, and hearing from your brother that you are about to go to Torquay, I lose no time in making you the offer. The poor dear animal is still alive—although it has had no appreciable means of sustenance for a month—and I am most anxious to get rid of the responsibility of starving it longer. In your hands it will thrive and have a fair chance of being developed without delay into some type of the Columbidæ—say a Pouter or a Tumbler.

My dear Darwin, I have been rambling through the north of Italy, and Germany lately. Everywhere have I heard your views and your admirable essay canvassed—the views of course often dissented from, according to the special bias of the speaker—but the work, its honesty of purpose, grandeur of conception, felicity of illustration, and

---

* Hugh Falconer, born 1809, died 1865. Chiefly known as a palæontologist, although employed as a botanist during his whole career in India, where he was a medical officer in the H. E. I. C. Service.

courageous exposition, always referred to in terms of the highest admiration. And among your warmest friends no one rejoiced more heartily in the just appreciation of Charles Darwin than did,

Yours very truly.

My father replied :—

Down [June 24, 1861].

My dear Falconer,—I have just received your note, and by good luck a day earlier than properly, and I lose not a moment in answering you, and thanking you heartily for your offer of the valuable specimen ; but I have no aquarium and shall soon start for Torquay, so that it would be a thousand pities that I should have it. Yet I should certainly much like to see it, but I fear it is impossible. Would not the Zoological Society be the best place? and then the interest which many would take in this extraordinary animal would repay you for your trouble.

Kind as you have been in taking this trouble and offering me this specimen, to tell the truth I value your note more than the specimen. I shall keep your note amongst a very few precious letters. Your kindness has quite touched me.

Yours affectionately and gratefully.

My father, who had the strongest belief in the value of Asa Gray's help, was anxious that his evolutionary writings should be more widely known in England. In the autumn of 1860, and the early part of 1861, he had a good deal of correspondence with him as to the publication, in the form of a pamphlet, of Gray's three articles in the July, August, and October numbers of the *Atlantic Monthly*, 1860.

The reader will find these articles republished in Dr. Gray's *Darwiniana*, p. 87, under the title "Natural Selection not inconsistent with Natural Theology." The pamphlet found many admirers, and my father believed that it was of much value in lessening opposition, and making converts to Evolution. His high opinion of it is shown not only in his letters, but by the fact that he inserted a special notice of it in a prominent place in the third edition of the *Origin*. Lyell, among others, recognised its value as an antidote to the kind of criticism from which the cause of Evolution suffered. Thus my father wrote to Dr. Gray: "Just to exemplify the use of your pamphlet, the Bishop of London was asking Lyell what he thought of the review in the *Quarterly*, and Lyell answered, 'Read Asa Gray in the *Atlantic*.' "

On the same subject he wrote to Gray in the following year :—

"I believe that your pamphlet has done my book *great* good ; and I thank you from my heart for myself : and believing that the views are in large part true, I must think that you have done natural science a good turn. Natural Selection seems to be making a little progress in England and on the Continent ; a new German edition is called for, and a French one has just appeared."

The following may serve as an example of the form assumed between these friends of the animosity at that time so strong between England and America * :—

"Talking of books, I am in the middle of one which pleases me, though it is very innocent food, viz. Miss Cooper's *Journal of a Naturalist*. Who is she ? She seems a very clever woman, and gives a capital account of the battle between *our* and *your* weeds.† Does it not hurt your Yankee pride that we thrash you so confoundedly ? I am sure Mrs. Gray will stick up for your own weeds. Ask her whether they are not more honest, downright good sort of weeds. The book gives an extremely pretty picture of one of your villages ; but I see your autumn, though so much more gorgeous than ours, comes on sooner, and that is one comfort."

A question constantly recurring in the letters to Gray is that of design. For instance :—

"Your question what would convince me of design is a poser. If I saw an angel come down to teach us good, and I was convinced from others seeing him that I was not mad, I should believe in design. If I could be convinced thoroughly that life and mind was in an unknown way a

---

* In his letters to Gray there are also numerous references to the American war. I give a single passage. "I never knew the newspapers so profoundly interesting. North America does not do England justice ; I have not seen or heard of a soul who is not with the North. Some few, and I am one of them, even wish to God, though at the loss of millions of lives, that the North would proclaim a crusade against slavery. In the long-run, a million horrid deaths would be amply repaid in the cause of humanity. What wonderful times we live in ! Massachusetts seems to show noble enthusiasm. Great God ! how I should like to see the greatest curse on earth—slavery—abolished !"

† This refers to the remarkable fact that many introduced European weeds have spread over large parts of the United States.

function of other imponderable force, I should be convinced. If man was made of brass or iron and no way connected with any other organism which had ever lived, I should perhaps be convinced. But this is childish writing.

" I have lately been corresponding with Lyell, who, I think, adopts your idea of the stream of variation having been led or designed. I have asked him (and he says he will hereafter reflect and answer me) whether he believes that the shape of my nose was designed. If he does I have nothing more to say. If not, seeing what Fanciers have done by selecting individual differences in the nasal bones of pigeons, I must think that it is illogical to suppose that the variations, which natural selection preserves for the good of any being, have been designed. But I know that I am in the same sort of muddle (as I have said before) as all the world seems to be in with respect to free will, yet with everything supposed to have been foreseen or preordained."

The shape of his nose would perhaps not have been used as an illustration, if he had remembered Fitz-Roy's objection to that feature (see *Autobiography*, p. 26). He should, too, have remembered the difficulty of predicting the value to an organism of an apparently unimportant character.

In England Professor Huxley was at work in the evolutionary cause. He gave, in 1862, two lectures at Edinburgh on *Man's Place in Nature*. My father wrote :—

" I am heartily glad of your success in the North. By Jove, you have attacked Bigotry in its stronghold. I thought you would have been mobbed. I am so glad that you will publish your Lectures. You seem to have kept a due medium between extreme boldness and caution. I am heartily glad that all went off so well."

A review,* by F. W. Hutton, afterwards Professor of Biology and Geology at Canterbury, N. Z., gave a hopeful note of the time not far off. when a broader view of the argument for Evolution would be accepted. My father wrote the author † :—

Down, April 20th, 1861.

DEAR SIR,—I hope that you will permit me to thank you for sending me a copy of your paper in the *Geologist*, and

---

* *Geologist*, 1861, p. 132.
† The letter is published in a lecture by Professor Hutton given before the Philosoph. Institute, Canterbury, N.Z., Sept. 12th, 1887.

at the same time to express my opinion that you have done the subject a real service by the highly original, striking, and condensed manner with which you have put the case. I am actually weary of telling people that I do not pretend to adduce direct evidence of one species changing into another, but that I believe that this view in the main is correct, because so many phenomena can be thus grouped together and explained.

But it is generally of no use, I cannot make persons see this. I generally throw in their teeth the universally admitted theory of the undulations of light—neither the undulations, nor the very existence of ether being proved—yet admitted because the views explain so much. You are one of the very few who have seen this, and have now put it most forcibly and clearly. I am much pleased to see how carefully you have read my book, and what is far more important, reflected on so many points with an independent spirit. As I am deeply interested in the subject (and I hope not exclusively under a personal point of view) I could not resist venturing to thank you for the right good service which you have done. Pray believe me, dear sir,
Yours faithfully and obliged.

It was a still more hopeful sign that work of the first rank in value, conceived on evolutionary principles, began to be published.

My father expressed this idea in a letter to the late Mr. Bates.*

"Under a general point of view, I am quite convinced (Hooker and Huxley took the same view some months ago) that a philosophic view of nature can solely be driven into naturalists by treating special subjects as you have done."

This refers to Mr. Bates' celebrated paper on mimicry, with which the following letter deals :—

---

* Mr. Bates is perhaps most widely known through his delightful *The Naturalist on the Amazons.* It was with regard to this book that my father wrote (April 1863) to the author:—"I have finished vol. i. My criticisms may be condensed into a single sentence, namely, that it is the best work of Natural History Travels ever published in England. Your style seems to me admirable. Nothing can be better than the discussion on the struggle for existence, and nothing better than the description of the Forest scenery. It is a grand book, and whether or not it sells quickly, it will last. You have spoken out boldly on Species; and boldness on the subject seems to get rarer and rarer. How beautifully illustrated it is."

Down, Nov. 20, [1862].

DEAR BATES,—I have just finished, after several reads,
your paper.* In my opinion it is one of the most remarkable and admirable papers I ever read in my life. The
mimetic cases are truly marvellous, and you connect excellently a host of analogous facts. The illustrations are beautiful, and seem very well chosen; but it would have saved
the reader not a little trouble, if the name of each had been
engraved below each separate figure. No doubt this would
have put the engraver into fits, as it would have destroyed
the beauty of the plate. I am not at all surprised at such a
paper having consumed much time. I am rejoiced that I
passed over the whole subject in the *Origin*, for I should
have made a precious mess of it. You have most clearly
stated and solved a wonderful problem. No doubt with
most people this will be the cream of the paper; but I am
not sure that all your facts and reasonings on variation, and
on the segregation of complete and semi-complete species,
is not really more, or at least as valuable a part. I never
conceived the process nearly so clearly before; one feels
present at the creation of new forms. I wish, however, you
had enlarged a little more on the pairing of similar varieties;

---

* Mr. Bates' paper, ' Contributions to an Insect Fauna of the Amazons
Valley ' (*Linn. Soc. Trans.* xxiii., 1862), in which the now familiar subject of
mimicry was founded. My father wrote a short review of it in the *Natural
History Review*, 1863, p. 219, parts of which occur almost verbatim in the later
editions of the *Origin of Species*. A striking passage occurs in the review,
showing the difficulties of the case from a creationist's point of view :—

"By what means, it may be asked, have so many butterflies of the Ama-
zonian region acquired their deceptive dress? Most naturalists will answer
that they were thus clothed from the hour of their creation—an answer which
will generally be so far triumphant that it can be met only by long-drawn
arguments; but it is made at the expense of putting an effectual bar to all
further inquiry. In this particular case, moreover, the creationist will meet
with special difficulties; for many of the mimicking forms of *Leptalis* can be
shown by a graduated series to be merely varieties of one species; other
mimickers are undoubtedly distinct species, or even distinct genera. So again,
some of the mimicked forms can be shown to be merely varieties; but the
greater number must be ranked as distinct species. Hence the creationist will
have to admit that some of these forms have become imitators, by means of
the laws of variation, whilst others he must look at as separately created under their present guise; he will further have to admit that some have been
created in imitation of forms not themselves created as we now see them, but
due to the laws of variation! Professor Agassiz, indeed, would think nothing
of this difficulty; for he believes that not only each species and each variety,
but that groups of individuals, though identically the same, when inhabiting
distinct countries, have been all separately created in due proportional numbers to the wants of each land. Not many naturalists will be content thus to
believe that varieties and individuals have been turned out all ready made,
almost as a manufacturer turns out toys according to the temporary demand
of the market."

a rather more numerous body of facts seems here wanted. Then, again, what a host of curious miscellaneous observations there are—as on related sexual and individual variability: these will some day, if I live, be a treasure to me.

With respect to mimetic resemblance being so common with insects, do you not think it may be connected with their small size; they cannot defend themselves; they cannot escape by flight, at least, from birds, therefore they escape by trickery and deception?

I have one serious criticism to make, and that is about the title of the paper; I cannot but think that you ought to have called prominent attention in it to the mimetic resemblances. Your paper is too good to be largely appreciated by the mob of naturalists without souls; but, rely on it, that it will have *lasting* value, and I cordially congratulate you on your first great work. You will find, I should think, that Wallace will appreciate it. How gets on your book? Keep your spirits up. A book is no light labour. I have been better lately, and working hard, but my health is very indifferent. How is your health? Believe me, dear Bates,

Yours very sincerely.

1863.

Although the battle* of Evolution was not yet won, the growth of belief was undoubtedly rapid. So that, for instance, Charles Kingsley could write to F. D. Maurice †:

"The state of the scientific mind is most curious; Darwin is conquering everywhere, and rushing in like a flood, by the mere force of truth and fact."

The change did not proceed without a certain amount of personal bitterness. My father wrote in February, 1863 :—

"What an accursed evil it is that there should be all this quarrelling within what ought to be the peaceful realms of science."

I do not desire to keep alive the memories of dead quarrels, but some of the burning questions of that day are too

---

* Mr. Huxley was as usual active in guiding and stimulating the growing tendency to tolerate or accept the views set forth in the *Origin of Species.* He gave a series of lectures to working men at the School of Mines in November, 1862. These were printed in 1863 from the shorthand notes of Mr. May, as six little blue books, price 4d. each, under the title, *Our Knowledge of the Causes of Organic Nature.*

† Kingsley's *Life*, vol. ii. p. 171.

important from the biographical point of view to be altogether omitted. Of this sort is the history of Lyell's conversion to Evolution. It led to no flaw in the friendship of
the two men principally concerned, but it shook and irritated
a number of smaller people. Lyell was like the Mississippi in
flood, and as he changed his course, the dwellers on the
banks were angered and frightened by the general upsetting
of landmarks.

C. D. to J. D. Hooker. Down, Feb. 24 [1863].

MY DEAR HOOKER,—I am astonished at your note. I
have not seen the *Athenæum*, * but I have sent for it, and
may get it to-morrow; and will then say what I think.

I have read Lyell's book. [*The Antiquity of Man.*] The
whole certainly struck me as a compilation, but of the highest
class, for when possible the facts have been verified on the
spot, making it almost an original work. The Glacial chapters seem to me best, and in parts magnificent. I could
hardly judge about Man, as all the gloss and novelty was
completely worn off. But certainly the aggregation of the
evidence produced a very striking effect on my mind. The
chapter comparing language and changes of species, seems
most ingenious and interesting. He has shown great skill
in picking out salient points in the argument for change of
species; but I am deeply disappointed (I do not mean personally) to find that his timidity prevents him from giving
any judgment. . . . From all my communications with him,
I must ever think that he has really entirely lost faith † in
the immutability of species; and yet one of his strongest
sentences is nearly as follows : " If it should *ever* ‡ be rendered highly probable that species change by variation and
natural selection," &c. &c. I had hoped he would have
guided the public as far as his own belief went. . . . One
thing does please me on this subject, that he seems to appre-

---

* In the *Antiquity of Man,* first edition, p. 480, Lyell criticised somewhat
severely Owen's account of the difference between the Human and Simian
brains. The number of the *Athenæum* here referred to (1863, p. 262) contains
a reply by Professor Owen to Lyell's strictures. The surprise expressed by
my father was at the revival of a controversy which every one believed to be
closed. Professor Huxley (*Medical Times,* Oct. 25th, 1862, quoted in *Man's
Place in Nature,* p. 117) spoke of the "two years during which this preposterous controversy has dragged its weary length." And this no doubt expressed
a very general feeling.

† This should obviously run, "that at one time he entirely had faith."

‡ The italics are not Lyell's.

ciate your work. No doubt the public or a part may be induced to think that, as he gives to us a larger space than to
Lamarck, he must think that there is something in our
views. When reading the brain chapter, it struck me forcibly that if he had said openly that he believed in change
of species, and as a consequence that man was derived from
some Quadrumanous animal, it would have been very proper
to have discussed by compilation the differences in the most
important organ, viz. the brain. As it is, the chapter seems
to me to come in rather by the head and shoulders. I do
not think (but then I am as prejudiced as Falconer and
Huxley, or more so) that it is too severe; it struck me as
given with judicial force. It might perhaps be said with
truth that he had no business to judge on a subject on
which he knows nothing; but compilers must do this to a
certain extent. (You know I value and rank high compilers,
being one myself!)

The Lyells are coming here on Sunday evening to stay
till Wednesday. I dread it, but I must say how much disappointed I am that he has not spoken out on species, still
less on man. And the best of the joke is that he thinks he
has acted with the courage of a martyr of old. I hope I
may have taken an exaggerated view of his timidity, and
shall *particularly* be glad of your opinion on this head.
When I got his book I turned over the pages, and saw he
had discussed the subject of species, and said that I thought
he would do more to convert the public than all of us, and
now (which makes the case worse for me) I must, in common honesty, retract. I wish to Heaven he had said not a
word on the subject.

C. D. to C. Lyell.   Down, March 6 [1863].

. . . I have been of course deeply interested by your
book.* I have hardly any remarks worth sending, but will
scribble a little on what most interested me. But I will
first get out what I hate saying, viz. that I have been greatly
disappointed that you have not given judgment and
spoken fairly out what you think about the derivation of
species. I should have been contented if you had boldly
said that species have not been separately created, and had
thrown as much doubt as you like on how far variation and

---

* *The Antiquity of Man.*

natural selection suffices. I hope to Heaven I am wrong
(and from what you say about Whewell it seems so), but I
cannot see how your chapters can do more good than an
extraordinary able review. I think the *Parthenon* is right,
that you will leave the public in a fog. No doubt they
may infer that as you give more space to myself, Wallace,
and Hooker, than to Lamarck, you think more of us. But
I had always thought that your judgment would have been
an epoch in the subject. All that is over with me, and I
will only think on the admirable skill with which you have
selected the striking points, and explained them. No
praise can be too strong, in my opinion, for the inimitable
chapter on language in comparison with species. . . .
  I know you will forgive me for writing with perfect
freedom, for you must know how deeply I respect you as my
old honoured guide and master. I heartily hope and expect that your book will have a gigantic circulation, and
may do in many ways as much good as it ought to do. I
am tired, so no more. I have written so briefly that you
will have to guess my meaning. I fear my remarks are
hardly worth sending. Farewell, with kindest remembrance
to Lady Lyell,

Ever yours.

  A letter from Lyell to Hooker (Mar. 9, 1863), published
in Lyell's *Life and Letters*, vol. ii. p. 361, shows what was
his feeling at the time :—
  " He [Darwin] seems much disappointed that I do not
go farther with him, or do not speak out more. I can only
say that I have spoken out to the full extent of my present
convictions, and even beyond my state of *feeling* as to man's
unbroken descent from the brutes, and I find I am half con-
verting not a few who were in arms against Darwin, and
are even now against Huxley." Lyell speaks, too, of having
had to abandon " old and long cherished ideas, which constituted the charm to me of the theoretical part of the
science in my earlier days, when I believed with Pascal in
the theory, as Hallam terms it, of ' the archangel ruined.' "

C. D. to C. Lyell. Down, 12th [March, 1863].

  MY DEAR LYELL,—I thank you for your very interesting and kind, I may say, charming letter. I feared you
might be huffed for a little time with me. I know some

men would have been so. . . . As you say that you have
gone as far as you believe on the species question, I have
not a word to say; but I must feel convinced that at times,
judging from conversation, expressions, letters, &c., you
have as completely given up belief in immutability of
specific forms as I have done. I must still think a clear
expression from you, *if you could have given it*, would have
been potent with the public, and all the more so, as you
formerly held opposite opinions. The more I work, the
more satisfied I become with variation and natural selection,
but that part of the case I look at as less important, though
more interesting to me personally. As you ask for criticisms
on this head (and believe me that I should not have made
them unasked), I may specify (pp. 412, 413) that such
words as "Mr. D. labours to show," "is believed by the
author to throw light," would lead a common reader to
think that you yourself do *not* at all agree, but merely
think it fair to give my opinion. Lastly, you refer repeatedly to my view as a modification of Lamarck's doctrine of development and progression. If this is your deliberate ópinion there is nothing to be said, but it does not
seem so to me. Plato, Buffon, my grandfather before
Lamarck, and others, propounded the *obvious* view that if
species were not created separately they must have descended from other species, and I can see nothing else in
common between the *Origin* and Lamarck. I believe this
way of putting the case is very injurious to its acceptance,
as it implies necessary progression, and closely connects
Wallace's and my views with what I consider, after two
deliberate readings, as a wretched book, and one from
which (I well remember my surprise) I gained nothing.
But I know you rank it higher, which is curious, as it did
not in the least shake your belief. But enough, and more
than enough. Please remember you have brought it all
down on yourself !!

I am very sorry to hear about Falconer's "reclamation." *
I hate the very word, and have a sincere affection for
him.

Did you ever read anything so wretched as the *Athe-*

---

* " Falconer, whom I [Lyell] referred to oftener than to any other author,
says I have not done justice to the part he took in resuscitating the cave question, and says he shall come out with a separate paper to prove it. I offered
to alter anything in the new edition, but this he declined."—C. Lyell to C.
Darwin, March 11, 1863; Lyell's *Life*, vol. ii. p. 364.

nœum reviews of you, and of Huxley * especially. Your *object* to make man old, and Huxley's *object* to degrade him. The wretched writer has not a glimpse of what the discovery of scientific truth means. How splendid some pages are in Huxley, but I fear the book will not be popular. . . .

In the *Athenæum*, Mar. 28, 1862, p. 417, appeared a notice of Dr. Carpenter's book on 'Foraminifera,' which led to more skirmishing in the same journal. The article was remarkable for upholding spontaneous generation.

My father wrote, Mar. 29, 1863 :—

" Many thanks for *Athenæum*, received this morning, and to be returned to-morrow morning. Who would have ever thought of the old stupid *Athenæum* taking to Oken-like transcendental philosophy written in Owenian style !

" It will be some time before we see 'slime, protoplasm, &c.' generating a new animal. But I have long regretted that I truckled to public opinion, and used the Pentateuchal term of creation, † by which I really meant 'appeared' by some wholly unknown process. It is mere rubbish, thinking at present of the origin of life; one might as well think of the origin of matter."

The *Athenæum* continued to be a scientific battle-ground. On April 4, 1863, Falconer wrote a severe article on Lyell. And my father wrote (*Athenæum*, 1863, p. 554), under the cloak of attacking spontaneous generation, to defend Evolution. In reply, an article appeared in the same Journal (May 2nd, 1863, p. 586), accusing my father of claiming for his views the exclusive merit of " connecting by an intelligible thread of reasoning " a number of facts in morphology, &c. The writer remarks that, " The different generalisations cited by Mr. Darwin as being connected by an intelligible thread of reasoning exclusively through his attempt to explain specific transmutation are in fact related to it in this wise, that they have prepared the minds of naturalists for a better reception of such attempts to explain the way of the origin of species from species."

---

* *Man's Place in Nature*, 1863.
† This refers to a passage in which the reviewer of Dr. Carpenter's book speaks of " an operation of force," or " a concurrence of forces which have now no place in nature," as being, " a creative force, in fact, which Darwin could only express in Pentateuchal terms as the primordial form 'into which life was first breathed.'" The conception of expressing a creative force as a primordial form is the reviewer's.

To this my father replied as follows in the *Athenæum* of May 9th, 1863 :—

Down, May 5 [1863].

I hope that you will grant me space to own that your reviewer is quite correct when he states that any theory of descent will connect, " by an intelligible thread of reasoning," the several generalizations before specified. I ought to have made this admission expressly ; with the reservation, however, that, as far as I can judge, no theory so well explains or connects these several generalizations (more especially the formation of domestic races in comparison with natural species, the principles of classification, embryonic resemblance, &c.) as the theory, or hypothesis, or guess, if the reviewer so likes to call it, of Natural Selection. Nor has any other satisfactory explanation been ever offered of the almost perfect adaptation of all organic beings to each other, and to their physical conditions of life. Whether the naturalist believes in the views given by Lamarck, by Geoffroy St. Hilaire, by the author of the *Vestiges*, by Mr. Wallace and myself, or in any other such view, signifies extremely little in comparison with the admission that species have descended from other species, and have not been created immutable ; for he who admits this as a great truth has a wide field opened to him for further inquiry. I believe, however, from what I see of the progress of opinion on the Continent, and in this country, that the theory of Natural Selection will ultimately be adopted, with, no doubt, many subordinate modifications and improvements.

CHARLES DARWIN.

In the following, he refers to the above letter to the *Athenæum* :—

C. D. to J. D. Hooker.   Saturday [May 11, 1863].

MY DEAR HOOKER,—You give good advice about not writing in newspapers; I have been gnashing my teeth at my own folly ; and this not caused by ——'s sneers, which were so good that I almost enjoyed them. I have written once again to own to a certain extent of truth in what he says, and then if I am ever such a fool again, have no mercy on me. I have read the squib in *Public Opinion ;* * it is capital; if there is more, and you have a

---

* *Public Opinion*, April 23, 1863. A lively account of a police case, in

copy, do lend it. It shows well that a scientific man had
better be trampled in dirt than squabble.

In the following year (1864) he received the greatest
honour which a scientific man can receive in this country,
the Copley Medal of the Royal Society. It is presented at
the Anniversary Meeting on St. Andrew's Day (Nov. 30),
the medalist being usually present to receive it, but this the
state of my father's health prevented. He wrote to Mr.
Fox :—

"I was glad to see your hand-writing. The Copley,
being open to all sciences and all the world, is reckoned a
great honour ; but excepting from several kind letters, such
things make little difference to me. It shows, however,
that Natural Selection is making some progress in this
country, and that pleases me. The subject, however, is safe
in foreign lands."

The presentation of the Copley Medal is of interest in
connection with what has gone before, inasmuch as it led to
Sir C. Lyell making in his after-dinner speech, a " confes-
sion of faith as to the *Origin*." He wrote to my father
(*Life of Sir C. Lyell*, vol. ii. p. 384), "I said I had been
forced to give up my old faith without thoroughly seeing
my way to a new one. But I think you would have been
satisfied with the length I went."

Lyell's acceptance of Evolution was made public in the
tenth edition of the *Principles*, published in 1867 and 1868.
It was a sign of improvement, " a great triumph," as my
father called it, that an evolutionary article by Wallace,
dealing with Lyell's book, should have appeared in the
Quarterly Review (April, 1869). Mr. Wallace wrote :—

---

which the quarrels of scientific men are satirised. Mr. John Bull gives evidence that—

"The whole neighbourhood was unsettled by their disputes ; Huxley
quarrelled with Owen, Owen with Darwin, Lyell with Owen, Falconer and
Prestwich with Lyell, and Gray the menagerie man with everybody. He
had pleasure, however, in stating that Darwin was the quietest of the set.
They were always picking bones with each other and fighting over their
gains. If either of the gravel sifters or stone breakers found anything, he
was obliged to conceal it immediately, or one of the old bone collectors would
be sure to appropriate it first and deny the theft afterwards, and the consequent wrangling and disputes were as endless as they were wearisome.

" Lord Mayor.—Probably the clergyman of the parish might exert some
influence over them ?

" The gentleman smiled, shook his head, and stated that he regretted to
say that no class of men paid so little attention to the opinions of the clergy
as that to which these unhappy men belonged."

"The history of science hardly presents so striking an instance of youthfulness of mind in advanced life as is shown by this abandonment of opinions so long held and so powerfully advocated; and if we bear in mind the extreme caution, combined with the ardent love of truth which characterise every work which our author has produced, we shall be convinced that so great a change was not decided on without long and anxious deliberation, and that the views now adopted must indeed be supported by arguments of overwhelming force. If for no other reason than that Sir Charles Lyell in his tenth edition has adopted it, the theory of Mr. Darwin deserves an attentive and respectful consideration from every earnest seeker after truth."

The incident of the Copley Medal is interesting as giving an index of the state of the scientific mind at the time.

My father wrote: "some of the old members of the Royal are quite shocked at my having the Copley." In the *Reader*, December 3, 1864, General Sabine's presidential address at the Anniversary Meeting is reported at some length. Special weight was laid on my father's work in Geology, Zoology, and Botany, but the *Origin of Species* was praised chiefly as containing a "mass of observations," &c. It is curious that as in the case of his election to the French Institute, so in this case, he was honoured not for the great work of his life, but for his less important work in special lines.

I believe I am right in saying that no little dissatisfaction at the President's manner of allusion to the *Origin* was felt by some Fellows of the Society.

My father spoke justly when he said that the subject was "safe in foreign lands." In telling Lyell of the progress of opinion, he wrote (March, 1863) :—

"A first-rate German naturalist * (I now forget the name!), who has lately published a grand folio, has spoken out to the utmost extent on the *Origin*. De Candolle, in a very good paper on 'Oaks,' goes, in Asa Gray's opinion, as far as he himself does; but De Candolle, in writing to me, says *we*, 'we think this and that;' so that I infer he really goes to the full extent with me, and tells me of a French good botanical palæontologist † (name forgotten), who

---

* No doubt Haeckel, whose monograph on the Radiolaria was published in 1862.

† The Marquis de Saporta.

writes to De Candolle that he is sure that my views will ultimately prevail. But I did not intend to have written all
this. It satisfies me with the final results, but this result, I
begin to see, will take two or three life-times. The entomologists are enough to keep the subject back for half a
century."

The official attitude of French science was not very
hopeful. The Secrétaire Perpétuel of the Académie published an *Examen du livre de M. Darwin*, on which my
father remarks :

" A great gun, Flourens, has written a little dull book *
against me, which pleases me much, for it is plain that our
good work is spreading in France."

Mr. Huxley, who reviewed the book, † quotes the following passage from Flourens :—

" M. Darwin continue: Aucune distinction absolue n'a
été et ne peut être établie entre les espèces et les variétés !
Je vous ai déjà dit que vous vous trompiez ; une distinction
absolue sépare les variétés d'avec les espèces." Mr. Huxley
remarks on this, " Being devoid of the blessings of an
Academy in England, we are unaccustomed to see our
ablest men treated in this way even by a Perpetual Secretary." After demonstrating M. Flourens' misapprehension
of Natural Selection, Mr. Huxley says, " How one knows it
all by heart, and with what relief one reads at p. 65, ' Je
laisse M. Darwin.' "

The deterrent effect of the Académie on the spread of
Evolution in France has been most striking. Even at the
present day a member of the Institute does not feel quite
happy in owning to a belief in Darwinism. We may indeed be thankful that we are " devoid of such a blessing."

Among the Germans, he was fast gaining supporters.
In 1865 he began a correspondence with the distinguished
Naturalist, Fritz Müller, then, as now, resident in Brazil.
They never met, but the correspondence with Müller, which
continued to the close of my father's life, was a source of
very great pleasure to him. My impression is that of all
his unseen friends Fritz Müller was the one for whom he
had the strongest regard. Fritz Müller is the brother of
another distinguished man, the late Hermann Müller, the

---

* *Examen du livre de M. Darwin sur l'origine des espèces.* Par P. Flourens. 8vo. Paris, 1864.
† *Lay Sermons*, p. 328.

author of *Die Befruchtung der Blumen* (The Fertilisation of Flowers), and of much other valuable work.

The occasion of writing to Fritz Müller was the latter's book, *Für Darwin*, which was afterwards translated by Mr. Dallas at my father's suggestion, under the title *Facts and Arguments for Darwin.*

Shortly afterwards, in 1866, began his connection with Professor Victor Carus, of Leipzig, who undertook the translation of the 4th edition of the *Origin.* From this time forward Professor Carus continued to translate my father's books into German. The conscientious care with which this work was done was of material service, and I well remember the admiration (mingled with a tinge of vexation at his own shortcomings) with which my father used to receive the lists of oversights, &c., which Professor Carus discovered in the course of translation. The connection was not a mere business one, but was cemented by warm feelings of regard on both sides.

About this time, too, he came in contact with Professor Ernst Haeckel, whose influence on German science has been so powerful.

The earliest letter which I have seen from my father to Professor Haeckel, was written in 1865, and from that time forward they corresponded (though not, I think, with any regularity) up to the end of my father's life. His friendship with Haeckel was not merely the growth of correspondence, as was the case with some others, for instance, Fritz Müller. Haeckel paid more than one visit to Down, and these were thoroughly enjoyed by my father. The following letter will serve to show the strong feeling of regard which he entertained for his correspondent—a feeling which I have often heard him emphatically express, and which was warmly returned. The book referred to is Haeckel's *Generelle Morphologie,* published in 1866, a copy of which my father received from the author in January, 1867.

Dr. E. Krause * has given a good account of Professor Haeckel's services in the cause of Evolution. After speaking of the lukewarm reception which the *Origin* met with in Germany on its first publication, he goes on to describe the first adherents of the new faith as more or less popular writers, not especially likely to advance its acceptance with the professorial or purely scientific world. And he claims

---

* *Charles Darwin und sein Verhältniss zu Deutschland,* 1885.

for Haeckel that it was his advocacy of Evolution in his
Radiolaria (1862), and at the "Versammlung" of Naturalists at Stettin in 1863, that placed the Darwinian question
for the first time publicly before the forum of German
science, and his enthusiastic propagandism that chiefly contributed to its success.

Mr. Huxley, writing in 1869, paid a high tribute to Professor Haeckel as the Coryphæus of the Darwinian movement in Germany. Of his *Generelle Morphologie*, "an
attempt to work out the practical applications" of the
doctrine of Evolution to their final results, he says that it
has the "force and suggestiveness, and . . . systematising
power of Oken without his extravagance." Mr. Huxley
also testifies to the value of Haeckel's *Schöpfungs-Geschichte*
as an exposition of the *Generelle Morphologie* "for an educated public."

Again, in his *Evolution in Biology,** Mr. Huxley wrote :
" Whatever hesitation may not unfrequently be felt by less
daring minds, in following Haeckel in many of his speculations, his attempt to systematise the doctrine of Evolution
and to exhibit its influence as the central thought of modern
biology, cannot fail to have a far-reaching influence on the
progress of science.".

In the following letter my father alludes to the somewhat fierce manner in which Professor Haeckel fought the
battle of ' Darwinismus,' and on this subject Dr. Krause
has some good remarks (p. 162). He asks whether much
that happened in the heat of the conflict might not well
have been otherwise, and adds that Haeckel himself is the
last man to deny this. Nevertheless he thinks that even
these things may have worked well for the cause of Evolution, inasmuch as Haeckel "concentrated on himself by
his *Ursprung des Menschen-Geschlechts*, his *Generelle Morphologie*, and *Schöpfungs-Geschichte*, all the hatred and
bitterness which Evolution excited in certain quarters," so
that, "in a surprisingly short time it became the fashion
in Germany that Haeckel alone should be abused, while
Darwin was held up as the ideal of forethought and moderation."

---

* An article in the *Encyclopædia Britannica*, 9th edit. reprinted in *Science
and Culture*, 1881, p. 298.

C. D. to E. Haeckel.   Down, May 21, 1867.

DEAR HAECKEL,—Your letter of the 18th has given me
great pleasure, for you have received what I said in the most
kind and cordial manner.   You have in part taken what I
said much stronger than I had intended.   It never occurred
to me for a moment to doubt that your work, with the whole
subject so admirably and clearly arranged, as well as forti-
fied by so many new facts and arguments, would not advance
our common object in the highest degree.   All that I think
is that you will excite anger, and that anger so completely
blinds every one that your arguments would have no chance
of influencing those who are already opposed to our views.
Moreover, I do not at all like that you, towards whom I feel
so much friendship, should unnecessarily make enemies,
and there is pain and vexation enough in the world without
more being caused.   But I repeat that I can feel no doubt
that your work will greatly advance our subject, and I
heartily wish it could be translated into English, for my
own sake and that of others.   With respect to what you say
about my advancing too strongly objections against my own
views, some of my English friends think that I have erred
on this side ; but truth compelled me to write what I did,
and I am inclined to think it was good policy.   The belief
in the descent theory is slowly spreading in England,* even
amongst those who can give no reason for their belief.   No
body of men were at first so much opposed to my views as
the members of the London Entomological Society, but now
I am assured that, with the exception of two or three old
men, all the members concur with me to a certain extent.
It has been a great disappointment to me that I have never
received your long letter written to me from the Canary
Islands.   I am rejoiced to hear that your tour, which seems
to have been a most interesting one, has done your health
much good.

. . . . I am very glad to hear that there is some chance
of your visiting England this autumn, and all in this house
will be delighted to see you here.

Believe me, my dear Haeckel, yours very sincerely.

---

* In October, 1867, he wrote to Mr. Wallace :—" Mr. Warrington has lately
read an excellent and spirited abstract of the *Origin* before the Victoria Institute, and as this is a most orthodox body, he has gained the name of the Devil's Advocate.   The discussion which followed during three consecutive
meetings is very rich from the nonsense talked."

I place here an extract from a letter of later date (Nov. 1868), which refers to one of Haeckel's later works.*

" Your chapters on the affinities and genealogy of the animal kingdom strike me as admirable and full of original thought. Your boldness, however, sometimes makes me tremble, but as Huxley remarked, some one must be bold enough to make a beginning in drawing up tables of descent. Although you fully admit the imperfection of the geological record, yet Huxley agreed with me in thinking that you are sometimes rather rash in venturing to say at what periods the several groups first appeared. I have this advantage over you, that I remember how wonderfully different any statement on this subject made 20 years ago, would have been to what would now be the case, and I expect the next 20 years will make quite as great a difference."

The following extract from a letter to Professor W. Preyer, a well-known physiologist, shows that he estimated at its true value the help he was to receive from the scientific workers of Germany :—

March 31, 1868.

. . . . I am delighted to hear that you uphold the doctrine of the Modification of Species, and defend my views. The support which I receive from Germany is my chief ground for hoping that our views will ultimately prevail. To the present day I am continually abused or treated with contempt by writers of my own country ; but the younger naturalists are almost all on my side, and sooner or later the public must follow those who make the subject their special study. The abuse and contempt of ignorant writers hurts me very little. . . .

I must now pass on to the publication, in 1868, of his book on *The Variation of Animals and Plants under Domestication.* It was began two days after the appearance of the second edition of the *Origin*, on Jan. 9, 1860, and it may, I think, be reckoned that about half of the eight years that elapsed between its commencement and completion was spent on it. The book did not escape adverse criticism : it was said, for instance, that the public had been patiently waiting for Mr. Darwin's *pièces justicatives*, and

---

* *Die natürliche Schöpfungs-Geschichte*, 1868. It was translated and published in 1876, under the title, *The History of Creation.*

that after eight years of expectation, all they got was a mass of detail about pigeons, rabbits and silkworms. But the true critics welcomed it as an expansion with unrivalled wealth of illustration of a section of the *Origin*. Variation under the influence of man was the only subject (except the question of man's origin) which he was able to deal with in detail so as to utilise his full stores of knowledge. When we remember how important for his argument is a knowledge of the action of artificial selection, we may well rejoice that this subject was chosen by him for amplification.

In 1864, he wrote to Sir Joseph Hooker:

"I have begun looking over my old MS., and it is as fresh as if I had never written it; parts are astonishingly dull, but yet worth printing, I think; and other parts strike me as very good. I am a complete millionaire in odd and curious little facts, and I have been really astounded at my own industry whilst reading my chapters on Inheritance and Selection. God knows when the book will ever be completed, for I find that I am very weak, and on my best days cannot do more than one or one and a half hours' work. It is a good deal harder than writing about my dear climbing plants."

In Aug. 1867, when Lyell was reading the proofs of the book, my father wrote :—

"I thank you cordially for your last two letters. The former one did me *real* good, for I had got so wearied with the subject that I could hardly bear to correct the proofs, and you gave me fresh heart. I remember thinking that when you came to the Pigeon chapter you would pass it over as quite unreadable. I have been particularly pleased that you have noticed Pangenesis. I do not know whether you ever had the feeling of having thought so much over a subject that you had lost all power of judging it. This is my case with Pangenesis (which is 26 or 27 years old), but I am inclined to think that if it be admitted as a probable hypothesis it will be a somewhat important step in Biology."

His theory of Pangenesis, by which he attempted to explain " how the characters of the parents are 'photographed' on the child, by means of material atoms derived from each cell in both parents, and developed in the child," has never met with much acceptance. Nevertheless, some of his contemporaries felt with him about it. Thus in February 1868, he wrote to Hooker :—

"I heard yesterday from Wallace, who says (excuse horrid vanity), 'I can hardly tell you how much I admire the chapter on *Pangenesis*. It is a *positive comfort* to me to have any feasible explanation of a difficulty that has always been haunting me, and I shall never be able to give it up till a better one supplies its place, and that I think hardly possible.' Now his foregoing [italicised] words express my sentiments exactly and fully : though perhaps I feel the relief extra strongly from having during many years vainly attempted to form some hypothesis. When you or Huxley say that a single cell of a plant, or the stump of an amputated limb, has the 'potentiality' of reproducing the whole —or 'diffuses an influence,' these words give me no positive idea ;—but when it is said that the cells of a plant, or stump, include atoms derived from every other cell of the whole organism and capable of development, I gain a distinct idea."

Immediately after the publication of the book, he wrote :

Down, February 10 [1868].

My dear Hooker,—What is the good of having a friend, if one may not boast of him? I heard yesterday that Murray has sold in a week the whole edition of 1500 copies of my book, and the sale so pressing that he has agreed with Clowes to get another edition in fourteen days ! This has done me a world of good, for I had got into a sort of dogged hatred of my book. And now there has appeared a review in the *Pall Mall* which has pleased me excessively, more perhaps than is reasonable. I am quite content, and do not care how much I may be pitched into. If by any chance you should hear who wrote the article in the *Pall Mall*, do please tell me ; it is some one who writes capitally, and who knows the subject. I went to luncheon on Sunday, to Lubbock's, partly in hopes of seeing you, and, be hanged to you, you were not there.

Your cock-a-hoop friend,
C. D.

Independently of the favourable tone of the able series of notices in the *Pall Mall Gazette* (Feb. 10, 15, 17, 1868), my father may well have been gratified by the following passages :—

"We must call attention to the rare and noble calmness with which he expounds his own views, undisturbed by the heats of polemical agitation which those views have excited, and persistently refusing to retort on his antagonists by ridicule, by indignation, or by contempt. Considering the amount of vituperation and insinuation which has come from the other side, this forbearance is supremely dignified."

And again in the third notice, Feb. 17 :—

" Nowhere has the author a word that could wound the most sensitive self-love of an antagonist; nowhere does he, in text or note, expose the fallacies and mistakes of brother investigators . . . but while abstaining from impertinent censure, he is lavish in acknowledging the smallest debts he may owe ; and his book will make many men happy."

I am indebted to Messrs. Smith and Elder for the information that these articles were written by Mr. G. H. Lewes.

The following extract from a letter (Feb. 1870) to his friend Professor Newton, the well-known ornithologist, shows how much he valued the appreciation of his colleagues.

" I suppose it would be universally held extremely wrong for a defendant to write to a Judge to express his satisfaction at a judgment in his favour ; and yet I am going thus to act. I have just read what you have said in the ' Record ' * about my pigeon chapters, and it has gratified me beyond measure. I have sometimes felt a little disappointed that the labour of so many years seemed to be almost thrown away, for you are the first man capable of forming a judgment (excepting partly Quatrefages), who seems to have thought anything of this part of my work. The amount of labour, correspondence, and care, which the subject cost me, is more than you could well suppose. I thought the article in the *Athenæum* was very unjust; but now I feel amply repaid, and I cordially thank you for your sympathy and too warm praise."

WORK ON MAN.

In February 1867, when the manuscript of *Animals and Plants* had been sent to Messrs. Clowes to be printed, and before the proofs began to come in, he had an interval of

---

* *Zoological Record.* The volume for 1868, published December, 1869.

spare time, and began a " Chapter on Man," but he soon
found it growing under his hands, and determined to publish it separately as a " very small volume."

It is remarkable that only four years before this date,
namely in 1864, he had given up hope of being able to work
out this subject. He wrote to Mr. Wallace :—

" I have collected a few notes on man, but I do not suppose that I shall ever use them. Do you intend to follow
out your views, and if so, would you like at some future time
to have my few references and notes? I am sure I hardly
know whether they are of any value, and they are at present
in a state of chaos. There is much more that I should like
to write, but I have not strength." But this was at a period of ill-health ; not long before, in 1863, he had written
in the same depressed tone about his future work generally :—

" I have been so steadily going downhill, I cannot help
doubting whether I can ever crawl a little uphill again.
Unless I can, enough to work a little, I hope my life may be
very short, for to lie on a sofa all day and do nothing but
give trouble to the best and kindest of wives and good dear
children is dreadful."

The " Chapter on Man," which afterwards grew into
the *Descent of Man*, was interrupted by the necessity of
correcting the proofs of *Animals and Plants*, and by some
botanical work, but was resumed with unremitting industry
on the first available day in the following year. He could
not rest, and he recognised with regret the gradual change
in his mind that rendered continuous work more and more
necessary to him as he grew older. This is expressed in a
letter to Sir J. D. Hooker, June 17, 1868, which repeats to
some extent what is given in the *Autobiography* :—

" I am glad you were at the *Messiah*, it is the one thing
that I should like to hear again, but I dare say I should
find my soul too dried up to appreciate it as in old days ;
and then I should feel very flat, for it is a horrid bore to
feel as I constantly do, that I am a withered leaf for every
subject except Science. It sometimes makes me hate Science, though God knows I ought to be thankful for such a
perennial interest, which makes me forget for some hours
every day my accursed stomach."

The Descent of Man (and this is indicated on its title-
page) consists of two separate books, namely on the pedigree of mankind, and on sexual selection in the animal

kingdom generally. In studying this latter part of the subject he had to take into consideration the whole subject of
colour. I give the two following characteristic letters, in
which the reader is as it were present at the birth of a
theory.

C. D. to A. R. Wallace. Down, February 23 [1867].

DEAR WALLACE,—I much regretted that I was unable
to call on you, but after Monday I was unable even to leave
the house. On Monday evening I called on Bates, and put
a difficulty before him, which he could not answer, and, as
on some former similar occasion, his first suggestion was,
" You had better ask Wallace." My difficulty is, why are
caterpillars sometimes so beautifully and artistically coloured? Seeing that many are coloured to escape danger, I
can hardly attribute their bright colour in other cases to
mere physical conditions. Bates says the most gaudy cater-
pillar he ever saw in Amazonia (of a sphinx) was conspicuous at the distance of yards, from its black and red colours,
whilst feeding on large green leaves. If any one objected to
male butterflies having been made beautiful by sexual selection, and asked why should they not have been made beautiful as well as their caterpillars, what would you answer?
I could not answer, but should maintain my ground. Will
you think over this, and some time, either by letter or when
we meet, tell me what you think? . . .

He seems to have received an explanation by return of
post, for a day or two afterwards he could write to Wallace :—
" Bates was quite right; you are the man to apply to in
a difficulty. I never heard anything more ingenious than
your suggestion, and I hope you may be able to prove it
true. That is a splendid fact about the white moths; it
warms one's very blood to see a theory thus almost proved
to be true."

Mr. Wallace's suggestion was that conspicuous caterpillars or perfect insects (*e. g.* white butterflies), which are
distasteful to birds, benefit by being promptly recognised
and therefore easily avoided.*

---

* Mr. Jenner Weir's observations published in the *Transactions of the
Entomological Society* (1869 and 1870) give strong support to the theory in
question.

The letter from Darwin to Wallace goes on : " The reason of my being so much interested just at present about sexual selection is, that I have almost resolved to publish a little essay on the origin of Mankind, and I still strongly think (though I failed to convince you, and this, to me, is the heaviest blow possible) that sexual selection has been the main agent in forming the races of man.

" By the way, there is another subject which I shall introduce in my essay, namely, expression of countenance. Now, do you happen to know by any odd chance a very good-natured and acute observer in the Malay Archipelago, who you think would make a few easy observations for me on the expression of the Malays when excited by various emotions ? "

The reference to the subject of expression in the above letter is explained by the fact, that my father's original intention was to give his essay on this subject as a chapter in the *Descent of Man*, which in its turn grew, as we have seen, out of a proposed chapter in *Animals and Plants*.

He got much valuable help from Dr. Günther, of the Natural History Museum, to whom he wrote in May 1870 :—

" As I crawl on with the successive classes I am astonished to find how similar the rules are about the nuptial or ' wedding dress ' of all animals. The subject has begun to interest me in an extraordinary degree ; but I must try not to fall into my common error of being too speculative. But a drunkard might as well say he would drink a little and not too much ! My essay, as far as fishes, batrachians and reptiles are concerned, will be in fact yours, only written by me."

The last revise of the *Descent of Man* was corrected on January 15th, 1871, so that the book occupied him for about three years. He wrote to Sir J. Hooker : " I finished the last proofs of my book a few days ago ; the work half-killed me, and I have not the most remote idea whether the book is worth publishing."

He also wrote to Dr. Gray :—

" I have finished my book on the *Descent of Man*, &c., and its publication is delayed only by the Index : when published, I will send you a copy, but I do not know that you will care about it. Parts, as on the moral sense, will, I dare say, aggravate you, and if I hear from you, I shall probably receive a few stabs from your polished stiletto of a pen."

The book was published on February 24, 1871. 2500 copies were printed at first, and 5000 more before the end of the year. My father notes that he received for this edition £1470.

Nothing can give a better idea (in a small compass) of the growth of Evolutionism, and its position at this time, than a quotation from Mr. Huxley * :—

"The gradual lapse of time has now separated us by more than a decade from the date of the publication of the *Origin of Species ;* and whatever may be thought or said about Mr. Darwin's doctrines, or the manner in which he has propounded them, this much is certain, that in a dozen years the *Origin of Species* has worked as complete a revolution in Biological Science as the *Principia* did in Astronomy;" and it had done so, "because in the words of Helmholtz, it contains 'an essentially new creative thought.' And, as time has slipped by, a happy change has come over Mr. Darwin's critics. The mixture of ignorance and insolence which at first characterised a large proportion of the attacks with which he was assailed, is no longer the sad distinction of anti-Darwinian criticism."

A passage in the Introduction to the *Descent of Man* shows that the author recognised clearly this improvement in the position of Evolutionism. "When a naturalist like Carl Vogt ventures to say in his address, as President of the National Institution of Geneva (1869), 'personne, en Europe au moins, n'ose plus soutenir la création indépendante et de toutes pièces, des espèces,' it is manifest that at least a large number of naturalists must admit that species are the modified descendants of other species; and this especially holds good with the younger and rising naturalists. . . . Of the older and honoured chiefs in natural science, many, unfortunately, are still opposed to Evolution in every form."

In Mr. James Hague's pleasantly written article, "A Reminiscence of Mr. Darwin" (*Harper's Magazine*, October, 1884), he describes a visit to my father "early in 1871," shortly after the publication of the *Descent of Man*. Mr. Hague represents my father as "much impressed by the general assent with which his views had been received," and as remarking that "everybody is talking about it without being shocked."

Later in the year the reception of the book is described

---

in different language in the *Edinburgh Review :* " On every side it is raising a storm of mingled wrath, wonder and admiration."

Haeckel seems to have been one of the first to write to my father about the *Descent of Man.* I quote from Darwin's reply :—

" I must send you a few words to thank you for your interesting, and I may truly say, charming letter. I am delighted that you approve of my book, as far as you have read it. I felt very great difficulty and doubt how often I ought to allude to what you have published ; strictly speaking every idea, although occurring independently to me, if published by you previously ought to have appeared as if taken from your works, but this would have made my book very dull reading ; and I hoped that a full acknowledgment at the beginning would suffice.* I cannot tell you how glad I am to find that I have expressed my high admiration of your labours with sufficient clearness ; I am sure that I have not expressed it too strongly."

In March he wrote to Professor Ray Lankester :—

" I think you will be glad to hear, as a proof of the increasing liberality of England, that my book has sold wonderfully . . . . and as yet no abuse (though some, no doubt, will come, strong enough), and only contempt even in the poor old *Athenæum.*"

About the same time he wrote to Mr. Murray :—

" Many thanks for the *Nonconformist* [March 8, 1871]. I like to see all that is written, and it is of some real use. If you hear of reviewers in out-of-the-way papers, especially the religious, as *Record, Guardian, Tablet,* kindly inform me. It is wonderful that there has been no abuse as yet. On the whole, the reviews have been highly favourable."

The following extract from a letter to Mr. Murray (April 13, 1871) refers to a review in the *Times* † :—

" I have no idea who wrote the *Times'* review. He has no knowledge of science, and seems to me a wind-bag full of metaphysics and classics, so that I do not much regard

---

* In the introduction to the *Descent of Man* the author wrote :—" This last naturalist [Haeckel] . . . has recently . . . published his *Natürliche Schöpf-ungs-Geschichte,* in which he fully discusses the genealogy of man. If this work had appeared before my essay had been written, I should probably never have completed it. Almost all the conclusions at which I have arrived, I find confirmed by this naturalist, whose knowledge on many points is much fuller than mine."

† April 7 and 8, 1871.

his adverse judgment, though I suppose it will injure the sale.".

A striking review appeared in the *Saturday Review* (March 4 and 11, 1871) in which the position of Evolution is well stated.

" He claims to have brought man himself, his origin and constitution, within that unity which he had previously sought to trace through all lower animal forms. The growth of opinion in the interval, due in chief measure to his own intermediate works, has placed the discussion of this problem in a position very much in advance of that held by it fifteen years ago. The problem of Evolution is hardly any longer to be treated as one of first principles; nor has Mr. Darwin to do battle for a first hearing of his central hypothesis, upborne as it is by a phalanx of names full of distinction and promise in either hemisphere."

We must now return to the history of the general principle of Evolution. At the beginning of 1869 * he was at work on the fifth edition of the *Origin.* The most important alterations were suggested by a remarkable paper in the *North British Review* (June, 1867) written by the late Fleeming Jenkin.

It is not a little remarkable that the criticisms, which my father, as I believe, felt to be the most valuable ever made on his views should have come, not from a professed naturalist but from a Professor of Engineering.

The point on which Fleeming Jenkin convinced my father is the extreme difficulty of believing that *single individuals* which differ from their fellows in the possession of some useful character can be the starting point of a new variety. Thus the origin of a new variety is more likely to

---

* His holiday this year was at Caerdeon, on the north shore of the beautiful Barmouth estuary, and pleasantly placed in being close to wild hill country behind, as well as to the picturesque wooded "hummocks," between the steeper hills and the river. My father was ill and somewhat depressed throughout this visit, and I think felt imprisoned and saddened by his inability to reach the hills over which he had once wandered for days together.

He wrote from Caerdeon to Sir J. D. Hooker (June 22nd):—

" We have been here for ten days, how I wish it was possible for you to pay us a visit here; we have a beautiful house with a terraced garden, and a really magnificent view of Cader, right opposite. Old Cader is a grand fellow, and shows himself off superbly with every changing light. We remain here till the end of July, when the H. Wedgwoods have the house. I have been as yet in a very poor way; it seems as soon as the stimulus of mental work stops, my whole strength gives way. As yet I have hardly crawled half a mile from the house, and then have been fearfully fatigued. It is enough to make one wish oneself quiet in a comfortable tomb."

be found in a species which presents the incipient character in a large number of its individuals. This point of view was of course perfectly familiar to him, it was this that induced him to study " unconscious selection," where a breed is formed by the long-continued preservation by Man of all those individuals which are best adapted to his needs : not as in the art of the professed breeder, where a single individual is picked out to breed from.

It is impossible to give in a short compass an account of Fleeming Jenkin's argument. My father's copy of the paper (ripped out of the volume as usual, and tied with a bit of string) is annotated in pencil in many places. I quote a passage opposite which my father has written " good sneers "—but it should be remembered that he used the word " sneer " in rather a special sense, not as necessarily implying a feeling of bitterness in the critic, but rather in the sense of " banter." Speaking of the " true believer," Fleeming Jenkin says, p. 293 :—

" He can invent trains of ancestors of whose existence there is no evidence ; he can marshal hosts of equally imaginary foes ; he can call up continents, floods, and peculiar atmospheres ; he can dry up oceans, split islands, and parcel out eternity at will; surely with these advantages he must be a dull fellow if he cannot scheme some series of animals and circumstances explaining our assumed difficulty quite naturally. Feeling the difficulty of dealing with adversaries who command so huge a domain of fancy, we will abandon these arguments, and trust to those which at least cannot be assailed by mere efforts of imagination."

In the fifth edition of the *Origin*, my father altered a passage in the Historical Sketch (fourth edition, p. xviii.). He thus practically gave up the difficult task of understanding whether or not Sir R. Owen claims to have discovered the principle of Natural Selection. Adding, " As far as the mere enunciation of the principle of Natural Selection is concerned, it is quite immaterial whether or not Professor Owen preceded me, for both of us . . . were long ago preceded by Dr. Wells and Mr. Matthew."

The desire that his views might spread in France was always strong with my father, and he was therefore justly annoyed to find that in 1869 the publisher of the French edition had brought out a third edition without consulting the author. He was accordingly glad to enter into an arrangement for a French translation of the fifth edition ;

this was undertaken by M. Reinwald, with whom he continued to have pleasant relations as the publisher of many of his books into French.

He wrote to Sir J. D. Hooker :—

" I must enjoy myself and tell you about Mdlle. C. Royer, who translated the *Origin* into French, and for whose second edition I took infinite trouble. She has just now brought out a third edition without informing me, so that all the corrections, &c., in the fourth and fifth English editions are lost. Besides her enormously long preface to the first edition, she has added a second preface abusing me like a pickpocket for Pangenesis, which of course has no relation to the *Origin*. So I wrote to Paris ; and Reinwald agrees to bring out at once a new translation from the fifth English edition, in competition with her third edition. . . . This fact shows that ' evolution of species ' must at last be spreading in France."

It will be well perhaps to place here all that remains to be said about the *Origin of Species*. The sixth or final edition was published in January 1872 in a smaller and cheaper form than its predecessors. The chief addition was a discussion suggested by Mr. Mivart's *Genesis of Species*, which appeared in 1871, before the publication of the *Descent of Man*. The following quotation from a letter to Wallace (July 9, 1871) may serve to show the spirit and method in which Mr. Mivart dealt with the subject. " I grieve to see the omission of the words by Mivart, detected by Wright.* I complained to Mivart that in two cases he quotes only the commencement of sentences by me, and thus modifies my meaning ; but I never supposed he would have omitted words. There are other cases of what I consider unfair treatment."

My father continues, with his usual charity and moderation :—

" I conclude with sorrow that though he means to be honourable, he is so bigoted that he cannot act fairly."

In July 1871, my father wrote to Mr. Wallace :—

" I feel very doubtful how far I shall succeed in answering Mivart, it is so difficult to answer objections to doubtful

---

* The late Chauncey Wright, in an article published in the *North American Review*, vol. cxiii. pp. 83, 84. Wright points out that the words omitted are "essential to the point on which he [Mr. Mivart] cites Mr. Darwin's authority." It should be mentioned that the passage from which words are omitted is not given within inverted commas by Mr. Mivart.

points, and make the discussion readable. I shall make
only a selection. The worst of it is, that I cannot possibly
hunt through all my references for isolated points, it would
take me three weeks of intolerably hard work. I wish I
had your power of arguing clearly. At present I feel sick
of everything, and if I could occupy my time and forget my
daily discomforts, or rather miseries, I would never publish
another word. But I shall cheer up, I dare say, soon, having only just got over a bad attack. Farewell; God knows
why I bother you about myself. I can say nothing more
about missing-links than what I have said. I should rely
much on pre-silurian times; but then comes Sir W. Thomson like an odious spectre.* Farewell.

"... There is a most cutting review of me in the
[July] *Quarterly ;* I have only read a few pages. The skill
and style make me think of Mivart. I shall soon be
viewed as the most despicable of men. This *Quarterly Review* tempts me to republish Ch. Wright,† even if not
read by any one, just to show some one will say a word
against Mivart, and that his (*i. e.* Mivart's) remarks ought
not to be swallowed without some reflection. . . . God knows
whether my strength and spirit will last out to write a
chapter versus Mivart and others; I do so hate controversy
and feel I shall do it so badly."

The *Quarterly* review was the subject of an article by
Mr. Huxley in the November number of the *Contemporary
Review.* Here, also, are discussed Mr. Wallace's *Contribution to the Theory of Natural Selection,* and the second edition of Mr. Mivart's *Genesis of Species.* What follows is
taken from Mr. Huxley's article. The *Quarterly* reviewer,
though to some extent an evolutionist, believes that Man
" differs more from an elephant or a gorilla, than do these
from the dust of the earth on which they tread." The reviewer also declares that Darwin has " with needless opposition, set at naught the first principles of both philosophy
and religion." Mr. Huxley passes from the *Quarterly* reviewer's further statement, that there is no necessary opposition between evolution and religion, to the more definite
position taken by Mr. Mivart, that the orthodox authorities
of the Roman Catholic Church agree in distinctly asserting

---

* My father, as an Evolutionist, felt that he required more time than Sir
W. Thomson's estimate of the age of the world allows.

† Chauncey Wright's review was published as a pamphlet in the autumn
of 1871.

derivative creation, so that "their teachings harmonize with all that modern science can possibly require." Here Mr. Huxley felt the want of that "study of Christian philosophy" (at any rate in its Jesuitic garb), which Mr. Mivart speaks of, and it was a want he at once set to work to fill up. He was then staying at St. Andrews, whence he wrote to my father :—

"By great good luck there is an excellent library here, with a good copy of Suarez,* in a dozen big folios. Among these I dived, to the great astonishment of the librarian, and looking into them 'as careful robins eye the delver's toil' (*vide Idylls*), I carried off the two venerable clasped volumes which were most promising." Even those who know Mr. Huxley's unrivalled power of tearing the heart out of a book must marvel at the skill with which he has made Suarez speak on his side. "So I have come out," he wrote, "in the new character of a defender of Catholic orthodoxy, and upset Mivart out of the mouth of his own prophet."

The remainder of Mr. Huxley's critique is largely occupied with a dissection of the *Quarterly* reviewer's psychology, and his ethical views. He deals, too, with Mr. Wallace's objections to the doctrine of Evolution by natural causes when applied to the mental faculties of Man. Finally, he devotes a couple of pages to justifying his description of the *Quarterly* reviewer's treatment of Mr. Darwin as alike "unjust and unbecoming." †

In the sixth edition my father also referred to the "direct action of the conditions of life" as a subordinate cause of modification in living things : On this subject he wrote to Dr. Moritz Wagner (Oct. 13, 1876) : "In my opinion the greatest error which I have committed, has been not allowing sufficient weight to the direct action of the environment, *i. e.* food, climate, &c., independently of natural selection.

---

* The learned Jesuit on whom Mr. Mivart mainly relies.

† The same words may be applied to Mr. Mivart's treatment of my father. The following extract from a letter to Mr. Wallace (June 17th, 1874) refers to Mr. Mivart's statement (*Lessons from Nature*, p. 144) that Mr. Darwin at first studiously disguised his views as to the "bestiality of man" :—

"I have only just heard of and procured your two articles in the *Academy*. I thank you most cordially for your generous defence of me against Mr. Mivart. In the *Origin* I did not discuss the derivation of any one species ; but that I might not be accused of concealing my opinion, I went out of my way, and inserted a sentence which seemed to me (and still so seems) to disclose plainly my belief. This was quoted in my *Descent of Man*. Therefore it is very unjust . . . of Mr. Mivart to accuse me of base fraudulent concealment."

Modifications thus caused, which are neither of advantage nor disadvantage to the modified organism, would be especially favoured, as I can now see chiefly through your observations, by isolation, in a small area, where only a few individuals lived under nearly uniform conditions."

It has been supposed that such statements indicate a serious change of front on my father's part. As a matter of fact the first edition of the *Origin* contains the words, " I am convinced that natural selection has been the main but not the exclusive means of modification." Moreover, any alteration that his views may have undergone was due not to a change of opinion, but to change in the materials on which a judgment was to be formed. Thus he wrote to Wagner in the above quoted letter :—

" When I wrote the *Origin*, and for some years afterwards, I could find little good evidence of the direct action of the environment; now there is a large body of evidence."

With the possibility of such action of the environment he had of course been familiar for many years. Thus he wrote to Mr. Davidson in 1861 :—

" My greatest trouble is, not being able to weigh the direct effects of the long-continued action of changed conditions of life without any selection, with the action of selection on mere accidental (so to speak) variability. I oscillate much on this head, but generally return to my belief that the direct action of the conditions of life has not been great. At least this direct action can have played an extremely small part in producing all the numberless and beautiful adaptations in every living creature."

And to Sir Joseph Hooker in the following year :—

" I hardly know why I am a little sorry, but my present work is leading me to believe rather more in the direct action of physical conditions. I presume I regret it, because it lessens the glory of Natural Selection, and is so confoundedly doubtful. Perhaps I shall change again when I get all my facts under one point of view, and a pretty hard job this will be."

Reference has already been made to the growth of his book on the *Expression of the Emotions* out of a projected chapter in the *Descent of Man*.

It was published in the autumn of 1872. The edition consisted of 7000, and of these 5267 copies were sold at Mr. Murray's sale in November. Two thousand were printed at the end of the year, and this proved a misfortune, as they did not afterwards sell so rapidly, and thus a mass of notes collected by the author was never employed for a second edition during his lifetime.*

As usual he had no belief in the possibility of the book being generally successful. The following passage in a letter to Haeckel serves to show that he had felt the writing of this book as a somewhat severe strain :—

"I have finished my little book on Expression, and when it is published in November I will of course send you a copy, in case you would like to read it for amusement. I have resumed some old botanical work, and perhaps I shall never again attempt to discuss theoretical views.

"I am growing old and weak, and no man can tell when his intellectual powers begin to fail. Long life and happiness to you for your own sake and for that of science."

A good review by Mr. Wallace appeared in the *Quarterly Journal of Science*, Jan. 1873. Mr. Wallace truly remarks that the book exhibits certain "characteristics of the author's mind in an eminent degree," namely, "the insatiable longing to discover the causes of the varied and complex phenomena presented by living things." He adds that in the case of the author "the restless curiosity of the child to know the 'what for?' the 'why?' and the 'how?' of everything" seems "never to have abated its force."

The publication of the Expression book was the occasion of the following letter to one of his oldest friends, the late Mrs. Haliburton, who was the daughter of a Shropshire neighbour, Mr. Owen of Woodhouse, and became the wife of the author of *Sam Slick*.

Nov. 1, 1872.

MY DEAR MRS. HALIBURTON,—I dare say you will be surprised to hear from me. My object in writing now is to say that I have just published a book on the *Expression of the Emotions in Man and Animals ;* and it has occurred to me that you might possibly like to read some parts of it; and I can hardly think that this would have been the case with any of the books which I have already published. So I send by this post my present book. Although I have had no communication with you or the other members of your family for so long a time, no scenes in my whole life pass so frequently or so vividly before my mind as those which re-

---

* They were utilised to some extent in the 2nd edition, edited by me, and published in 1890.—F. D.

late to happy old days spent at Woodhouse. I should very much like to hear a little news about yourself and the other members of your family, if you will take the trouble to write to me. Formerly I used to glean some news about you from my sisters.

I have had many years of bad health and have not been able to visit anywhere; and now I feel very old. As long as I pass a perfectly uniform life, I am able to do some daily work in Natural History, which is still my passion, as it was in old days, when you used to laugh at me for collecting beetles with such zeal at Woodhouse. Excepting from my continued ill-health, which has excluded me from society, my life has been a very happy one; the greatest drawback being that several of my children have inherited from me feeble health. I hope with all my heart that you retain, at least to a large extent, the famous " Owen constitution." With sincere feelings of gratitude and affection for all bearing the name of Owen, I venture to sign myself,

Yours affectionately,
CHARLES DARWIN.

CHAPTER XV.

In 1874 a second edition of his *Coral Reefs* was published, which need not specially concern us. It was not until some time afterwards that the criticisms of my father's theory appeared, which have attracted a good deal of attention.

The following interesting account of the subject is taken from Professor's Judd's " Critical Introduction " to Messrs. Ward, Lock and Co's. edition of *Coral Reefs* and *Volcanic Islands, &c.**

" The first serious note of dissent to the generally accepted theory was heard in 1863, when a distinguished German naturalist, Dr. Karl Semper, declared that his study of the Pelew Islands showed that uninterrupted subsidence could not have been going on in that region. Dr. Semper's objections were very carefully considered by Mr. Darwin, and a reply to them appeared in the second and revised edition of his *Coral Reefs*, which was published in 1874. With characteristic frankness and freedom from prejudices, Darwin admitted that the facts brought forward by Dr. Semper proved that in certain specified cases, subsidence could not have played the chief part in originating the peculiar forms of the coral islands. But while making this admission, he firmly maintained that exceptional cases, like those described in the Pelew Islands, were not sufficient to invalidate the theory of subsidence as applied to the widely spread atolls, encircling reefs, and barrier-reefs of the Pacific and Indian Oceans. It is worthy of note that to the end of his life Darwin maintained a friendly correspondence with Semper concerning the points on which they were at issue.

---

* *The Minerva Library of Famous Books*, 1890, edited by G. T. Bettany.

" After the appearance of Semper's work, Dr. J. J. Rein
published an account of the Bermudas, in which he opposed
the interpretation of the structure of the islands given by
Nelson and other authors, and maintained that the facts observed in them are opposed to the views of Darwin. Although so far as I am aware, Darwin had no opportunity of
studying and considering these particular objections, it may
be mentioned that two American geologists have since carefully re-examined the district—Professor W. N. Rice in
1884 and Professor A. Heilprin in 1889—and they have independently arrived at the conclusion that Dr. Rein's objections cannot be maintained.

" The most serious objection to Darwin's coral-reef theory, however, was that which developed itself after the return of H. M. S. *Challenger* from her famous voyage. Mr.
John Murray, one of the staff of naturalists on board that
vessel, propounded a new theory of coral-reefs, and maintained that the view that they were formed by subsidence
was one that was no longer tenable; these objections have
been supported by Professor Alexander Agassiz in the
United States, and by Dr. A. Geikie, and Dr. H. B. Guppy
in this country.

" Although Mr. Darwin did not live to bring out a third
edition of his *Coral Reefs*, I know from several conversations with him that he had given the most patient and
thoughtful consideration to Mr. Murray's paper on the subject. He admitted to me that had he known, when he
wrote his work, of the abundant deposition of the remains
of calcareous organisms on the sea floor, he might have regarded this cause as sufficient in a few cases to raise the
summit of submerged volcanoes or other mountains to a level
at which reef-forming corals can commence to flourish.
But he did not think that the admission that under certain
favourable conditions, atolls might be thus formed without
subsidence, necessitated an abandonment of his theory in
the case of the innumerable examples of the kind which
stud the Indian and Pacific Oceans.

" A letter written by Darwin to Professor Alexander
Agassiz in May 1881, shows exactly the attitude which careful consideration of the subject led him to maintain towards
the theory propounded by Mr. Murray:—

" ' You will have seen,' he writes, ' Mr. Murray's views
on the formation of atolls and barrier reefs. Before publishing my book, I thought long over the same view, but

only as far as ordinary marine organisms are concerned, for
at that time little was known of the multitude of minute
oceanic organisms. I rejected this view, as from the few
dredgings made in the *Beagle*, in the south temperate regions, I concluded that shells, the smaller corals, &c., decayed, and were dissolved, when not protected by the deposition of sediment, and sediment could not accumulate in
the open ocean. Certainly, shells, &c., were in several cases
completely rotten, and crumbled into mud between my fingers; but you will know well whether this is in any degree
common. I have expressly said that a bank at the proper
depth would give rise to an atoll, which could not be distinguished from one formed during subsidence. I can,
however, hardly believe in the former presence of as many
banks (there having been no subsidence) as there are atolls
in the great oceans, within a reasonable depth, on which
minute oceanic organisms could have accumulated to the
thickness of many hundred feet.

" Darwin's concluding words in the same letter written
within a year of his death, are a striking proof of the candour and openness of mind which he preserved so well to
, the end, in this as in other controversies.

" ' If I am wrong, the sooner I am knocked on the head
and annihilated so much the better. It still seems to me a
marvellous thing that there should not have been much,
and long continued, subsidence in the beds of the great
oceans. I wish that·some doubly rich millionaire would
take it into his head to have borings made in some of the
Pacific and Indian atolls, and bring home cores for slicing
from a depth of 500 or 600 feet.'

" It is noteworthy that the objections to Darwin's theory
have for the most part proceeded from zoologists, while
those who have fully appreciated the geological aspect of
the question have been the staunchest supporters of the theory of subsidence. The desirability of such boring operations in atolls has been insisted upon by several geologists,
and it may be hoped that before many years have passed
away, Darwin's hopes may be realised, either with or without the intervention of the ' doubly rich millionaire.'

" Three years after the death of Darwin, the veteran
Professor Dana re-entered the lists and contributed a powerful defence of the theory of subsidence in the form of a reply to an essay written by the ablest exponent of the anti-
Darwinian views on this subject, Dr. A. Geikie. While

pointing out that the Darwinian position had been to a great extent misunderstood by its opponents, he showed that the rival theory presented even greater difficulties than those which it professed to remove.

"During the last five years, the whole question of the origin of coral-reefs and islands has been re-opened, and a controversy has arisen, into which, unfortunately, acrimonious elements have been very unnecessarily introduced. Those who desire it, will find clear and impartial statements of the varied and often mutually destructive views put forward by different authors, in three works which have made their appearance within the last year—*The Bermuda Islands*, by Professor Angelo Heilprin : *Corals and Coral Islands*, new edition by Professor J. D. Dana ; and the third edition of Darwin's *Coral-Reefs*, with Notes and Appendix by Professor T. G. Bonney.

"Most readers will, I think, rise from the perusal of these works with the conviction that, while on certain points of detail it is clear that, through the want of knowledge concerning the action of marine organisms in the open ocean, Darwin was betrayed into some grave errors, yet the main foundations of his argument have not been seriously impaired by the new facts observed in the deep-sea researches, or by the severe criticisms to which his theory has been subjected during the last ten years. On the other hand, I think it will appear that much misapprehension has been exhibited by some of Darwin's critics, as to what his views and arguments really were ; so that the reprint and wide circulation of the book in its original form is greatly to be desired, and cannot but be attended with advantage to all those who will have the fairness to acquaint themselves with Darwin's views at first hand, before attempting to reply to them."

The only important geological work of my father's later years is embodied in his book on earthworms (1881), which may therefore be conveniently considered in this place. This subject was one which had interested him many years before this date, and in 1838 a paper on the formation of mould was published in the *Proceedings of the Geological Society*.

Here he showed that " fragments of burnt marl, cinders, &c., which had been thickly strewed over the surface of several meadows were found after a few years lying at a depth of some inches beneath the turf, but still forming a layer."

For the explanation of this fact, which forms the central idea of the geological part of the book, he was indebted to his uncle Josiah Wedgwood, who suggested that worms, by bringing earth to the surface in their castings, must undermine any objects lying on the surface and cause an apparent sinking.

In the book of 1881 he extended his observations on this burying action, and devised a number of different ways of checking his estimates as to the amount of work done. He also added a mass of observations on the natural history and intelligence of worms, a part of the work which added greatly to its popularity.

In 1877 Sir Thomas Farrer had discovered close to his garden the remains of a building of Roman-British times, and thus gave my father the opportunity of seeing for himself the effects produced by earthworms on the old concrete floors, walls, &c. On his return he wrote to Sir Thomas Farrer :—

" I cannot remember a more delightful week than the last. I know very well that E. will not believe me, but the worms were by no means the sole charm."

In the autumn of 1880, when the *Power of Movement in Plants* was nearly finished, he began once more on the subject. He wrote to Professor Carus (September 21) :—

" In the intervals of correcting the press, I am writing a very little book, and have done nearly half of it. Its title will be (as at present designed), *The Formation of Vegetable Mould through the Action of Worms.** As far as I can judge, it will be a curious little book."

The manuscript was sent to the printers in April 1881, and when the proof-sheets were coming in he wrote to Professor Carus : " The subject has been to me a hobby-horse, and I have perhaps treated it in foolish detail."

It was published on October 10, and 2000 copies were sold at once. He wrote to Sir J. D. Hooker, " I am glad that you approve of the *Worms.* When in old days I used to tell you whatever I was doing, if you were at all interested, I always felt as most men do when their work is finally published."

To Mr. Mellard Reade he wrote (November 8) : " It has been a complete surprise to me how many persons have

---

* The full title is *The Formation of Vegetable Mould through the Action of Worms, with Observations on their Habits,* 1881.

cared for the subject." And to Mr. Dyer (in November) :
" My book has been received with almost laughable enthusiasm, and 3500 copies have been sold ! ! ! " Again to his
friend Mr. Anthony Rich, he wrote on February 4, 1882,
" I have been plagued with an endless stream of letters on
the subject; most of them very foolish and enthusiastic;
but some containing good facts which I have used in correcting yesterday the *Sixth Thousand.*" The popularity of
the book may be roughly estimated by the fact that, in the
three years following its publication, 8500 copies were sold
—a sale relatively greater than that of the *Origin of Species.*

It is not difficult to account for its success with the non-
scientific public. Conclusions so wide and so novel, and so
easily understood, drawn from the study of creatures so
familiar, and treated with unabated vigour and freshness,
may well have attracted many readers. A reviewer remarks :
" In the eyes of most men . . . the earthworm is a mere
blind, dumb, senseless, and unpleasantly slimy annelid. Mr.
Darwin undertakes to rehabilitate his character, and the
earthworm steps forth at once as an intelligent and beneficent personage, a worker of vast geological changes, a planer
down of mountain sides . . . a friend of man . . . and an
ally of the Society for the preservation of ancient monuments." The *St. James's Gazette,* of October 17th, 1881,
pointed out that the teaching of the cumulative importance
of the infinitely little is the point of contact between this
book and the author's previous work.

One more book remains to be noticed, the *Life of Erasmus Darwin.*

In February 1879 an essay by Dr. Ernst Krause, on the
scientific work of Erasmus Darwin, appeared in the evolutionary journal, *Kosmos.* The number of *Kosmos* in question was a " Gratulationsheft," * or special congratulatory
issue in honour of my father's birthday, so that Dr. Krause's
essay, glorifying the older evolutionist, was quite in its
place. He wrote to Dr. Krause, thanking him cordially for
the honour paid to Erasmus, and asking his permission to
publish an English translation of the Essay.

His chief reason for writing a notice of his grandfather's
life was " to contradict flatly some calumnies by Miss Sew-

---

* The same number contains a good biographical sketch of my father of
which the material was to a large extent supplied by him to the writer, Professor Preyer of Jena. The article contains an excellent list of my father's
publications.

ard." This appears from a letter of March 27, 1879, to his
cousin Reginald Darwin, in which he asks for any docu-
ments and letters which might throw light on the character of Erasmus. This led to Mr. Reginald Darwin placing
in my father's hands a quantity of valuable material, including a curious folio common-place book, of which he
wrote : " I have been deeply interested by the great book,
. . . reading and looking at it is like having communion
with the dead . . . [it] has taught me a good deal about
the occupations and tastes of our grandfather."

Dr. Krause's contribution formed the second part of the
Life of Erasmus Darwin, my father supplying a " preliminary notice." This expression on the title-page is somewhat misleading; my father's contribution is more than
half the book, and should have been described as a biography. Work of this kind was new to him, and he wrote
doubtfully to Mr. Thiselton Dyer, June 18th : " God only
knows what I shall make of his life, it is such a new kind
of work to me." The strong interest he felt about his fore-
bears helped to give zest to the work, which became a decided enjoyment to him. With the general public the book
was not markedly successful, but many of his friends recognised its merits. Sir J. D. Hooker was one of these, and to
him my father wrote, " Your praise of the Life of Dr. D. has
pleased me exceedingly, for I despised my work, and thought
myself a perfect fool to have undertaken such a job."

To Mr. Galton, too, he wrote, November 14 :—

" I am extremely glad that you approve of the little *Life*
of our grandfather, for I have been repenting that I ever
undertook it, as the work was quite beyond my tether."

THE VIVISECTION QUESTION.

Something has already been said of my father's strong
feeling with regard to suffering * both in man and beast. It

---

* He once made an attempt to free a patient in a mad-house, who (as he
wrongly supposed) was sane. He was in correspondence with the gardener
at the asylum, and on one occasion he found a letter from the patient enclosed
with one from the gardener. The letter was rational in tone and declared that
the writer was sane and wrongfully confined.

My father wrote to the Lunacy Commissioners (without explaining the
source of his information) and in due time heard that the man had been visited by the Commissioners, and that he was certainly insane. Some time
afterward the patient was discharged, and wrote to thank my father for his
interference, adding that he had undoubtedly been insane when he wrote his
former letter.

was indeed one of the strongest feelings in his nature, and was exemplified in matters small and great, in his sympathy with the educational miseries of dancing dogs, or his horror at the sufferings of slaves.

The remembrance of screams, or other sounds heard in Brazil, when he was powerless to interfere with what he believed to be the torture of a slave, haunted him for years, especially at night. In smaller matters, where he could interfere, he did so vigorously. He returned one day from his walk pale and faint from having seen a horse ill-used, and from the agitation of violently remonstrating with the man. On another occasion he saw a horse-breaker teaching his son to ride; the little boy was frightened and the man was rough; my father stopped, and jumping out of the carriage reproved the man in no measured terms.

One other little incident may be mentioned, showing that his humanity to animals was well known in his own neighbourhood. A visitor, driving from Orpington to Down, told the cabman to go faster. "Why," said the man, "if I had whipped the horse *this* much, driving Mr. Darwin, he would have got out of the carriage and abused me well."

With respect to the special point under consideration,— the sufferings of animals subjected to experiment,—nothing could show a stronger feeling than the following words from a letter to Professor Ray Lankester (March 22, 1871):—

"You ask about my opinion on vivisection. I quite agree that it is justifiable for real investigations on physiology; but not for mere damnable and detestable curiosity. It is a subject which makes me sick with horror, so I will not say another word about it, else I shall not sleep tonight."

The Anti-Vivisection agitation, to which the following letters refer, seems to have become specially active in 1874, as may be seen, *e.g.* by the index to *Nature* for that year, in which the word "Vivisection" suddenly comes into prominence. But before that date the subject had received the earnest attention of biologists. Thus at the Liverpool Meeting of the British Association in 1870, a Committee was appointed, whose report defined the circumstances and conditions under which, in the opinion of the signatories, experiments on living animals were justifiable. In the spring of 1875, Lord Hartismere introduced a bill into the Upper House to regulate the course of physiological

research. Shortly afterwards a Bill more just towards
science in its provisions was introduced to the House of
Commons by Messrs. Lyon Playfair, Walpole, and Ashley.
It was, however, withdrawn on the appointment of a Royal
Commission to inquire into the whole question. The Com-
missioners were Lords Cardwell and Winmarleigh, Mr. W.
E. Forster, Sir J. B. Karslake, Mr. Huxley, Professor
Erichssen, and Mr. R. H. Hutton : they commenced their
inquiry in July, 1875, and the Report was published early
in the following year.

In the early summer of 1876, Lord Carnarvon's Bill,
entitled, " An Act to amend the Law relating to Cruelty to
Animals," was introduced. The framers of this Bill, yield-
ing to the unreasonable clamour of the public, went far
beyond the recommendations of the Royal Commission. As
a correspondent writes in *Nature* (1876, p. 248), " the evidence on the strength of which legislation was recommended went beyond the facts, the Report went beyond the
evidence, the Recommendations beyond the Report; and
the Bill can hardly be said to have gone beyond the Recommendations ; but rather to have contradicted them."

The legislation which my father worked for, was practically what was introduced as Dr. Lyon Playfair's Bill.

The following letter appeared in the *Times*, April 18th,
1881) :—

*C. D. to Frithiof Holmgren.** Down, April 14, 1881.

DEAR SIR,—In answer to your courteous letter of April
7, I have no objection to express my opinion with respect
to the right of experimenting on living animals. I use this
latter expression as more correct and comprehensive than
that of vivisection. You are at liberty to make any use of
this letter which you may think fit, but if published I should
wish the whole to appear. I have all my life been a strong
advocate for humanity to animals, and have done what I
could in my writings to enforce this duty. Several years
ago, when the agitation against physiologists commenced in
England, it was asserted that inhumanity was here prac-
tised, and useless suffering caused to animals; and I was
led to think that it might be advisable to have an Act of
Parliament on the subject. I then took an active part in

---

* Professor of Physiology at Upsala.

trying to get a Bill passed, such as would have removed all just cause of complaint, and at the same time have left physiologists free to pursue their researches—a Bill very different from the Act which has since been passed. It is right to add that the investigation of the matter by a Royal Commission proved that the accusations made against our English physiologists were false. From all that I have heard, however, I fear that in some parts of Europe little regard is paid to the sufferings of animals, and if this be the case, I should be glad to hear of legislation against inhumanity in any such country. On the other hand, I know that physiology cannot possibly progress except by means of experiments on living animals, and I feel the deepest conviction that he who retards the progress of physiology commits a crime against mankind. Any one who remembers, as I can, the state of this science half a century ago must admit that it has made immense progress, and it is now progressing at an ever-increasing rate. What improvements in medical practice may be directly attributed to physiological research is a question which can be properly discussed only by those physiologists and medical practitioners who have studied the history of their subjects; but, as far as I can learn, the benefits are already great. However this may be, no one, unless he is grossly ignorant of what science has done for mankind, can entertain any doubt of the incalculable benefits which will hereafter be derived from physiology, not only by man, but by the lower animals. Look for instance at Pasteur's results in modifying the germs of the most malignant diseases, from which, as it happens, animals will in the first place receive more relief than man. Let it be remembered how many lives and what a fearful amount of suffering have been saved by the knowledge gained of parasitic worms through the experiments of Virchow and others on living animals. In the future every one will be astonished at the ingratitude shown, at least in England, to these benefactors of mankind. As for myself, permit me to assure you that I honour, and shall always honour, every one who advances the noble science of physiology.

Dear Sir, yours faithfully.

In the *Times* of the following day appeared a letter headed " Mr. Darwin and Vivisection," signed by Miss Frances Power Cobbe. To this my father replied in the

Times of April 22, 1881. On the same day he wrote to Mr. Romanes :—

"As I have a fair opportunity, I sent a letter to the *Times* on Vivisection, which is printed to-day. I thought it fair to bear my share of the abuse poured in so atrocious a manner on all physiologists."

C. D. to the Editor of the Times.

SIR,—I do not wish to discuss the views expressed by Miss Cobbe in the letter which appeared in the *Times* of the 19th inst.; but as she asserts that I have "misinformed" my correspondent in Sweden in saying that "the investigation of the matter by a Royal Commission proved that the accusations made against our English physiologists were false," I will merely ask leave to refer to some other sentences from the report of the Commission.

(1.) The sentence—"It is not to be doubted that inhumanity may be found in persons of very high position as physiologists," which Miss Cobbe quotes from page 17 of the report, and which, in her opinion, "can necessarily concern English physiologists alone and not foreigners," is immediately followed by the words "We have seen that it was so in Magendie." Magendie was a French physiologist who became notorious some half century ago for his cruel experiments on living animals.

(2.) The Commissioners, after speaking of the "general sentiment of humanity" prevailing in this country, say (p. 10) :—

"This principle is accepted generally by the very highly educated men whose lives are devoted either to scientific investigation and education or to the mitigation or the removal of the sufferings of their fellow-creatures; though differences of degree in regard to its practical application will be easily discernible by those who study the evidence as it has been laid before us."

Again, according to the Commissioners (p. 10):—

"The secretary of the Royal Society for the Prevention of Cruelty to Animals, when asked whether the general tendency of the scientific world in this country is at variance with humanity, says he believes it to be very different indeed from that of foreign physiologists; and while giving it as the opinion of the society that experiments are performed which are in their nature beyond any legitimate province of

science, and that the pain which they inflict is pain which
it is not justifiable to inflict even for the scientific object in
view, he readily acknowledges that he does not know a single
case of wanton cruelty, and that in general the English
physiologists have used anæsthetics where they think they
can do so with safety to the experiment."

                    I am, Sir, your obedient servant.
      April 21.

During the later years of my father's life there was a
growing tendency in the public to do him honour.* The
honours which he valued most highly were those which
united the sympathy of friends with a mark of recognition
of his scientific colleagues. Of this type was the article
" Charles Darwin," published in *Nature*, June 4, 1874, and
written by Asa Gray. This admirable estimate of my
father's work in science is given in the form of a comparison
and contrast between Robert Brown and Charles Darwin.

To Gray he wrote :—

" I wrote yesterday and cannot remember exactly what I
said, and now cannot be easy without again telling you how
profoundly I have been gratified. Every one, I suppose,
occasionally thinks that he has worked in vain, and when
one of these fits overtakes me, I will think of your article,
and if that does not dispel the evil spirit, I shall know that
I am at the time a little bit insane, as we all are occasionally.

" What you say about Teleology † pleases me especially,
and I do not think any one else has ever noticed the point.
I have always said you were the man to hit the nail on the
head."

In 1877 he received the honorary degree of LL.D. from
the University of Cambridge. The degree was conferred
on November 17, and with the customary Latin speech from
the Public Orator, concluding with the words : " Tu vero,
qui leges naturæ tam docte illustraveris, legum doctor nobis
esto."

The honorary degree led to a movement being set on
foot in the University to obtain some permanent memorial

---

* In 1867 he had received a distinguished honour from Germany,—the order " Pour le Mérite."

† "Let us recognise Darwin's great service to Natural Science in bringing
back to it Teleology ; so that instead of Morphology *versus* Teleology, we
shall have Morphology wedded to Teleology." Similar remarks had been
previously made by Mr. Huxley. See *Critiques and Addresses*, p. 305.

of my father. In June 1879 he sat to Mr. W. Richmond for the portrait in the possession of the University, now placed in the Library of the Philosophical Society at Cambridge.

A similar wish on the part of the Linnean Society—with which my father was so closely associated—led to his sitting in August, 1881, to Mr. John Collier, for the portrait now in the possession of the Society. The portrait represents him standing facing the observer in the loose cloak so familiar to those who knew him with his slouch hat in his hand Many of those who knew his face most intimately, think that Mr. Collier's picture is the best of the portraits, and in this judgment the sitter himself was inclined to agree. According to my feeling it is not so simple or strong a representation of him as that given by Mr. Ouless. The last-named portrait was painted at Down in 1875; it is in the possession of the family,* and is known to many through Rajon's fine etching. Of Mr. Ouless's picture my father wrote to Sir J. D. Hooker:

"I look a very venerable, acute, melancholy old dog; whether I really look so I do not know."

Besides the Cambridge degree, he received about the same time honours of an academic kind from some foreign societies.

On August 5, 1878, he was elected a Corresponding Member of the French Institute in the Botanical Section,† and wrote to Dr. Asa Gray:—

"I see that we are both elected Corresponding Members of the Institute. It is rather a good joke that I should be elected in the Botanical Section, as the extent of my knowledge is little more than that a daisy is a Compositous plant and a pea a Leguminous one."

He valued very highly two photographic albums con-

---

* A *replica* by the artist hangs alongside of the portraits of Milton and Paley in the hall of Christ's College, Cambridge.

† He received twenty-six votes out of a possible thirty-nine, five blank papers were sent in, and eight votes were recorded for the other candidates. In 1872 an attempt had been made to elect him in the Section of Zoology, when, however, he only received fifteen out of forty-eight votes, and Lovén was chosen for the vacant place. It appears (*Nature*, August 1st, 1872) that an eminent member of the Academy wrote to *Les Mondes* to the following effect:—

"What has closed the doors of the Academy to Mr. Darwin is that the science of those of his books which have made his chief title to fame—the *Origin of Species*, and still more the *Descent of Man*, is not science, but a mass of assertions and absolutely gratuitous hypotheses, often evidently fallacious. This kind of publication and these theories are a bad example, which a body that respects itself cannot encourage."

taining portraits of a large number of scientific men in Germany and Holland, which he received as birthday gifts in 1877.

In the year 1878 my father received a singular mark of recognition in the form of a letter from a stranger, announcing that the writer intended to leave to him the reversion of the greater part of his fortune. Mr. Anthony Rich, who desired thus to mark his sense of my father's services to science, was the author of a *Dictionary of Roman and Greek Antiquities*, said to be the best book of the kind. It has been translated into French, German, and Italian, and has, in English, gone through several editions. Mr. Rich lived a great part of his life in Italy, painting, and collecting books and engravings. He finally settled, many years ago, at Worthing (then a small village), where he was a friend of Byron's Trelawny. My father visited Mr. Rich at Worthing, more than once, and gained a cordial liking and respect for him.

Mr. Rich died in April, 1891, having arranged that his bequest * should not lapse in consequence of the predecease of my father.

In 1879 he received from the Royal Academy of Turin the *Bressa* Prize for the years 1875–78, amounting to the sum of 12,000 francs. He refers to this in a letter to Dr. Dohrn (February 15th, 1880) :—

" Perhaps you saw in the papers that the Turin Society honoured me to an extraordinary degree by awarding me the *Bressa* Prize. Now it occurred to me that if your station wanted some piece of apparatus, of about the value of £100, I should very much like to be allowed to pay for it. Will you be so kind as to keep this in mind, and if any want should occur to you, I would send you a cheque at any time."

I find from my father's accounts that £100 was presented to the Naples Station.

Two years before my father's death, and twenty-one years after the publication of his greatest work, a lecture was given (April 9, 1880) at the Royal Institution by Mr. Huxley † which was aptly named " The Coming of Age of the Origin of Species." The following characteristic letter, referring to this subject, may fitly close the present chapter.

---

* Mr. Rich leaves a single near relative, to whom is bequeathed the life-interest in his property.

† Published in *Science and Culture*, p. 310.

Abinger Hall, Dorking, Sunday, April 11, 1880.

MY DEAR HUXLEY,—I wished much to attend your Lecture, but I have had a bad cough, and we have come here to see whether a change would do me good, as it has done. What a magnificent success your lecture seems to have been, as I judge from the reports in the *Standard* and *Daily News*, and more especially from the accounts given me by three of my children. I suppose that you have not written out your lecture, so I fear there is no chance of its being printed *in extenso*. You appear to have piled, as on so many other occasions, honours high and thick on my old head. But I well know how great a part you have played in establishing and spreading the belief in the descent-theory, ever since that grand review in the *Times* and the battle royal at Oxford up to the present day.

Ever, my dear Huxley,
Yours sincerely and gratefully,
CHARLES DARWIN.

P. S.—It was absurdly stupid in me, but I had read the announcement of your Lecture, and thought that you meant the maturity of the subject, until my wife one day remarked, " it is almost twenty-one years since the *Origin* appeared," and then for the first time the meaning of your words flashed on me.

BOTANICAL WORK.

"I have been making some little trifling observations which have interested and perplexed me much."

From a letter of June, 1860.

CHAPTER XVI.

THE botanical work which my father accomplished by the guidance of the light cast on the study of natural history by his own work on evolution remains to be noticed. In a letter to Mr. Murray, September 24th, 1861, speaking of his book the *Fertilisation of Orchids*, he says: "It will perhaps serve to illustrate how Natural History may be worked under the belief of the modification of species." This remark gives a suggestion as to the value and interest of his botanical work, and it might be expressed in far more emphatic language without danger of exaggeration.

In the same letter to Mr. Murray, he says: "I think this little volume will do good to the *Origin*, as it will show that I have worked hard at details." It is true that his botanical work added a mass of corroborative detail to the case for Evolution, but the chief support given to his doctrines by these researches was of another kind. They supplied an argument against those critics who have so freely dogmatised as to the uselessness of particular structures, and as to the consequent impossibility of their having been developed by means of natural selection. His observations on Orchids enabled him to say: "I can show the meaning of some of the apparently meaningless ridges and horns; who will now venture to say that this or that structure is useless?" A kindred point is expressed in a letter to Sir J. D. Hooker (May 14th, 1862):—

"When many parts of structure, as in the woodpecker, show distinct adaptation to external bodies, it is preposterous to attribute them to the effects of climate, &c., but when a single point alone, as a hooked seed, it is conceivable it may thus have arisen. I have found the study of Orchids eminently useful in showing me how nearly all parts of the flower are co-adapted for fertilisation by insects, and therefore the results of natural selection,—even the most trifling details of structure."

One of the greatest services rendered by my father to the study of Natural History is the revival of Teleology. The evolutionist studies the purpose or meaning of organs with the zeal of the older Teleologist, but with far wider and more coherent purpose. He has the invigorating knowledge that he is gaining not isolated conceptions of the economy of the present, but a coherent view of both past and present. And even where he fails to discover the use of any part, he may, by a knowledge of its structure, unravel the history of the past vicissitudes in the life of the species. In this way a vigour and unity is given to the study of the forms of organised beings, which before it lacked. Mr. Huxley has well remarked : * " Perhaps the most remarkable service to the philosophy of Biology rendered by Mr. Darwin is the reconciliation of Teleology and Morphology, and the explanation of the facts of both, which his views offer. The teleology which supposes that the eye, such as we see it in man, or one of the higher vertebrata, was made with the precise structure it exhibits, for the purpose of enabling the animal which possesses it to see, has undoubtedly received its death-blow. Nevertheless, it is necessary to remember that there is a wider teleology which is not touched by the doctrine of Evolution, but is actually based upon the fundamental proposition of Evolution."

The point which here especially concerns us is to recognise that this " great service to natural science," as Dr. Gray describes it, was effected almost as much by Darwin's special botanical work as by the *Origin of Species.*

For a statement of the scope and influence of my father's botanical work, I may refer to Mr. Thiselton Dyer's article in ' Charles Darwin,' one of the *Nature Series.* Mr. Dyer's wide knowledge, his friendship with my father, and his power of sympathising with the work of others, combine to give this essay a permanent value. The following passage (p. 43) gives a true picture :—

" Notwithstanding the extent and variety of his botanical work, Mr. Darwin always disclaimed any right to be regarded as a professed botanist. He turned his attention to plants, doubtless because they were convenient objects for

---

* The " Genealogy of Animals" (*The Academy,* 1869), reprinted in *Critiques and Addresses.*

organic phenomena in their least complicated
d this point of view, which, if one may use the ex-
without disrespect, had something of the amateur
was in itself of the greatest importance. For, from
g, till he took up any point, familiar with the liter-
aring on it, his mind was absolutely free from any
ssion. He was never afraid of his facts, or of fram-
hypothesis, however startling, which seemed to ex-
hem. . . . In any one else such an attitude would
roduced much work that was crude and rash. But
arwin—if one may venture on language which will
no one who had conversed with him as over-strained
med by gentle persuasion to have penetrated that re-
of nature which baffles smaller men. In other words,
ng experience had given him a kind of instinctive in-
into the method of attack of any biological problem,
ever unfamiliar to him, while he rigidly controlled the
lity of his mind in hypothetical explanations by the no
fertility of ingeniously devised experiment."
To form any just idea of the greatness of the revolution
rked by my father's researches in the study of the fertili-
ion of flowers, it is necessary to know from what a con-
tion this branch of knowledge has emerged. It should
remembered that it was only during the early years of
he present century that the idea of sex, as applied to plants,
ecame firmly established. Sachs, in his *History of Botany**
1875), has given some striking illustrations of the remark-
ble slowness with which its acceptance gained ground. He
remarks that when we consider the experimental proofs given
by Camerarius (1694), and by Kölreuter (1761–66), it appears incredible that doubts should afterwards have been
to the sexuality of plants. Yet he shows that such
edly crop up. These adverse criti-
less experiments 1820.

---

of my father. In June 1879 he sat to Mr. W. Richmond for
the portrait in the possession of the University, now placed
in the Library of the Philosophical Society at Cambridge.

A similar wish on the part of the Linnean Society—with
which my father was so closely associated—led to his sitting
in August, 1881, to Mr. John Collier, for the portrait now
in the possession of the Society. The portrait represents
him standing facing the observer in the loose cloak so familiar to those who knew him with his slouch hat in his
hand Many of those who knew his face most intimately,
think that Mr. Collier's picture is the best of the portraits,
and in this judgment the sitter himself was inclined to
agree. According to my feeling it is not so simple or strong
a representation of him as that given by Mr. Ouless.. The
last-named portrait was painted at Down in 1875; it is in
the possession of the family,* and is known to many through
Rajon's fine etching. Of Mr. Ouless's picture my father
wrote to Sir J. D. Hooker:

"I look a very venerable, acute, melancholy old dog;
whether I really look so I do not know."

Besides the Cambridge degree, he received about the
same time honours of an academic kind from some foreign
societies.

On August 5, 1878, he was elected a Corresponding
Member of the French Institute in the Botanical Section,†
and wrote to Dr. Asa Gray:—

"I see that we are both elected Corresponding Members
of the Institute. It is rather a good joke that I should be
elected in the Botanical Section, as the extent of my knowledge is little more than that a daisy is a Compositous plant
and a pea a Leguminous one."

He valued very highly two photographic albums con-

---

* A *replica* by the artist hangs alongside of the portraits of Milton and
Paley in the hall of Christ's College, Cambridge.

† He received twenty-six votes out of a possible thirty-nine, five blank papers were sent in, and eight votes were recorded for the other candidates. In
1872 an attempt had been made to elect him in the Section of Zoology, when,
however, he only received fifteen out of forty-eight votes, and Lovén was
chosen for the vacant place. It appears (*Nature*, August 1st, 1872) that an
eminent member of the Academy wrote to *Les Mondes* to the following
effect:—

"What has closed the doors of the Academy to Mr. Darwin is that the science of those of his books which have made his chief title to fame—the *Origin of Species*, and still more the *Descent of Man*, is not science, but a mass of
assertions and absolutely gratuitous hypotheses, often evidently fallacious.
This kind of publication and these theories are a bad example, which a body
that respects itself cannot encourage."

When the belief in the sexuality of plants ha[d]
established as an incontrovertible piece of kno[wl]
weight of misconception remained, weighing d[own]
rational view of the subject. Camerarius* believ[ed natu-]
rally enough in his day) that hermaphrodite† flo[wers]
necessarily self-fertilised. He had the wit to be as[tonished]
at this, a degree of intelligence which, as Sachs po[ints out]
the majority of his successors did not attain to.

The following extracts from a note-book show th[at]
point occurred to my father as early as 1837:

"Do not plants which have male and female org[ans to-]
gether [i.e. in the same flower] yet receive influence[s]
other plants? Does not Lyell give some argument[s]
varieties being difficult to keep [true] on account of [crossing]
from other plants? Because this may be applied to [show]
all plants do receive intermixture."

Sprengel,‡ indeed, understood that the hermaphro[dite]
structure of flowers by no means necessarily leads to [self-]
fertilisation. But although he discovered that in many c[ases]
pollen is of necessity carried to the stigma of another flo[wer]
he did not understand that in the advantage gained by
intercrossing of distinct plants lies the key to the wh[ole]
question. Hermann Müller# has well remarked that th[is]
"omission was for several generations fatal to Sprenge[l's]
work. . . . For both at the time and subsequently, bot[a-]
nists felt above all the weakness of his theory, and they se[t it]
aside, along with his defective ideas, the rich store of hi[s]
patient and acute observations and his comprehensive and
accurate interpretations." It remained for my father to
convince the world that the meaning hidden in the structure
of flowers was to be found by seeking light in the same
direction in which Sprengel, seventy years before, had
laboured. Robert Brown was the connecting link between
them, for it was at his recommendation that my father in
1841 read Sprengel's now celebrated Secret of Nature Dis-
played.||

The book impressed him as being "full of truth," al-

* Sachs, Geschichte d. Botanik, p. 419.
† That is to say, flowers possessing both stamens, or male organs, and pistils or female organs.
‡ Christian Conrad Sprengel, born 1750, died 1816.
# Fertilisation of Flowers (Eng. Trans.) 1883, p. 3.
|| Das entdeckte Geheimniss der Natur im Baue und in der Befruchtung
der Blumen. Berlin, 1793.

in English, g[...]
a great part of his li[...]
books and engravings. He finally se[...]
at Worthing (then a small village), where he was [...]
of Byron's Trelawny. My father visited Mr. Rich at Worthing, more than once, and gained a cordial liking and respect for him.

Mr. Rich died in April, 1891, having arranged that his
bequest* should not lapse in consequence of the predecease
of my father.

In 1879 he received from the Royal Academy of Turin
the *Bressa* Prize for the years 1875–78, amounting to the
sum of 12,000 francs. He refers to this in a letter to Dr.
Dohrn (February 15th, 1880):—

"Perhaps you saw in the papers that the Turin Society
honoured me to an extraordinary degree by awarding me
the *Bressa* Prize. Now it occurred to me that if your
station wanted some piece of apparatus, of about the value
of £100, I should very much like to be allowed to pay for it.
Will you be so kind as to keep this in mind, and if any want
should occur to you, I would send you a cheque at any time."

I find from my father's accounts that £100 was presented
to the Naples Station.

Two years before my father's death, and twenty-one
years after the publication of his greatest work, a lecture
was given (April 9, 1880) at the Royal Institution by Mr.
Huxley† which was aptly named "The Coming of Age of
the Origin of Species." The following characteristic letter,
referring to this subject, may fitly close the present chapter.

* Mr. Rich leaves a single near relative, to whom is bequeathed the life-
interest in his property.
† Published in *Science and Culture*, p. 310.

though "with some little nonsense." It not only encouraged him in kindred speculation, but guided him in his work, for in 1844 he speaks of verifying Sprengel's observations. It may be doubted whether Robert Brown ever planted a more fruitful seed than in putting such a book into such hands.

A passage in the *Autobiography* (p. 44) shows how it was that my father was attracted to the subject of fertilisation : " During the summer of 1839, and I believe during the previous summer, I was led to attend to the cross-fertilisation of flowers by the aid of insects, from having come to the conclusion in my speculations on the origin of species, that crossing played an important part in keeping specific forms constant."

The original connection between the study of flowers and the problem of evolution is curious, and could hardly have been predicted. Moreover, it was not a permanent bond. My father proved by a long series of laborious experiments, that when a plant is fertilised and sets seeds under the influence of pollen from a distinct individual, the offspring so produced are superior in vigour to the offspring of self-fertilisation, *i.e.* of the union of the male and female elements of a single plant. When this fact was established, it was possible to understand the *raison d'être* of the machinery which insures cross-fertilisation in so many flowers; and to understand how natural selection can act on, and mould, the floral structure.

Asa Gray has well remarked with regard to this central idea (*Nature*, June 4, 1874) :—" The aphorism, ' Nature abhors a vacuum,' is a characteristic specimen of the science of the middle ages. The aphorism, ' Nature abhors close fertilisation,' and the demonstration of the principle, belong to our age and to Mr. Darwin. To have originated this, and also the principle of Natural Selection . . . and to have applied these principles to the system of nature, in such a manner as to make, within a dozen years, a deeper impression upon natural history than has been made since Linnæus, is ample title for one man's fame."

The flowers of the Papilionaceæ * attracted his attention early, and were the subject of his first paper on fertilisation.† The following extract from an undated letter to Asa

---

* The order to which the pea and bean belong.
† *Gardeners' Chronicle*, 1857, p. 725. It appears that this paper was a piece

Gray seems to have been written before the publication of this paper, probably in 1856 or 1857 :—

". . . What you say on Papilionaceous flowers is very true; and I have no facts to show that varieties are crossed; but yet (and the same remark is applicable in a beautiful way to Fumaria and Dielytra, as I noticed many years ago), I must believe that the flowers are constructed partly in direct relation to the visits of insects; and how insects can avoid bringing pollen from other individuals I cannot understand. It is really pretty to watch the action of a humble-bee on the scarlet kidney bean, and in this genus (and in *Lathyrus grandiflorus*) * the honey is so placed that the bee invariably alights on that *one* side of the flower towards which the spiral pistil is protruded (bringing out with it pollen), and by the depression of the wing-petal is forced against the bee's side all dusted with pollen. In the broom the pistil is rubbed on the centre of the back of the bee. I suspect there is something to be made out about the Leguminosæ, which will bring the case within *our* theory; though I have failed to do so. Our theory will explain why in the vegetable . . . kingdom the act of fertilisation even in hermaphrodites usually takes place *sub jove*, though thus exposed to *great* injury from damp and rain."

A letter to Dr. Asa Gray (September 5th, 1857) gives the substance of the paper in the *Gardeners' Chronicle:*—

" Lately I was led to examine buds of kidney bean with the pollen shed; but I was led to believe that the pollen could *hardly* get on the stigma by wind or otherwise, except by bees visiting [the flower] and moving the wing petals: hence I included a small bunch of flowers in two bottles in every way treated the same: the flowers in one I daily just momentarily moved, as if by a bee; these set three fine pods, the other *not one*. Of course this little experiment must be tried again, and this year in England it is too late, as the flowers seem now seldom to set. If bees are necessary to this flower's self-fertilisation, bees must almost cross them, as their dusted right-side of head and right legs constantly touch the stigma.

" I have, also, lately been reobserving daily *Lobelia ful-*

---

of "over-time" work. He wrote to a friend, "that confounded Leguminous paper was done in the afternoon, and the consequence was I had to go to Moor Park for a week."

* The sweet pea and everlasting pea belong to the genus Lathyrus.

gens—this in my garden is never visited by insects, and never sets seeds, without pollen be put on the stigma (whereas the small blue Lobelia is visited by bees and does set seed) ; I mention this because there are such beautiful contrivances to prevent the stigma ever getting its own pollen ; which seems only explicable on the doctrine of the advantage of crosses."

The paper was supplemented by a second in 1858.[*] The chief object of these publications seems to have been to obtain information as to the possibility of growing varieties of Leguminous plants near each other, and yet keeping them true. It is curious that the Papilionaceæ should not only have been the first flowers which attracted his attention by their obvious adaptation to the visits of insects, but should also have constituted one of his sorest puzzles. The common pea and the sweet pea gave him much difficulty, because, although they are as obviously fitted for insect-visits as the rest of the order, yet their varieties keep true. The fact is that neither of these plants being indigenous, they are not perfectly adapted for fertilisation by British insects. He could not, at this stage of his observations, know that the co-ordination between a flower and the particular insect which fertilises it may be as delicate as that between a lock and its key, so that this explanation was not likely to occur to him.

Besides observing the Leguminosæ, he had already begun, as shown in the foregoing extracts, to attend to the structure of other flowers in relation to insects. At the beginning of 1860 he worked at Leschenaultia,[†] which at first puzzled him, but was ultimately made out. A passage in a letter chiefly relating to Leschenaultia seems to show that it was only in the spring of 1860 that he began widely to apply his knowledge to the relation of insects to other flowers. This is somewhat surprising, when we remember that he had read Sprengel many years before. He wrote (May 14) :—

"I should look at this curious contrivance as specially related to visits of insects; as I begin to think is almost universally the case."

Even in July 1862 he wrote to Asa Gray :

---

* *Gardeners' Chronicle*, 1858, p. 828.
† He published a short paper on the manner of fertilisation of this flower, in the *Gardeners' Chronicle* 1871, p. 1166.

" There is no end to the adaptations. Ought not these cases to make one very cautious when one doubts about the use of all parts ? I fully believe that the structure of all irregular flowers is governed in relation to insects. Insects are the Lords of the floral (to quote the witty *Athenæum*) world."

This idea has been worked out by H. Müller, who has written on insects in the character of flower-breeders or flower-fanciers, showing how the habits and structure of the visitors are reflected in the forms and colours of the flowers visited.

He was probably attracted to the study of Orchids by the fact that several kinds are common near Down. The letters of 1860 show that these plants occupied a good deal of his attention ; and in 1861 he gave part of the summer and all the autumn to the subject. He evidently considered himself idle for wasting time on Orchids which ought to have been given to *Variation under Domestication*. Thus he wrote :—

" There is to me incomparably more interest in observing than in writing ; but I feel quite guilty in trespassing on these subjects, and not sticking to varieties of the confounded cocks, hens and ducks. I hear that Lyell is savage at me."

It was in the summer of 1860 that he made out one of the most striking and familiar facts in the Orchid-book, namely, the manner in which the pollen masses are adapted for removal by insects. He wrote to Sir J. D. Hooker, July 12 :—

" I have been examining *Orchis pyramidalis*, and it almost equals, perhaps even beats, your Listera case ; the sticky glands are congenitally united into a saddle-shaped organ, which has great power of movement, and seizes hold of a bristle (or proboscis) in an admirable manner, and then another movement takes place in the pollen masses, by which they are beautifully adapted to leave pollen on the two lateral stigmatic surfaces. I never saw anything so beautiful."

In June of the same year he wrote :—

" You speak of adaptation being rarely visible, though present in plants. I have just recently been looking at the common Orchis, and I declare I think its adaptations in every part of the flower quite as beautiful and plain, or even more beautiful than in the woodpecker." *

---

* The woodpecker was one of his stock examples of adaptation.

He wrote also to Dr. Gray, June 8, 1860 :—

"Talking of adaptation, I have lately been looking at our common orchids, and I dare say the facts are as old and well-known as the hills, but I have been so struck with admiration at the contrivances, that I have sent a notice to the *Gardeners' Chronicle*."

Besides attending to the fertilisation of the flowers he was already, in 1860, busy with the homologies of the parts, a subject of which he made good use in the Orchid book. He wrote to Sir Joseph Hooker (July):—

"It is a real good joke my discussing homologies of Orchids with you, after examining only three or four genera; and this very fact makes me feel positive I am right! I do not quite understand some of your terms; but sometime I must get you to explain the homologies; for I am intensely interested in the subject, just as at a game of chess."

This work was valuable from a systematic point of view. In 1880 he wrote to Mr. Bentham :—

"It was very kind in you to write to me about the Orchideæ, for it has pleased me to an extreme degree that I could have been of the *least* use to you about the nature of the parts."

The pleasure which his early observations on Orchids gave him is shown in such passages as the following from a letter to Sir J. D. Hooker (July 27, 1861) :—

"You cannot conceive how the Orchids have delighted me. They came safe, but box rather smashed; cylindrical old cocoa- or snuff-canister much safer. I enclose postage. As an account of the movement, I shall allude to what I suppose is Oncidium, to make *certain*,—is the enclosed flower with crumpled petals this genus? Also I most specially want to know what the enclosed little globular brown Orchid is. I have only seen pollen of a Cattleya on a bee, but surely have you not unintentionally sent me what I wanted most (after Catasetum or Mormodes), viz., one of the Epidendreæ ? ! I *particularly* want (and will presently tell you why) another spike of this little Orchid, with older flowers, some even almost withered."

His delight in observation is again shown in a letter to Dr. Gray (1863). Referring to Crüger's letters from Trinidad, he wrote :—" Happy man, he has actually seen crowds of bees flying round Catasetum, with the pollinia sticking to their backs ! "

The following extracts of letters to Sir J. D. Hooker

illustrate further the interest which his work excited in
him :—

"Veitch sent me a grand lot this morning. What
wonderful structures!

"I have now seen enough, and you must not send me
more, for though I enjoy looking at them *much*, and it has
been very useful to me, seeing so many different forms, it is
idleness. For my object each species requires studying for
days. I wish you had time to take up the group. I would
give a good deal to know what the rostellum is, of which I
have traced so many curious modifications. I suppose it cannot be one of the stigmas,* there seems a great tendency for
two lateral stigmas to appear. My paper, though touching
on only subordinate points will run, I fear, to 100 MS. folio
pages! The beauty of the adaptation of parts seems to me
unparalleled. I should think or guess waxy pollen was most
differentiated. In Cypripedium which seems least modified,
and a much exterminated group, the grains are single. In
all others, as far as I have seen, they are in packets of four;
and these packets cohere into many wedge-formed masses
in Orchis; into eight, four, and finally two. It seems curious that a flower should exist, which could *at most* fertilise
only two other flowers, seeing how abundant pollen generally
is; this fact I look at as explaining the perfection of the
contrivance by which the pollen, so important from its few-
ness, is carried from flower to flower " † (1861).

"I was thinking of writing to you to-day, when your
note with the Orchids came. What frightful trouble you
have taken about Vanilla; you really must not take an atom
more; for the Orchids are more play than real work. I
have been much interested by Epidendrum, and have worked
all morning at them; for Heaven's sake, do not corrupt me
by any more" (August 30, 1861).

He originally intended to publish his notes on Orchids
as a paper in the Linnean Society's *Journal*, but it soon became evident that a separate volume would be a more suitable form of publication. In a letter to Sir J. D. Hooker,
Sept. 24, 1861, he writes :—

---

* It is a modification of the upper stigma.
† This rather obscure statement may be paraphrased thus :—
The machinery is so perfect that the plant can afford to minimise the
amount of pollen produced. Where the machinery for pollen distribution is
of a cruder sort, for instance where it is carried by the wind, enormous quan-
tities are produced, *e. g.* in the fir tree.

"I have been acting, I fear that you will think, like a goose; and perhaps in truth I have. When I finished a few days ago my Orchis paper, which turns out one hundred and forty folio pages !! and thought of the expense of woodcuts, I said to myself, I will offer the Linnean Society to withdraw it, and publish it in a pamphlet. It then flashed on me that perhaps Murray would publish it, so I gave him a cautious description, and offered to share risks and profits. This morning he writes that he will publish and take all risks, and share profits and pay for all illustrations. It is a risk, and Heaven knows whether it will not be a dead failure, but I have not deceived Murray, and [have] told him that it would interest those alone who cared much for natural history. I hope I do not exaggerate the curiosity of the many special contrivances."

And again on September 28th :—

" What a good soul you are not to sneer at me, but to pat me on the back. I have the greatest doubt whether I am not going to do, in publishing my paper, a most ridiculous thing. It would annoy me much, but only for Murray's sake, if the publication were a dead failure."

There was still much work to be done, and in October he was still receiving Orchids from Kew, and wrote to Hooker :—

" It is impossible to thank you enough. I was almost mad at the wealth of Orchids." And again—

" Mr. Veitch most generously has sent me two splendid buds of Mormodes, which will be capital for dissection, but I fear will never be irritable ; so for the sake of charity and love of heaven do, I beseech you, observe what movement takes place in Cychnoches, and what part must be touched. Mr. V. has also sent me one splendid flower of Catasetum, the most wonderful Orchid I have seen."

On October 13 he wrote to Sir Joseph Hooker :—

" It seems that I cannot exhaust your good nature. I have had the hardest day's work at Catasetum and buds of Mormodes, and believe I understand at last the mechanism of movements and the functions. Catasetum is a beautiful case of slight modification of structure leading to new functions. I never was more interested in any subject in all my life than in this of Orchids. I owe very much to you."

Again to the same friend, November 1, 1861 :—

" If you really can spare another Catasetum, when nearly ready, I shall be most grateful ; had I not better send for it? The case is truly marvellous; the (so-called) sensation, or stimulus from a light touch is certainly transmitted through the antennæ for more than one inch *instantaneously.* . . . A cursed insect or something let my last flower off last night."

Professor de Candolle has remarked * of my father, " Ce n'est pas lui qui aurait demandé de construire des palais pour y loger des laboratoires." This was singularly true of his orchid work, or rather it would be nearer the truth to say that he had no laboratory, for it was only after the publication of the *Fertilisation of Orchids,* that he built himself a green-house. He wrote to Sir J. D. Hooker (December 24th, 1862) :—

"And now I am going to tell you a *most* important piece of news!! I have almost resolved to build a small hothouse ; my neighbour's really first-rate gardener has suggested it, and offered to make me plans, and see that it is well done, and he is really a clever fellow, who wins lots of prizes, and is very observant. He believes that we should succeed with a little patience; it will be a grand amusement for me to experiment with plants."

Again he wrote (February 15th, 1863) :—

" I write now because the new hot-house is ready, and I long to stock it, just like a schoolboy. Could you tell me pretty soon what plants you can give me ; and then I shall know what to order? And do advise me how I had better get such plants as you can *spare.* Would it do to send my tax-cart early in the morning, on a day that was not frosty, lining the cart with mats, and arriving here before night? I have no idea whether this degree of exposure (and of course the cart would be cold) could injure stove-plants ; they would be about five hours (with bait) on the journey home."

A week later he wrote :—

" You cannot imagine what pleasure your plants give me (far more than your dead Wedgwood-ware can give you) ; H. and I go and gloat over them, but we privately confessed to each other, that if they were not our own, perhaps we should not see such transcendant beauty in each leaf."

And in March, when he was extremely unwell, he wrote :—

---

* " Darwin considéré, &c.," *Archives des Sciences Physiques et Naturelles,* 3ème période. Tome vii. 481, 1882.

"A few words about the stove-plants; they do so amuse me. I have crawled to see them two or three times. Will you correct and answer, and return enclosed. I have hunted in all my books and cannot find these names, and I like much to know the family." His difficulty with regard to the names of plants is illustrated, with regard to a Lupine on which he was at work, in an extract from a letter (July 21, 1866) to Sir J. D. Hooker: "I sent to the nursery garden, whence I bought the seed, and could only hear that it was 'the common blue Lupine,' the man saying 'he was no scholard, and did not know Latin, and that parties who make experiments ought to find out the names.'"

The book was published May 15th, 1862. Of its reception he writes to Mr. Murray, June 13th and 18th :—

"The Botanists praise my Orchid-book to the skies. Some one sent me (perhaps you) the *Parthenon*, with a good review. The *Athenæum* * treats me with very kind pity and contempt; but the reviewer knew nothing of his subject."

"There is a superb, but I fear exaggerated, review in the *London Review*.† But I have not been a fool, as I thought I was, to publish; for Asa Gray, about the most competent judge in the world, thinks almost as highly of the book as does the *London Review*. The *Athenæum* will hinder the sale greatly."

The Rev. M. J. Berkeley was the author of the notice in the *London Review*, as my father learned from Sir J. D. Hooker, who added, "I thought it very well done indeed. I have read a good deal of the Orchid-book, and echo all he says."

To this my father replied (June 30th, 1862) :—

"My dear old friend,—You speak of my warming the cockles of your heart, but you will never know how often you have warmed mine. It is not your approbation of my scientific work (though I care for that more than for any one's) : it is something deeper. To this day I remember keenly a letter you wrote to me from Oxford, when I was at the Water-cure, and how it cheered me when I was utterly weary of life. Well, my Orchid-book is a success (but I do not know whether it sells)."

In another letter to the same friend, he wrote :—

"You have pleased me much by what you say in regard

to Bentham and Oliver approving of my book; for I had got a sort of nervousness, and doubted whether I had not made an egregious fool of myself, and concocted pleasant little stinging remarks for reviews, such as ' Mr. Darwin's head seems to have been turned by a certain degree of success, and he thinks that the most trifling observations are worth publication.' "

He wrote too, to Asa Gray :—

" Your generous sympathy makes you over-estimate what you have read of my Orchid-book. But your letter of May 18th and 26th has given me an almost foolish amount of satisfaction. The subject interested me, I knew, beyond its real value; but I had lately got to think that I had made myself a complete fool by publishing in a semi-popular form. Now I shall confidently defy the world. . . . No doubt my volume contains much error: how curiously difficult it is to be accurate, though I try my utmost. Your notes have interested me beyond measure. I can now afford to d— my critics with ineffable complacency of mind. Cordial thanks for this benefit."

Sir Joseph Hooker reviewed the book in the *Gardeners' Chronicle*, writing in a successful imitation of the style of Lindley, the Editor. My father wrote to Sir Joseph (Nov. 12, 1862) :—

" So you did write the review in the *Gardeners' Chronicle*. Once or twice I doubted whether it was Lindley; but when I came to a little slap at R. Brown, I doubted no longer. You arch-rogue! I do not wonder you have deceived others also. Perhaps I am a conceited dog; but if so, you have much to answer for ; I never received so much praise, and coming from you I value it much more than from any other."

With regard to botanical opinion generally, he wrote to Dr. Gray, " I am fairly astonished at the success of my book with botanists." Among naturalists who were not botanists, Lyell was pre-eminent in his appreciation of the book. I have no means of knowing when he read it, but in later life, as I learn from Professor Judd, he was enthusiastic in praise of the *Fertilisation of Orchids*, which he considered " next to the *Origin*, as the most valuable of all Darwin's works." Among the general public the author did not at first hear of many disciples, thus he wrote to his cousin Fox in September 1862 : " Hardly any one not a botanist, except yourself, as far as I know, has cared for it."

If we examine the literature relating to the fertilisation of flowers, we do not find that this new branch of study showed any great activity immediately after the publication of the Orchid-book. There a few papers by Asa Gray, in 1862 and 1863, by Hildebrand in 1864, and by Moggridge in 1865, but the great mass of work by Axell, Delpino, Hildebrand, and the Müllers, did not begin to appear until about 1867. The period during which the new views were being assimilated, and before they became thoroughly fruitful, was, however, surprisingly short. The later activity in this department may be roughly gauged by the fact that the valuable 'Bibliography,' given by Professor D'Arcy Thompson in his translation of Müller's *Befruchtung* (1883),* contains references to 814 papers.

In 1877 a second edition of the *Fertilisation of Orchids* was published, the first edition having been for some time out of print. The new edition was remodelled and almost rewritten, and a large amount of new matter added, much of which the author owed to his friend Fritz Müller.

With regard to this edition he wrote to Dr. Gray :—

" I do not suppose I shall ever again touch the book. After much doubt I have resolved to act in this way with all my books for the future; that is to correct them once and never touch them again, so as to use the small quantity of work left in me for new matter."

One of the latest references to his Orchid-work occurs in a letter to Mr. Bentham, February 16, 1880. It shows the amount of pleasure which this subject gave to my father, and (what is characteristic of him) that his reminiscence of the work was one of delight in the observations which preceded its publication, not to the applause which followed it :—

" They are wonderful creatures, these Orchids, and I sometimes think with a glow of pleasure, when I remember making out some little point in their method of fertilisation."

The Effect of Cross- and Self-fertilisation in the Vegetable Kingdom. Different Forms of Flowers on Plants of the same Species.

Two other books bearing on the problem of sex in plants require a brief notice. *The Effects of Cross- and*

* My father's " Prefatory Notice " to this work is dated February 6th, 1882, and is therefore almost the last of his writings.

Self-Fertilisation, published in 1876, is one of his most important works, and at the same time one of the most unreadable to any but the professed naturalist. Its value lies in the proof it offers of the increased vigour given to the offspring by the act of cross-fertilisation. It is the complement of the Orchid book because it makes us understand the advantage gained by the mechanisms for insuring cross-fertilisation described in that work.

The book is also valuable in another respect, because it throws light on the difficult problems of the origin of sexuality. The increased vigour resulting from cross-fertilisation is allied in the closest manner to the advantage gained by change of conditions. So strongly is this the case, that in some instances cross-fertilisation gives no advantage to the offspring, unless the parents have lived under slightly different conditions. So that the really important thing is not that two individuals of different *blood* shall unite, but two individuals which have been subjected to different conditions. We are thus led to believe that sexuality is a means for infusing vigour into the offspring by the coalescence of differentiated elements, an advantage which could not accompany asexual reproductions.

It is remarkable that this book, the result of eleven years of experimental work, owed its origin to a chance observation. My father had raised two beds of *Linaria vulgaris* —one set being the offspring of cross and the other of self-fertilisation. The plants were grown for the sake of some observations on inheritance, and not with any view to crossbreeding, and he was astonished to observe that the offspring of self-fertilisation were clearly less vigorous than the others. It seemed incredible to him that this result could be due to a single act of self-fertilisation, and it was only in the following year, when precisely the same result occurred in the case of a similar experiment on inheritance in carnations, tha this attention was "thoroughly aroused," and that he determined to make a series of experiments specially directed to the question.

The volume on *Forms of Flowers* was published in 1877, and was dedicated by the author to Professor Asa Gray, "as a small tribute of respect and affection." It consists of certain earlier papers re-edited, with the addition of a quantity of new matter. The subjects treated in the book are:—

(i.) Heterostyled Plants.

(ii.) Polygamous, Diœcious, and Gynodiœcious Plants.
(iii.) Cleistogamic Flowers.

The nature of heterostyled plants may be illustrated in the primrose, one of the best known examples of the class. If a number of primroses be gathered, it will be found that some plants yield nothing but " pin-eyed " flowers, in which the style (or organ for the transmission of the pollen to the ovule) is long, while the others yield only " thrum-eyed " flowers with short styles. Thus primroses are divided into two sets or castes differing structurally from each other. My father showed that they also differ sexually, and that in fact the bond between the two castes more nearly resembles that between separate sexes than any other known relationship. Thus for example a long-styled primrose, though it can be fertilised by its own pollen, is not *fully* fertile unless it is impregnated by the pollen of a short-styled flower. Heterostyled plants are comparable to hermaphrodite animals, such as snails, which require the concourse of two individuals, although each possesses both the sexual elements. The difference is that in the case of the primrose it is *perfect fertility*, and not simply *fertility*, that depends on the mutual action of the two sets of individuals.

The work on heterostyled plants has a special bearing, to which the author attached much importance, on the problem of the origin of species.*

He found that a wonderfully close parallelism exists between hybridisation (*i.e.* crosses between distinct species), and certain forms of fertilisation among heterostyled plants. So that it is hardly an exaggeration to say that the " illegitimately " reared seedlings are hybrids, although both their parents belong to identically the same species. In a letter to Professor Huxley, given in the second volume of the *Life and Letters* (p. 384), my father writes as if his researches on heterostyled plants tended to make him believe that sterility is a selected or acquired quality. But in his later publications, e.g. in the sixth edition of the *Origin*, he adheres to the belief that sterility is an incidental † rather than a selected quality. The result of his work on

---

* See *Autobiography*, p. 48.

† The pollen or fertilising element is in each species adapted to produce a certain change in the egg-cell (or female element), just as a key is adapted to a lock. If a key opens a lock for which it was never intended it is an incidental result. In the same way if the pollen of species of A. proves to be capable of fertilising the egg-cell of species B. we may call it incidental.

heterostyled plants is of importance as showing that sterility is no test of specific distinctness, and that it depends
on differentiation of the sexual elements which is independent of any racial difference. I imagine that it was his
instinctive love of making out a difficulty which to a great
extent kept him at work so patiently on the heterostyled
plants. But it was the fact that general conclusions of the
above character could be drawn from his results which made
him think his results worthy of publication.

CHAPTER XVII.

Climbing Plants ; Power of Movement in Plants ; Insectivorous Plants ; Kew Index of Plant Names.

My father mentions in his *Autobiography* (p. 45) that he was led to take up the subject of climbing plants by reading Dr. Gray's paper, "Note on the Coiling of the Tendrils of Plants." [*] This essay seems to have been read in 1862, but I am only able to guess at the date of the letter in which he asks for a reference to it, so that the precise date of his beginning this work cannot be determined.

In June 1863, he was certainly at work, and wrote to Sir J. D. Hooker for information as to previous publications on the subject, being then in ignorance of Palm's and H. v. Mohl's works on climbing plants, both of which were published in 1827.

C. Darwin to Asa Gray. Down, August 4 [1863].

My present hobby-horse I owe to you, viz. the tendrils : their irritability is beautiful, as beautiful in all its modifications as anything in Orchids. About the *spontaneous* movement (independent of touch) of the tendrils and upper internodes, I am rather taken aback by your saying, "is it not well known?" I can find nothing in any book which I have. . . . The spontaneous movement of the tendrils is independent of the movement of the upper internodes, but both work harmoniously together in sweeping a circle for the tendrils to grasp a stick. So with all climbing plants (without tendrils) as yet examined, the upper internodes go on night and day sweeping a circle in one fixed direction. It is surprising to watch the Apocyneæ with shoots 18 inches long (beyond the supporting stick), steadily searching for something to climb up. When the shoot meets a

---

[*] *Proc. Amer. Acad. of Arts and Sciences*, 1858.

stick, the motion at that point is arrested, but in the upper part is continued ; so that the climbing of all plants yet examined is the simple result of the spontaneous circulatory movement of the upper internodes.* Pray tell me whether anything has been published on this subject? I hate publishing what is old ; but I shall hardly regret my work if it is old, as it has much amused me. . . .

He soon found that his observations were not entirely novel, and wrote to Hooker : " I have now read two German books, and all I believe that has been written on climbers, and it has stirred me up to find that I have a good deal of new matter. It is strange, but I really think no one has explained simple twining plants. These books have stirred me up, and made me wish for plants specified in them."

He continued his observations on climbing plants during the prolonged illness from which he suffered in the autumn of 1863, and in the following spring. He wrote to Sir J. D. Hooker, apparently in March 1864 :—

" The hot-house is such an amusement to me, and my amusement I owe to you, as my delight is to look at the many odd leaves and plants from Kew. . . . The only approach to work which I can do is to look at tendrils and climbers, this does not distress my weakened brain. Ask Oliver to look over the enclosed queries (and do you look) and amuse a broken-down brother naturalist by answering any which he can. If you ever lounge through your houses, remember me and climbing plants."

A letter to Dr. Gray, April 9, 1865, has a word or two on the subject :—

" I have began correcting proofs of my paper on Climbing Plants. I suppose I shall be able to send you a copy in four or five weeks. I think it contains a good deal new, and some curious points, but it is so fearfully long, that no one will ever read it. If, however, you do not *skim* through it, you will be an unnatural parent, for it is your child."

Dr. Gray not only read it but approved of it, to my father's great satisfaction, as the following extracts show :—

" I was much pleased to get your letter of July 24th. Now that I can do nothing, I maunder over old subjects, and your approbation of my climbing paper gives me *very* great satisfaction. I made my observations when I could

---

* This view is rejected by some botanists.

do nothing else and much enjoyed it, but always doubted whether they were worth publishing. . . .

"I received yesterday your article* on climbers, and it has pleased me in an extraordinary and even silly manner. You pay me a superb compliment, and as I have just said to my wife, I think my friends must perceive that I like praise, they give me such hearty doses. I always admire your skill in reviews or abstracts, and you have done this article excellently and given the whole essence of my paper. . . . I have had a letter from a good zoologist in S. Brazil, F. Müller, who has been stirred up to observe climbers, and gives me some curious cases of *branch*-climbers, in which branches are converted into tendrils, and then continue to grow and throw out leaves and new branches, and then lose their tendril character."

The paper on Climbing Plants was republished in 1875, as a separate book. The author had been unable to give his customary amount of care to the style of the original essay, owing to the fact that it was written during a period of continued ill-health, and it was now found to require a great deal of alteration. He wrote to Sir J. D. Hooker (March 3, 1875) : "It is lucky for authors in general that they do not require such dreadful work in merely licking what they write into shape." And to Mr. Murray, in September, he wrote : "The corrections are heavy in *Climbing Plants*, and yet I deliberately went over the MS. and old sheets three times." The book was published in September 1875, an edition of 1500 copies was struck off ; the edition sold fairly well, and 500 additional copies were printed in June of the following year.

The Power of Movement in Plants. 1880.

The few sentences in the autobiographical chapter give with sufficient clearness the connection between the *Power of Movement* and the book on Climbing Plants. The central idea of the book is that the movements of plants in relation to light, gravitation, &c., are modifications of a spontaneous tendency to revolve or circumnutate, which is widely inherent in the growing parts of plants. This conception has not been generally adopted, and has not taken

---

* In the September number of *Silliman's Journal*, concluded in the January number, 1866.

a place among the canons of orthodox physiology. The book has been treated by Professor Sachs with a few words of professorial contempt; and by Professor Wiesner it has been honoured by careful and generously expressed criticism.

Mr. Thiselton Dyer * has well said : " Whether this masterly conception of the unity of what has hitherto seemed a chaos of unrelated phenomena will be sustained, time alone will show. But no one can doubt the importance of what Mr. Darwin has done, in showing that for the future the phenomena of plant movement can and indeed must be studied from a single point of view."

The work was begun in the summer of 1877, after the publication of *Different Forms of Flowers*, and by the autumn his enthusiasm for the subject was thoroughly established, and he wrote to Mr. Dyer : " I am all on fire at the work." At this time he was studying the movements of cotyledons, in which the sleep of plants is to be observed in its simplest form ; in the following spring he was trying to discover what useful purpose these sleep-movements could serve, and wrote to Sir Joseph Hooker (March 25th, 1878):—

" I think we have *proved* that the sleep of plants is to lessen the injury to the leaves from radiation. This has interested me much, and has cost us great labour, as it has been a problem since the time of Linnæus. But we have killed or badly injured a multitude of plants. N.B.—*Oxalis carnosa* was most valuable, but last night was killed."

The book was published on November 6, 1880, and 1500 copies were disposed of at Mr. Murray's sale. With regard to it he wrote to Sir J. D. Hooker (November 23) :—

" Your note has pleased me much, for I did not expect that you would have had time to read *any* of it. Read the last chapter, and you will know the whole result, but without the evidence. The case, however, of radicles bending after exposure for an hour to geotropism, with their tips (or brains) cut off is, I think worth your reading (bottom of p. 525) ; it astounded me. But I will bother you no more about my book. The sensitiveness of seedlings to light is marvellous."

To another friend, Mr. Thiselton Dyer, he wrote (November 28, 1880) :

" Very many thanks for your most kind note, but you think too highly of our work, not but what this is very

---

* *Charles Darwin, Nature* Series, p. 41.

pleasant. . . . Many of the Germans are very contemptuous about making out the use of organs ; but they may sneer the souls out of their bodies, and I for one shall think it the most interesting part of Natural History. Indeed you are greatly mistaken if you doubt for one moment on the very great value of your constant and most kind assistance to us."

The book was widely reviewed, and excited much interest among the general public. The following letter refers to a leading article in the *Times*, November 20, 1880 :

*C. D. to Mrs. Haliburton.** Down, November 22, 1880.

MY DEAR SARAH,—You see how audaciously I begin ; but I have always loved and shall ever love this name. Your letter has done more than please me, for its kindness has touched my heart. I often think of old days and of the delight of my visits to Woodhouse, and of the deep debt of gratitude which I owe to your father. It was very good of you to write. I had quite forgotten my old ambition about the Shrewsbury newspaper ;† but I remember the pride which I felt when I saw in a book about beetles the impressive words " captured by C. Darwin." Captured sounded so grand compared with caught. This seemed to me glory enough for any man ! I do not know in the least what made the *Times* glorify me, for it has sometimes pitched into me ferociously.

I should very much like to see you again, but you would find a visit here very dull, for we feel very old and have no amusement, and lead a solitary life. But we intend in a few weeks to spend a few days in London, and then if you have anything else to do in London, you would perhaps come and lunch with us.

Believe me, my dear Sarah,
Yours gratefully and affectionately.

The following letter was called forth by the publication of a volume devoted to the criticism of the *Power of Move-*

---

* Mrs. Haliburton was a daughter of my father's early friend, the late Mr. Owen, of Woodhouse.

† Mrs. Haliburton had reminded him of his saying as a boy that if Eddowes' newspaper ever alluded to him as " our deserving fellow-townsman," his ambition would be amply gratified.

ment in Plants by an accomplished botanist, Dr. Julius
Wiesner, Professor of Botany in the University of Vienna:

C. D. to Julius Wiesner.   Down, October 25th, 1881.

MY DEAR SIR,—I have now finished your book,* and
have understood the whole except a very few passages. In
the first place, let me thank you cordially for the manner in
which you have everywhere treated me. You have shown
how a man may differ from another in the most decided
manner, and yet express his difference with the most perfect
courtesy. Not a few English and German naturalists might
learn a useful lesson from your example; for the coarse language often used by scientific men towards each other does
no good, and only degrades science.

I have been profoundly interested by your book, and
some of your experiments are so beautiful, that I actually
felt pleasure while being vivisected. It would take up too
much space to discuss all the important topics in your book.
I fear that you have quite upset the interpretation which I
have given of the effects of cutting off the tips of horizon-
tally extended roots, and of those laterally exposed to moisture; but I cannot persuade myself that the horizontal position of lateral branches and roots is due simply to their
lessened power of growth. Nor when I think of my experiments with the cotyledons of *Phalaris*, can I give up the
belief of the transmission of some stimulus due to light from
the upper to the lower part. At p. 60 you have misunderstood my meaning, when you say that I believe that the
effects from light are transmitted to a part which is not
itself heliotropic. I never considered whether or not the
short part beneath the ground was heliotropic; but I believe
that with young seedlings the part which bends *near*, but
above the ground is heliotropic, and I believe so from this
part bending only moderately when the light is oblique, and
bending rectangularly when the light is horizontal. Nevertheless the bending of this lower part, as I conclude from
my experiments with opaque caps, is influenced by the action of light on the upper part. My opinion, however, on
the above and many other points, signifies very little, for I
have no doubt that your book will convince most botanists
that I am wrong in all the points on which we differ.

---

* *Das Bewegungsvermögen der Pflanzen.* Vienna, 1881.

Independently of the question of transmission, my mind is so full of facts leading me to believe that light, gravity, &c., act not in a direct manner on growth, but as stimuli, that I am quite unable to modify my judgment on this head. I could not understand the passage at p. 78, until I consulted my son George, who is a mathematician. He supposes that your objection is founded on the diffused light from the lamp illuminating both sides of the object, and not being reduced, with increasing distance in the same ratio as the direct light; but he doubts whether this *necessary* correction will account for the very little difference in the heliotropic curvature of the plants in the successive pots.

With respect to the sensitiveness of the tips of roots to contact, I cannot admit your view until it is proved that I am in error about bits of card attached by liquid gum causing movement; whereas no movement was caused if the card remained separated from the tip by a layer of the liquid gum. The fact also of thicker and thinner bits of card attached on opposite sides of the same root by shellac, causing movement in one direction, has to be explained. You often speak of the tip having been injured ; but externally there was no sign of injury : and when the tip was plainly injured, the extreme part became curved *towards* the injured side. I can no more believe that the tip was injured by the bits of card, at least when attached by gum-water, than that the glands of Drosera are injured by a particle of thread or hair placed on it, or that the human tongue is so when it feels any such object.

About the most important subject in my book, namely, circumnutation, I can only say that I feel utterly bewildered at the difference in our conclusions; but I could not fully understand some parts which my son Francis will be able to translate to me when he returns home. The greater part of your book is beautifully clear.

Finally, I wish that I had enough strength and spirit to commence a fresh set of experiments, and publish the results, with a full recantation of my errors when convinced of them ; but I am too old for such an undertaking, nor do I suppose that I shall be able to do much, or any more, original work. I imagine that I see one possible source of error in your beautiful experiment of a plant rotating and exposed to a lateral light.

With high respect, and with sincere thanks for the kind

manner in which you have treated me and my mistakes, I
remain,

My dear Sir, yours sincerely.

Insectivorous Plants.

In the summer of 1860 he was staying at the house of
his sister-in-law, Miss Wedgwood, in Ashdown Forest,
whence he wrote (July 29, 1860) to Sir Joseph Hooker :—

"Latterly I have done nothing here; but at first I
amused myself with a few observations on the insect-catch-
ing power of Drosera : * and I must consult you some time
whether my 'twaddle' is worth communicating to the Linnean Society."

In August he wrote to the same friend :—

"I will gratefully send my notes on Drosera when copied
by my copier : the subject amused me when I had nothing
to do."

He has described in the *Autobiography* (p. 47), the general nature of these early experiments. He noticed insects
sticking to the leaves, and finding that flies, &c., placed on
the adhesive glands, were held fast and embraced, he sus-
pected that the captured prey was digested and absorbed by
the leaves. He therefore tried the effect on the leaves of various nitrogenous fluids—with results which, as far as they
went, verified his surmise. In September, 1860, he wrote
to Dr. Gray :—

"I have been infinitely amused by working at Drosera :
the movements are really curious; and the manner in which
the leaves detect certain nitrogenous compounds is marvellous. You will laugh; but it is, at present, my full belief
(after endless experiments) that they detect (and move in
consequence of) the $\frac{1}{2880}$ part of a single grain of nitrate of
ammonia; but the muriate and sulphate of ammonia bother
their chemical skill, and they cannot make anything of the
nitrogen in these salts!"

Later in the autumn he was again obliged to leave home
for Eastbourne, where he continued his work on Drosera.

On his return home he wrote to Lyell (November
1860) :—

"I will and must finish my Drosera MS., which will
take me a week, for, at the present moment, I care more

---

* The common sun-dew.

about Drosera than the origin of all the species in the world. But I will not publish on Drosera till next year, for I am frightened and astounded at my results. I declare it is a certain fact that one organ is so sensitive to touch, that a weight seventy-eight times less than that, viz., $\frac{1}{1000}$ of a grain, which will move the best chemical balance, suffices to cause a conspicuous movement. Is it not curious that a plant should be far more sensitive to the touch than any nerve in the human body? Yet I am perfectly sure that this is true. When I am on my hobby-horse, I never can resist telling my friends how well my hobby goes, so you must forgive the rider."

The work was continued, as a holiday task, at Bournemouth, where he stayed during the autumn of 1862.

A long break now ensued in his work on insectivorous plants, and it was not till 1872 that the subject seriously occupied him again. A passage in a letter to Dr. Asa Gray, written in 1863 or 1864, shows, however, that the question was not altogether absent from his mind in the interim :—

"Depend on it you are unjust on the merits of my beloved Drosera; it is a wonderful plant, or rather a most sagacious animal. I will stick up for Drosera to the day of my death. Heaven knows whether I shall ever publish my pile of experiments on it."

He notes in his diary that the last proof of the *Expression of the Emotions* was finished on August 22, 1872, and that he began to work on Drosera on the following day.

C. D. to Asa Gray.   [Sevenoaks], October 22 [1872].

. . . I have worked pretty hard for four or five weeks on Drosera, and then broke down ; so that we took a house near Sevenoaks for three weeks (where I now am) to get complete rest. I have very little power of working now, and must put off the rest of the work on Drosera till next spring, as my plants are dying. It is an endless subject, and I must cut it short, and for this reason shall not do much on Dionæa. The point which has interested me most is tracing the *nerves !* which follow the vascular bundles. By a prick with a sharp lancet at a certain point, I can paralyse one-half the leaf, so that a stimulus to the other half causes no movement. It is just like dividing the spinal marrow of a frog :—no stimulus can be sent from the brain or anterior part of the spine to the hind legs : but if these latter are

stimulated, they move by reflex action. I find my old results about the astonishing sensitiveness of the nervous system (!?) of Drosera to various stimulants fully confirmed
and extended. . . .

C. D. to Asa Gray. Down, June 3 [1874].

. . . I am now hard at work getting my book on Drosera
& Co. ready for the printers, but it will take some time, for I
am always finding out new points to observe. I think you
will be interested by my observations on the digestive process in Drosera ; the secretion contains an acid of the acetic
series, and some ferment closely analogous to, but not iden-
tical with, pepsine ; for I have been making a long series of
comparative trials. No human being will believe what I
shall publish about the smallness of the doses of phosphate
of ammonia which act. . . .

The manuscript of *Insectivorous Plants* was finished in
March 1875. He seems to have been more than usually op-
pressed by the writing of this book, thus he wrote to Sir J.
D. Hooker in February :—

" You ask about my book, and all that I can say is that
I am ready to commit suicide ; I thought it was decently
written, but find so much wants rewriting, that it will not
be ready to go to printers for two months, and then will
make a confoundedly big book. Murray will say that it is
no use publishing in the middle of summer, so I do not
know what will be the upshot ; but I begin to think that
every one who publishes a book is a fool."

The book was published on July 2nd, 1875, and 2700
copies were sold out of the edition of 3000.

The Kew Index of Plant-Names.

Some account of my father's connection with the *Index
of Plant-Names*, now (1892) being printed by the Claren-
don Press, will be found in Mr. B. Daydon Jackson's paper
in the *Journal of Botany*, 1887, p. 151. Mr. Jackson quotes
the following statement by Sir J. D. Hooker :—

" Shortly before his death, Mr. Charles Darwin informed
Sir Joseph Hooker that it was his intention to devote a considerable sum of money annually for some years in aid or
furtherance of some work or works of practical utility to
biological science, and to make provisions in his will in

the event of these not being completed during his lifetime.

" Amongst other objects connected with botanical science, Mr. Darwin regarded with especial interest the importance of a complete index to the names and authors of the
genera and species of plants known to botanists, together
with their native countries. Steudel's *Nomenclator* is the
only existing work of this nature, and although now nearly
half a century old, Mr. Darwin had found it of great aid in
his own researches. It has been indispensable to every botanical institution, whether as a list of all known flowering
plants, as an indication of their authors, or as a digest of
botanical geography."

Since 1840, when the *Nomenclator* was published, the
number of described plants may be said to have doubled, so
that Steudel is now seriously below the requirements of botanical work. To remedy this want, the *Nomenclator* has
been from time to time posted up in an interleaved copy in
the Herbarium at Kew, by the help of " funds supplied by
private liberality." *

My father, like other botanists, had, as Sir Joseph Hooker
points out, experienced the value of Steudel's work. He
obtained plants from all sorts of sources, which were often
incorrectly named, and he felt the necessity of adhering to
the accepted nomenclature, so that he might convey to other
workers precise indications as to the plants which he had
studied. It was also frequently a matter of importance to
him to know the native country of his experimental plants.
Thus it was natural that he should recognise the desirability
of completing and publishing the interleaved volume at
Kew. The wish to help in this object was heightened by
the admiration he felt for the results for which the world
has to thank the Royal Gardens at Kew, and by his gratitude for the invaluable aid which for so many years he received from its Director and his staff. He expressly stated
that it was his wish " to aid in some way the scientific work
carried on at the Royal Gardens " †—which induced him to
offer to supply funds for the completion of the Kew *Nomenclator*.

The following passage, for which I am indebted to Professor Judd, is of interest, as illustrating the motives that

---

* *Kew Gardens Report*, 1881, p 62.
† See *Nature*, January 5, 1882.

actuated my father in this matter. Professor Judd
writes :—

"On the occasion of my last visit to him, he told me
that his income having recently greatly increased, while his
wants remained the same, he was most anxious to devote
what he could spare to the advancement of Geology or
Biology. He dwelt in the most touching manner on the
fact that he owed so much happiness and fame to the natural history sciences, which had been the solace of what
might have been a painful existence ;—and he begged me,
if I knew if any research which could be aided by a grant
of a few hundreds of pounds, to let him know, as it would be
a delight to him to feel that he was helping in promoting the
progress of science. He informed me at the same time
that he was making the same suggestion to Sir Joseph
Hooker and Professor Huxley with respect to Botany and
Zoology respectively. I was much impressed by the earnestness, and, indeed, deep emotion, with which he spoke of his
indebtedness to Science, and his desire to promote its interests."

The plan of the proposed work having been carefully
considered, Sir Joseph Hooker was able to confide its elaboration in detail to Mr. B. Daydon Jackson, Secretary of
the Linnean Society, whose extensive knowledge of botanical literature qualifies him for the task. My father's original idea of producing a modern edition of Steudel's *Nomenclator* has been practically abandoned, the aim now kept in
view is rather to construct a list of genera and species (with
references) founded on Bentham and Hooker's *Genera
Plantarum.* Under Sir Joseph Hooker's supervision, the
work, carried out with admirable zeal by Mr. Jackson, goes
steadily forward. The colossal nature of the undertaking
may be estimated by the fact that the manuscript of the
Index is at the present time (1892) believed to weigh more
than a ton.

The Kew 'Index,' will be a fitting memorial of my
father : and his share in its completion illustrates a part of
his character—his ready sympathy with work outside his
own lines of investigation—and his respect for minute and
patient labour in all branches of science.

CHAPTER XVIII.

CONCLUSION.

SOME idea of the general course of my father's health may have been gathered from the letters given in the preceding pages. The subject of health appears more prominently than is often necessary in a Biography, because it was, unfortunately, so real an element in determining the outward form of his life.

My father was at one time in the hands of Dr. Bence Jones, from whose treatment he certainly derived benefit. In later years he became a patient of Sir Andrew Clark, under whose care he improved greatly in general health. It was not only for his generously rendered service that my father felt a debt of gratitude towards Sir Andrew Clark. He owed to his cheering personal influence an often-repeated encouragement, which latterly added something real to his happiness, and he found sincere pleasure in Sir Andrew's friendship and kindness towards himself and his children. During the last ten years of his life the state of his health was a cause of satisfaction and hope to his family. His condition showed signs of amendment in several particulars. He suffered less distress and discomfort, and was able to work more steadily.

Scattered through his letters are one or two references to pain or uneasiness felt in the region of the heart. How far these indicate that the heart was affected early in life, I cannot pretend to say; in any case it is certain that he had no serious or permanent trouble of this nature until shortly before his death. In spite of the general improvement in his health, which has been above alluded to, there was a certain loss of physical vigour occasionally apparent during the last few years of his life. This is illustrated by a sentence in a letter to his old friend Sir James Sulivan, written on January 10, 1879: "My scientific work tires me more than it used to do, but I have nothing else to do, and whether one is worn out a year or two sooner or later signifies but little."

A similar feeling is shown in a letter to Sir J. D. Hooker of June 15, 1881. My father was staying at Patterdale, and wrote : " I am rather despondent about myself. . . . I have not the heart or strength to begin any investigation lasting years, which is the only thing I enjoy, and I have no little jobs which I can do."

In July, 1881, he wrote to Mr. Wallace : " We have just returned home after spending five weeks on Ullswater ; the scenery is quite charming, but I cannot walk, and everything tires me, even seeing scenery. . . . What I shall do with my few remaining years of life I can hardly tell. I have everything to make me happy and contented, but life has become very wearisome to me." He was, however, able to do a good deal of work, and that of a trying sort,* during the autumn of 1881, but towards the end of the year, he was clearly in need of rest : and during the winter was in a lower condition than was usual with him.

On December 13, he went for a week to his daughter's house in Bryanston Street. During his stay in London he went to call on Mr. Romanes, and was seized when on the door-step with an attack apparently of the same kind as those which afterwards became so frequent. The rest of the incident, which I give in Mr. Romanes' words, is interesting too from a different point of view, as giving one more illustration of my father's scrupulous consideration for others :—

" I happened to be out, but my butler, observing that Mr. Darwin was ill, asked him to come in. He said he would prefer going home, and although the butler urged him to wait at least until a cab could be fetched, he said he would rather not give so much trouble. For the same reason he refused to allow the butler to accompany him. Accordingly he watched him walking with difficulty towards the direction in which cabs were to be met with, and saw that, when he had got about three hundred yards from the house, he staggered and caught hold of the park-railings as if to prevent himself from falling. The butler therefore hastened to his assistance, but after a few seconds saw him turn round with the evident purpose of retracing

---

* On the action of carbonate of ammonia on roots and leaves.

his steps to my house. However, after he had returned part of the way he seems to have felt better, for he again changed his mind, and proceeded to find a cab."

During the last week of February and in the beginning of March, attacks of pain in the region of the heart, with irregularity of the pulse, became frequent, coming on indeed nearly every afternoon. A seizure of this sort occurred about March 7, when he was walking alone at a short distance from the house; he got home with difficulty, and this was the last time that he was able to reach his favourite 'Sand-walk.' Shortly after this, his illness became obviously more serious and alarming, and he was seen by Sir Andrew Clark, whose treatment was continued by Dr. Norman Moore, of St. Bartholomew's Hospital, and Dr. Allfrey, at that time in practice at St. Mary Cray. He suffered from distressing sensations of exhaustion and faintness, and seemed to recognise with deep depression the fact that his working days were over. He gradually recovered from this condition, and became more cheerful and hopeful, as is shown in the following letter to Mr. Huxley, who was anxious that my father should have closer medical supervision than the existing arrangements allowed :—

Down, March 27, 1882.

MY DEAR HUXLEY,—Your most kind letter has been a real cordial to me. I have felt better to-day than for three weeks, and have felt as yet no pain. Your plan seems an excellent one, and I will probably act upon it, unless I get very much better. Dr. Clark's kindness is unbounded to me, but he is too busy to come here. Once again, accept my cordial thanks, my dear old friend. I wish to God there were more automata* in the world like you.

Ever yours,
CH. DARWIN.

The allusion to Sir Andrew Clark requires a word of explanation. Sir Andrew himself was ever ready to devote himself to my father, who however, could not endure the thought of sending for him, knowing how severely his great practice taxed his strength.

---

* The allusion is to Mr. Huxley's address, " On the hypothesis that animals are automata, and its history," given at the Belfast Meeting of the British Association, 1874, and republished in *Science and Culture*.

No especial change occurred during the beginning of April, but on Saturday 15th he was seized with giddiness while sitting at dinner in the evening, and fainted in an attempt to reach his sofa. On the 17th he was again better, and in my temporary absence recorded for me the progress of an experiment in which I was engaged. During the night of April 18th, about a quarter to twelve, he had a severe attack and passed into a faint, from which he was brought back to consciousness with great difficulty. He seemed to recognise the approach of death, and said, "I am not the least afraid to die." All the next morning he suffered from terrible nausea and faintness, and hardly rallied before the end came.

He died at about four o'clock on Wednesday, April 19th, 1882, in the 74th year of his age.

I close the record of my father's life with a few words of retrospect added to the manuscript of his *Autobiography* in 1879 :—

"As for myself, I believe that I have acted rightly in steadily following and devoting my life to Science. I feel no remorse from having committed any great sin, but have often and often regretted that I have not done more direct good to my fellow creatures."

APPENDIX I.

ON the Friday succeeding my father's death, the following letter, signed by twenty Members of Parliament, was addressed to Dr. Bradley, Dean of Westminster:—

HOUSE OF COMMONS, April 21, 1882.

VERY REV. SIR,—We hope you will not think we are taking a liberty if we venture to suggest that it would be acceptable to a very large number of our fellow-countryman of all classes and opinions that our illustrious countryman, Mr. Darwin, should be buried in Westminster Abbey.

We remain, your obedient servants,

JOHN LUBBOCK,	RICHARD B. MARTIN,
NEVIL STOREY MASKELYNE,	FRANCIS W. BUXTON,
A. J. MUNDELLA,	E. L. STANLEY,
G. O. TREVELYAN,	HENRY BROADHURST,
LYON PLAYFAIR,	JOHN BARRAN,
CHARLES W. DILKE,	J. F. CHEETHAM,
DAVID WEDDERBURN,	H. S. HOLLAND,
ARTHUR RUSSELL,	H. CAMPBELL-BANNERMAN,
HORACE DAVEY,	CHARLES BRUCE,
BENJAMIN ARMITAGE,	RICHARD FORT.

The Dean was abroad at the time, and telegraphed his cordial acquiescence:—

The family had desired that my father should be buried at Down: with regard to their wishes, Sir John Lubbock wrote:—

HOUSE OF COMMONS, April 25, 1882.

MY DEAR DARWIN,—I quite sympathise with your feeling, and personally I should greatly have preferred that your father should have rested in Down amongst us all. It is, I am sure, quite understood that the initiative was not taken by you. Still, from a national point of view, it is clearly right that he should be buried in the Abbey. I esteem it a great privilege to be allowed to accompany my dear master to the grave.

Believe me, yours most sincerely,

JOHN LUBBOCK.

W. E. DARWIN, ESQ.

The family gave up their first-formed plans, and the funeral took place in Westminster Abbey on April 26th. The pall-bearers were:—

SIR JOHN LUBBOCK,	CANON FARRAR,
MR. HUXLEY,	SIR JOSEPH HOOKER,
MR. JAMES RUSSELL LOWELL (American Minister),	MR. WILLIAM SPOTTISWOODE (President of the Royal Society),
MR. A. R. WALLACE,	The EARL OF DERBY,
The DUKE OF DEVONSHIRE,	The DUKE OF ARGYLL.

The funeral was attended by the representatives of France, Germany, Italy, Spain, Russia, and by those of the Universities and learned Societies, as well as by large numbers of personal friends and distinguished men.

The grave is in the north aisle of the Nave, close to the angle of the choir-screen, and a few feet from the grave of Sir Isaac Newton. The stone bears the inscription—

CHARLES ROBERT DARWIN.

Born 12 February, 1809.

Died 19 April, 1882.

APPENDIX II.

PORTRAITS.

Date.	Description.	Artist.	In the Possession of
1838	Water-colour . .	G. Richmond.	The Family.
1851	Lithograph . . .	Ipswich British Assn. Series.	
1853	Chalk Drawing .	Samuel Lawrence.	The Family.
1853 ?	Chalk Drawing * .	Samuel Lawrence.	Professor Hughes, Cambridge.
1869	Bust, marble . .	T. Woolner, R.A.	The Family.
1875	Oil Painting † . .	W. Ouless, R.A.	The Family.
	Etched by . . .	P. Rajon.	
1879	Oil Painting . .	W. B. Richmond.	The University of Cambridge.
1881	Oil Painting ‡ . .	Hon. John Collier.	The Linnean Society.
	Etched by . . .	Leopold Flameng.	

CHIEF PORTRAITS AND MEMORIALS NOT TAKEN FROM LIFE.

	Statue #	Joseph Boehm, R.A.	Museum, South Kensington.
	Bust	Chr. Lehr, Junr.	
	Plaque	T. Woolner, R.A., and Josiah Wedgwood and Sons.	Christ's College, in Charles Darwin's Room.
	Deep Medallion .	J. Boehm, R.A.	In Westminster Abbey.

* Probably a sketch made at one of the sittings for the last-mentioned.
† A *replica* by the artist is in the possession of Christ's College, Cambridge.
‡ A *replica* by the artist is in the possession of W. E. Darwin, Esq., Southampton.
# A cast from this work is now placed in the New Museums at Cambridge.

CHIEF ENGRAVINGS FROM PHOTOGRAPHS.

*1854 ? By Messrs. Maull and Fox, engraved on wood for *Harper's Magazine* (Oct., 1884). Frontispiece, *Life and Letters,* vol. i.

1868 By the late Mrs. Cameron, reproduced in heliogravure by the Cambridge Engraving Company for the present work.

*1870 ? By O. J. Rejlander, engraved on Steel by C. H. Jeens for *Nature* (June 4, 1874).

*1874 ? By Major Darwin, engraved on wood for the *Century Magazine* (Jan., 1883). Frontispiece, *Life and Letters,* vol. ii.

1881 By Messrs. Elliot and Fry, engraved on wood by G. Kruells, for vol. iii. of the *Life and Letters.*

---

* The dates of these photographs must, from various causes, remain uncertain. Owing to a loss of books by fire, Messrs. Maull and Fox can give only an approximate date. Mr. Rejlander died some years ago, and his business was broken up. My brother, Major Darwin, has no record of the date at which his photograph was taken.

INDEX.

THE END.

A CATALOGUE OF
SELECTED DOVER BOOKS
IN ALL FIELDS OF INTEREST

A CATALOGUE OF SELECTED DOVER
BOOKS IN ALL FIELDS OF INTEREST

RACKHAM'S COLOR ILLUSTRATIONS FOR WAGNER'S RING. Rackham's finest mature work—all 64 full-color watercolors in a faithful and lush interpretation of the *Ring*. Full-sized plates on coated stock of the paintings used by opera companies for authentic staging of Wagner. Captions aid in following complete Ring cycle. Introduction. 64 illustrations plus vignettes. 72pp. 8⅝ x 11¼. 23779-6 Pa. $6.00

CONTEMPORARY POLISH POSTERS IN FULL COLOR, edited by Joseph Czestochowski. 46 full-color examples of brilliant school of Polish graphic design, selected from world's first museum (near Warsaw) dedicated to poster art. Posters on circuses, films, plays, concerts all show cosmopolitan influences, free imagination. Introduction. 48pp. 9⅜ x 12¼. 23780-X Pa. $6.00

GRAPHIC WORKS OF EDVARD MUNCH, Edvard Munch. 90 haunting, evocative prints by first major Expressionist artist and one of the greatest graphic artists of his time: *The Scream, Anxiety, Death Chamber, The Kiss, Madonna,* etc. Introduction by Alfred Werner. 90pp. 9 x 12. 23765-6 Pa. $5.00

THE GOLDEN AGE OF THE POSTER, Hayward and Blanche Cirker. 70 extraordinary posters in full colors, from Maitres de l'Affiche, Mucha, Lautrec, Bradley, Cheret, Beardsley, many others. Total of 78pp. 9⅜ x 12¼. 22753-7 Pa. $5.95

THE NOTEBOOKS OF LEONARDO DA VINCI, edited by J. P. Richter. Extracts from manuscripts reveal great genius; on painting, sculpture, anatomy, sciences, geography, etc. Both Italian and English. 186 ms. pages reproduced, plus 500 additional drawings, including studies for *Last Supper,* Sforza monument, etc. 860pp. 7⅞ x 10¾. (Available in U.S. only) 22572-0, 22573-9 Pa., Two-vol. set $15.90

THE CODEX NUTTALL, as first edited by Zelia Nuttall. Only inexpensive edition, in full color, of a pre-Columbian Mexican (Mixtec) book. 88 color plates show kings, gods, heroes, temples, sacrifices. New explanatory, historical introduction by Arthur G. Miller. 96pp. 11⅜ x 8½. (Available in U.S. only) 23168-2 Pa. $7.95

UNE SEMAINE DE BONTÉ, A SURREALISTIC NOVEL IN COLLAGE, Max Ernst. Masterpiece created out of 19th-century periodical illustrations, explores worlds of terror and surprise. Some consider this Ernst's greatest work. 208pp. 8⅛ x 11. 23252-2 Pa. $6.00

DRAWINGS OF WILLIAM BLAKE, William Blake. 92 plates from Book of Job, *Divine Comedy, Paradise Lost,* visionary heads, mythological figures, Laocoon, etc. Selection, introduction, commentary by Sir Geoffrey Keynes. 178pp. 8⅛ x 11.   22303-5 Pa. $4.00

ENGRAVINGS OF HOGARTH, William Hogarth. 101 of Hogarth's greatest works: *Rake's Progress, Harlot's Progress, Illustrations for Hudibras, Before and After, Beer Street and Gin Lane,* many more. Full commentary. 256pp. 11 x 13¾.   22479-1 Pa. $12.95

DAUMIER: 120 GREAT LITHOGRAPHS, Honore Daumier. Wide-ranging collection of lithographs by the greatest caricaturist of the 19th century. Concentrates on eternally popular series on lawyers, on married life, on liberated women, etc. Selection, introduction, and notes on plates by Charles F. Ramus. Total of 158pp. 9⅜ x 12¼.   23512-2 Pa. $6.00

DRAWINGS OF MUCHA, Alphonse Maria Mucha. Work reveals draftsman of highest caliber: studies for famous posters and paintings, renderings for book illustrations and ads, etc. 70 works, 9 in color; including 6 items not drawings. Introduction. List of illustrations. 72pp. 9⅜ x 12¼. (Available in U.S. only)   23672-2 Pa. $4.00

GIOVANNI BATTISTA PIRANESI: DRAWINGS IN THE PIERPONT MORGAN LIBRARY, Giovanni Battista Piranesi. For first time ever all of Morgan Library's collection, world's largest. 167 illustrations of rare Piranesi drawings—archeological, architectural, decorative and visionary. Essay, detailed list of drawings, chronology, captions. Edited by Felice Stampfle. 144pp. 9⅜ x 12¼.   23714-1 Pa. $7.50

NEW YORK ETCHINGS (1905-1949), John Sloan. All of important American artist's N.Y. life etchings. 67 works include some of his best art; also lively historical record—Greenwich Village, tenement scenes. Edited by Sloan's widow. Introduction and captions. 79pp. 8⅜ x 11¼.
23651-X Pa. $4.00

CHINESE PAINTING AND CALLIGRAPHY: A PICTORIAL SURVEY, Wan-go Weng. 69 fine examples from John M. Crawford's matchless private collection: landscapes, birds, flowers, human figures, etc., plus calligraphy. Every basic form included: hanging scrolls, handscrolls, album leaves, fans, etc. 109 illustrations. Introduction. Captions. 192pp. 8⅞ x 11¾.
23707-9 Pa. $7.95

DRAWINGS OF REMBRANDT, edited by Seymour Slive. Updated Lippmann, Hofstede de Groot edition, with definitive scholarly apparatus. All portraits, biblical sketches, landscapes, nudes, Oriental figures, classical studies, together with selection of work by followers. 550 illustrations. Total of 630pp. 9⅛ x 12¼.   21485-0, 21486-9 Pa., Two-vol. set $15.00

THE DISASTERS OF WAR, Francisco Goya. 83 etchings record horrors of Napoleonic wars in Spain and war in general. Reprint of 1st edition, plus 3 additional plates. Introduction by Philip Hofer. 97pp. 9⅜ x 8¼.
21872-4 Pa. $4.00

THE EARLY WORK OF AUBREY BEARDSLEY, Aubrey Beardsley. 157 plates, 2 in color: *Manon Lescaut, Madame Bovary, Morte Darthur, Salome,* other. Introduction by H. Marillier. 182pp. 8⅛ x 11. 21816-3 Pa. $4.50

THE LATER WORK OF AUBREY BEARDSLEY, Aubrey Beardsley. Exotic masterpieces of full maturity: *Venus and Tannhauser, Lysistrata, Rape of the Lock, Volpone,* Savoy material, etc. 174 plates, 2 in color. 186pp. 8⅛ x 11. 21817-1 Pa. $5.95

THOMAS NAST'S CHRISTMAS DRAWINGS, Thomas Nast. Almost all Christmas drawings by creator of image of Santa Claus as we know it, and one of America's foremost illustrators and political cartoonists. 66 illustrations. 3 illustrations in color on covers. 96pp. 8⅜ x 11¼. 23660-9 Pa. $3.50

THE DORÉ ILLUSTRATIONS FOR DANTE'S DIVINE COMEDY, Gustave Doré. All 135 plates from Inferno, Purgatory, Paradise; fantastic tortures, infernal landscapes, celestial wonders. Each plate with appropriate (translated) verses. 141pp. 9 x 12. 23231-X Pa. $4.50

DORÉ'S ILLUSTRATIONS FOR RABELAIS, Gustave Doré. 252 striking illustrations of *Gargantua and Pantagruel* books by foremost 19th-century illustrator. Including 60 plates, 192 delightful smaller illustrations. 153pp. 9 x 12. 23656-0 Pa. $5.00

LONDON: A PILGRIMAGE, Gustave Doré, Blanchard Jerrold. Squalor, riches, misery, beauty of mid-Victorian metropolis; 55 wonderful plates, 125 other illustrations, full social, cultural text by Jerrold. 191pp. of text. 9⅜ x 12¼. 22306-X Pa. $7.00

THE RIME OF THE ANCIENT MARINER, Gustave Doré, S. T. Coleridge. Dore's finest work, 34 plates capture moods, subtleties of poem. Full text. Introduction by Millicent Rose. 77pp. 9¼ x 12. 22305-1 Pa. $3.50

THE DORE BIBLE ILLUSTRATIONS, Gustave Doré. All wonderful, detailed plates: Adam and Eve, Flood, Babylon, Life of Jesus, etc. Brief King James text with each plate. Introduction by Millicent Rose. 241 plates. 241pp. 9 x 12. 23004-X Pa. $6.00

THE COMPLETE ENGRAVINGS, ETCHINGS AND DRYPOINTS OF ALBRECHT DURER. "Knight, Death and Devil"; "Melencolia," and more—all Dürer's known works in all three media, including 6 works formerly attributed to him. 120 plates. 235pp. 8⅜ x 11¼. 22851-7 Pa. $6.50

MECHANICK EXERCISES ON THE WHOLE ART OF PRINTING, Joseph Moxon. First complete book (1683-4) ever written about typography, a compendium of everything known about printing at the latter part of 17th century. Reprint of 2nd (1962) Oxford Univ. Press edition. 74 illustrations. Total of 550pp. 6⅛ x 9¼. 23617-X Pa. $7.95

THE COMPLETE WOODCUTS OF ALBRECHT DURER, edited by Dr. W. Kurth. 346 in all: "Old Testament," "St. Jerome," "Passion," "Life of Virgin," Apocalypse," many others. Introduction by Campbell Dodgson. 285pp. 8½ x 12¼. 21097-9 Pa. $7.50

DRAWINGS OF ALBRECHT DURER, edited by Heinrich Wolfflin. 81 plates show development from youth to full style. Many favorites; many new. Introduction by Alfred Werner. 96pp. 8⅛ x 11. 22352-3 Pa. $5.00

THE HUMAN FIGURE, Albrecht Dürer. Experiments in various techniques—stereometric, progressive proportional, and others. Also life studies that rank among finest ever done. Complete reprinting of *Dresden Sketchbook.* 170 plates. 355pp. 8⅜ x 11¼. 21042-1 Pa. $7.95

OF THE JUST SHAPING OF LETTERS, Albrecht Dürer. Renaissance artist explains design of Roman majuscules by geometry, also Gothic lower and capitals. Grolier Club edition. 43pp. 7⅞ x 10¾ 21306-4 Pa. $3.00

TEN BOOKS ON ARCHITECTURE, Vitruvius. The most important book ever written on architecture. Early Roman aesthetics, technology, classical orders, site selection, all other aspects. Stands behind everything since. Morgan translation. 331pp. 5⅜ x 8½. 20645-9 Pa. $4.50

THE FOUR BOOKS OF ARCHITECTURE, Andrea Palladio. 16th-century classic responsible for Palladian movement and style. Covers classical architectural remains, Renaissance revivals, classical orders, etc. 1738 Ware English edition. Introduction by A. Placzek. 216 plates. 110pp. of text. 9½ x 12¾. 21308-0 Pa. $10.00

HORIZONS, Norman Bel Geddes. Great industrialist stage designer, "father of streamlining," on application of aesthetics to transportation, amusement, architecture, etc. 1932 prophetic account; function, theory, specific projects. 222 illustrations. 312pp. 7⅞ x 10¾. 23514-9 Pa. $6.95

FRANK LLOYD WRIGHT'S FALLINGWATER, Donald Hoffmann. Full, illustrated story of conception and building of Wright's masterwork at Bear Run, Pa. 100 photographs of site, construction, and details of completed structure. 112pp. 9¼ x 10. 23671-4 Pa. $5.50

THE ELEMENTS OF DRAWING, John Ruskin. Timeless classic by great Viltorian; starts with basic ideas, works through more difficult. Many practical exercises. 48 illustrations. Introduction by Lawrence Campbell. 228pp. 5⅜ x 8½. 22730-8 Pa. $3.75

GIST OF ART, John Sloan. Greatest modern American teacher, Art Students League, offers innumerable hints, instructions, guided comments to help you in painting. Not a formal course. 46 illustrations. Introduction by Helen Sloan. 200pp. 5⅜ x 8½. 23435-5 Pa. $4.00

THE ANATOMY OF THE HORSE, George Stubbs. Often considered the great masterpiece of animal anatomy. Full reproduction of 1766 edition, plus prospectus; original text and modernized text. 36 plates. Introduction by Eleanor Garvey. 121pp. 11 x 14¾. 23402-9 Pa. $6.00

BRIDGMAN'S LIFE DRAWING, George B. Bridgman. More than 500 illustrative drawings and text teach you to abstract the body into its major masses, use light and shade, proportion; as well as specific areas of anatomy, of which Bridgman is master. 192pp. 6½ x 9¼. (Available in U.S. only) 22710-3 Pa. $3.50

ART NOUVEAU DESIGNS IN COLOR, Alphonse Mucha, Maurice Verneuil, Georges Auriol. Full-color reproduction of *Combinaisons orne-mentales* (c. 1900) by Art Nouveau masters. Floral, animal, geometric, interlacings, swashes—borders, frames, spots—all incredibly beautiful. 60 plates, hundreds of designs. 9⅜ x 8-1/16. 22885-1 Pa. $4.00

FULL-COLOR FLORAL DESIGNS IN THE ART NOUVEAU STYLE, E. A. Seguy. 166 motifs, on 40 plates, from *Les fleurs et leurs applications decoratives* (1902): borders, circular designs, repeats, allovers, "spots." All in authentic Art Nouveau colors. 48pp. 9⅜ x 12¼.

23439-8 Pa. $5.00

A DIDEROT PICTORIAL ENCYCLOPEDIA OF TRADES AND IN-DUSTRY, edited by Charles C. Gillispie. 485 most interesting plates from the great French Encyclopedia of the 18th century show hundreds of working figures, artifacts, process, land and cityscapes; glassmaking, paper-making, metal extraction, construction, weaving, making furniture, clothing, wigs, dozens of other activities. Plates fully explained. 920pp. 9 x 12. 22284-5, 22285-3 Clothbd., Two-vol. set $40.00

HANDBOOK OF EARLY ADVERTISING ART, Clarence P. Hornung. Largest collection of copyright-free early and antique advertising art ever compiled. Over 6,000 illustrations, from Franklin's time to the 1890's for special effects, novelty. Valuable source, almost inexhaustible.
Pictorial Volume. Agriculture, the zodiac, animals, autos, birds, Christmas, fire engines, flowers, trees, musical instruments, ships, games and sports, much more. Arranged by subject matter and use. 237 plates. 288pp. 9 x 12. 20122-8 Clothbd. $14.50

Typographical Volume. Roman and Gothic faces ranging from 10 point to 300 point, "Barnum," German and Old English faces, script, logotypes, scrolls and flourishes, 1115 ornamental initials, 67 complete alphabets, more. 310 plates. 320pp. 9 x 12. 20123-6 Clothbd. $15.00

CALLIGRAPHY (CALLIGRAPHIA LATINA), J. G. Schwandner. High point of 18th-century ornamental calligraphy. Very ornate initials, scrolls, borders, cherubs, birds, lettered examples. 172pp. 9 x 13. 20475-8 Pa. $7.00

ART FORMS IN NATURE, Ernst Haeckel. Multitude of strangely beautiful natural forms: Radiolaria, Foraminifera, jellyfishes, fungi, turtles, bats, etc. All 100 plates of the 19th-century evolutionist's *Kunstformen der Natur* (1904). 100pp. 9⅜ x 12¼. 22987-4 Pa. $5.00

CHILDREN: A PICTORIAL ARCHIVE FROM NINETEENTH-CENTURY SOURCES, edited by Carol Belanger Grafton. 242 rare, copyright-free wood engravings for artists and designers. Widest such selection available. All illustrations in line. 119pp. 8⅜ x 11¼. 23694-3 Pa. $4.00

WOMEN: A PICTORIAL ARCHIVE FROM NINETEENTH-CENTURY SOURCES, edited by Jim Harter. 391 copyright-free wood engravings for artists and designers selected from rare periodicals. Most extensive such collection available. All illustrations in line. 128pp. 9 x 12. 23703-6 Pa. $4.50

ARABIC ART IN COLOR, Prisse d'Avennes. From the greatest ornamentalists of all time—50 plates in color, rarely seen outside the Near East, rich in suggestion and stimulus. Includes 4 plates on covers. 46pp. 9⅜ x 12¼. 23658-7 Pa. $6.00

AUTHENTIC ALGERIAN CARPET DESIGNS AND MOTIFS, edited by June Beveridge. Algerian carpets are world famous. Dozens of geometrical motifs are charted on grids, color-coded, for weavers, needleworkers, craftsmen, designers. 53 illustrations plus 4 in color. 48pp. 8¼ x 11. (Available in U.S. only) 23650-1 Pa. $1.75

DICTIONARY OF AMERICAN PORTRAITS, edited by Hayward and Blanche Cirker. 4000 important Americans, earliest times to 1905, mostly in clear line. Politicians, writers, soldiers, scientists, inventors, industrialists, Indians, Blacks, women, outlaws, etc. Identificatory information. 756pp. 9¼ x 12¾. 21823-6 Clothbd. $40.00

HOW THE OTHER HALF LIVES, Jacob A. Riis. Journalistic record of filth, degradation, upward drive in New York immigrant slums, shops, around 1900. New edition includes 100 original Riis photos, monuments of early photography. 233pp. 10 x 7⅞. 22012-5 Pa. $7.00

NEW YORK IN THE THIRTIES, Berenice Abbott. Noted photographer's fascinating study of city shows new buildings that have become famous and old sights that have disappeared forever. Insightful commentary. 97 photographs. 97pp. 11⅜ x 10. 22967-X Pa. $5.00

MEN AT WORK, Lewis W. Hine. Famous photographic studies of construction workers, railroad men, factory workers and coal miners. New supplement of 18 photos on Empire State building construction. New introduction by Jonathan L. Doherty. Total of 69 photos. 63pp. 8 x 10¾. 23475-4 Pa. $3.00

THE DEPRESSION YEARS AS PHOTOGRAPHED BY ARTHUR ROTH-STEIN, Arthur Rothstein. First collection devoted entirely to the work of outstanding 1930s photographer: famous dust storm photo, ragged children, unemployed, etc. 120 photographs. Captions. 119pp. 9¼ x 10¾.
23590-4 Pa. $5.00

CAMERA WORK: A PICTORIAL GUIDE, Alfred Stieglitz. All 559 illustrations and plates from the most important periodical in the history of art photography, Camera Work (1903-17). Presented four to a page, reduced in size but still clear, in strict chronological order, with complete captions. Three indexes. Glossary. Bibliography. 176pp. 8⅜ x 11¼.
23591-2 Pa. $6.95

ALVIN LANGDON COBURN, PHOTOGRAPHER, Alvin L. Coburn. Revealing autobiography by one of greatest photographers of 20th century gives insider's version of Photo-Secession, plus comments on his own work. 77 photographs by Coburn. Edited by Helmut and Alison Gernsheim. 160pp. 8⅛ x 11.
23685-4 Pa. $6.00

NEW YORK IN THE FORTIES, Andreas Feininger. 162 brilliant photographs by the well-known photographer, formerly with Life magazine, show commuters, shoppers, Times Square at night, Harlem nightclub, Lower East Side, etc. Introduction and full captions by John von Hartz. 181pp. 9¼ x 10¾.
23585-8 Pa. $6.95

GREAT NEWS PHOTOS AND THE STORIES BEHIND THEM, John Faber. Dramatic volume of 140 great news photos, 1855 through 1976, and revealing stories behind them, with both historical and technical information. Hindenburg disaster, shooting of Oswald, nomination of Jimmy Carter, etc. 160pp. 8¼ x 11.
23667-6 Pa. $5.00

THE ART OF THE CINEMATOGRAPHER, Leonard Maltin. Survey of American cinematography history and anecdotal interviews with 5 masters—Arthur Miller, Hal Mohr, Hal Rosson, Lucien Ballard, and Conrad Hall. Very large selection of behind-the-scenes production photos. 105 photographs. Filmographies. Index. Originally Behind the Camera. 144pp. 8¼ x 11.
23686-2 Pa. $5.00

DESIGNS FOR THE THREE-CORNERED HAT (LE TRICORNE), Pablo Picasso. 32 fabulously rare drawings—including 31 color illustrations of costumes and accessories—for 1919 production of famous ballet. Edited by Parmenia Migel, who has written new introduction. 48pp. 9⅜ x 12¼. (Available in U.S. only)
23709-5 Pa. $5.00

NOTES OF A FILM DIRECTOR, Sergei Eisenstein. Greatest Russian filmmaker explains montage, making of Alexander Nevsky, aesthetics; comments on self, associates, great rivals (Chaplin), similar material. 78 illustrations. 240pp. 5⅜ x 8½.
22392-2 Pa. $4.50

HOLLYWOOD GLAMOUR PORTRAITS, edited by John Kobal. 145 photos capture the stars from 1926-49, the high point in portrait photography. Gable, Harlow, Bogart, Bacall, Hedy Lamarr, Marlene Dietrich, Robert Montgomery, Marlon Brando, Veronica Lake; 94 stars in all. Full background on photographers, technical aspects, much more. Total of 160pp. 8⅜ x 11¼.   23352-9 Pa. $6.00

THE NEW YORK STAGE: FAMOUS PRODUCTIONS IN PHOTO-GRAPHS, edited by Stanley Appelbaum. 148 photographs from Museum of City of New York show 142 plays, 1883-1939. *Peter Pan, The Front Page, Dead End, Our Town,* O'Neill, hundreds of actors and actresses, etc. Full indexes. 154pp. 9½ x 10.   23241-7 Pa. $6.00

DIALOGUES CONCERNING TWO NEW SCIENCES, Galileo Galilei. Encompassing 30 years of experiment and thought, these dialogues deal with geometric demonstrations of fracture of solid bodies, cohesion, leverage, speed of light and sound, pendulums, falling bodies, accelerated motion, etc. 300pp. 5⅜ x 8½.   60099-8 Pa. $4.00

THE GREAT OPERA STARS IN HISTORIC PHOTOGRAPHS, edited by James Camner. 343 portraits from the 1850s to the 1940s: Tamburini, Mario, Caliapin, Jeritza, Melchior, Melba, Patti, Pinza, Schipa, Caruso, Farrar, Steber, Gobbi, and many more—270 performers in all. Index. 199pp. 8⅜ x 11¼.   23575-0 Pa. $7.50

J. S. BACH, Albert Schweitzer. Great full-length study of Bach, life, background to music, music, by foremost modern scholar. Ernest Newman translation. 650 musical examples. Total of 928pp. 5⅜ x 8½. (Available in U.S. only)   21631-4, 21632-2 Pa., Two-vol. set $11.00

COMPLETE PIANO SONATAS, Ludwig van Beethoven. All sonatas in the fine Schenker edition, with fingering, analytical material. One of best modern editions. Total of 615pp. 9 x 12. (Available in U.S. only)
                              23134-8, 23135-6 Pa., Two-vol. set $15.50

KEYBOARD MUSIC, J. S. Bach. Bach-Gesellschaft edition. For harpsichord, piano, other keyboard instruments. English Suites, French Suites, Six Partitas, Goldberg Variations, Two-Part Inventions, Three-Part Sinfonias. 312pp. 8⅛ x 11. (Available in U.S. only)   22360-4 Pa. $6.95

FOUR SYMPHONIES IN FULL SCORE, Franz Schubert. Schubert's four most popular symphonies: No. 4 in C Minor ("Tragic"); No. 5 in B-flat Major; No. 8 in B Minor ("Unfinished"); No. 9 in C Major ("Great"). Breitkopf & Hartel edition. Study score. 261pp. 9⅜ x 12¼.
                                                    23681-1 Pa. $6.50

THE AUTHENTIC GILBERT & SULLIVAN SONGBOOK, W. S. Gilbert, A. S. Sullivan. Largest selection available; 92 songs, uncut, original keys, in piano rendering approved by Sullivan. Favorites and lesser-known fine numbers. Edited with plot synopses by James Spero. 3 illustrations. 399pp. 9 x 12.   23482-7 Pa. $9.95

PRINCIPLES OF ORCHESTRATION, Nikolay Rimsky-Korsakov. Great classical orchestrator provides fundamentals of tonal resonance, progression of parts, voice and orchestra, tutti effects, much else in major document. 330pp. of musical excerpts. 489pp. 6½ x 9¼.   21266-1 Pa. $7.50

TRISTAN UND ISOLDE, Richard Wagner. Full orchestral score with complete instrumentation. Do not confuse with piano reduction. Commentary by Felix Mottl, great Wagnerian conductor and scholar. Study score. 655pp. 8⅛ x 11.   22915-7 Pa. $13.95

REQUIEM IN FULL SCORE, Giuseppe Verdi. Immensely popular with choral groups and music lovers. Republication of edition published by C. F. Peters, Leipzig, n. d. German frontmaker in English translation. Glossary. Text in Latin. Study score. 204pp. 9⅜ x 12¼.
23682-X Pa. $6.00

COMPLETE CHAMBER MUSIC FOR STRINGS, Felix Mendelssohn. All of Mendelssohn's chamber music: Octet, 2 Quintets, 6 Quartets, and Four Pieces for String Quartet. (Nothing with piano is included). Complete works edition (1874-7). Study score. 283 pp. 9⅜ x 12¼.
23679-X Pa. $7.50

POPULAR SONGS OF NINETEENTH-CENTURY AMERICA, edited by Richard Jackson. 64 most important songs: "Old Oaken Bucket," "Arkansas Traveler," "Yellow Rose of Texas," etc. Authentic original sheet music, full introduction and commentaries. 290pp. 9 x 12.   23270-0 Pa. $7.95

COLLECTED PIANO WORKS, Scott Joplin. Edited by Vera Brodsky Lawrence. Practically all of Joplin's piano works—rags, two-steps, marches, waltzes, etc., 51 works in all. Extensive introduction by Rudi Blesh. Total of 345pp. 9 x 12.   23106-2 Pa. $14.95

BASIC PRINCIPLES OF CLASSICAL BALLET, Agrippina Vaganova. Great Russian theoretician, teacher explains methods for teaching classical ballet; incorporates best from French, Italian, Russian schools. 118 illustrations. 175pp. 5⅜ x 8½.   22036-2 Pa. $2.50

CHINESE CHARACTERS, L. Wieger. Rich analysis of 2300 characters according to traditional systems into primitives. Historical-semantic analysis to phonetics (Classical Mandarin) and radicals. 820pp. 6⅛ x 9¼.
21321-8 Pa. $10.00

EGYPTIAN LANGUAGE: EASY LESSONS IN EGYPTIAN HIERO-GLYPHICS, E. A. Wallis Budge. Foremost Egyptologist offers Egyptian grammar, explanation of hieroglyphics, many reading texts, dictionary of symbols. 246pp. 5 x 7½. (Available in U.S. only)
21394-3 Clothbd. $7.50

AN ETYMOLOGICAL DICTIONARY OF MODERN ENGLISH, Ernest Weekley. Richest, fullest work, by foremost British lexicographer. Detailed word histories. Inexhaustible. Do not confuse this with *Concise Etymological Dictionary,* which is abridged. Total of 856pp. 6½ x 9¼.
21873-2, 21874-0 Pa., Two-vol. set $12.00

A MAYA GRAMMAR, Alfred M. Tozzer. Practical, useful English-language grammar by the Harvard anthropologist who was one of the three greatest American scholars in the area of Maya culture. Phonetics, grammatical processes, syntax, more. 301pp. 5⅜ x 8½.　　　23465-7 Pa. $4.00

THE JOURNAL OF HENRY D. THOREAU, edited by Bradford Torrey, F. H. Allen. Complete reprinting of 14 volumes, 1837-61, over two million words; the sourcebooks for *Walden*, etc. Definitive. All original sketches, plus 75 photographs. Introduction by Walter Harding. Total of 1804pp. 8½ x 12¼.　　　20312-3, 20313-1 Clothbd., Two-vol. set $70.00

CLASSIC GHOST STORIES, Charles Dickens and others. 18 wonderful stories you've wanted to reread: "The Monkey's Paw," "The House and the Brain," "The Upper Berth," "The Signalman," "Dracula's Guest," "The Tapestried Chamber," etc. Dickens, Scott, Mary Shelley, Stoker, etc. 330pp. 5⅜ x 8½.　　　20735-8 Pa. $4.50

SEVEN SCIENCE FICTION NOVELS, H. G. Wells. Full novels. *First Men in the Moon, Island of Dr. Moreau, War of the Worlds, Food of the Gods, Invisible Man, Time Machine, In the Days of the Comet*. A basic science-fiction library. 1015pp. 5⅜ x 8½. (Available in U.S. only)
　　　20264-X Clothbd. $8.95

ARMADALE, Wilkie Collins. Third great mystery novel by the author of *The Woman in White* and *The Moonstone*. Ingeniously plotted narrative shows an exceptional command of character, incident and mood. Original magazine version with 40 illustrations. 597pp. 5⅜ x 8½.
　　　23429-0 Pa. $6.00

MASTERS OF MYSTERY, H. Douglas Thomson. The first book in English (1931) devoted to history and aesthetics of detective story. Poe, Doyle, LeFanu, Dickens, many others, up to 1930. New introduction and notes by E. F. Bleiler. 288pp. 5⅜ x 8½. (Available in U.S. only)
　　　23606-4 Pa. $4.00

FLATLAND, E. A. Abbott. Science-fiction classic explores life of 2-D being in 3-D world. Read also as introduction to thought about hyperspace. Introduction by Banesh Hoffmann. 16 illustrations. 103pp. 5⅜ x 8½.
　　　20001-9 Pa. $2.00

THREE SUPERNATURAL NOVELS OF THE VICTORIAN PERIOD, edited, with an introduction, by E. F. Bleiler. Reprinted complete and unabridged, three great classics of the supernatural: *The Haunted Hotel* by Wilkie Collins, *The Haunted House at Latchford* by Mrs. J. H. Riddell, and *The Lost Stradivarious* by J. Meade Falkner. 325pp. 5⅜ x 8½.
　　　22571-2 Pa. $4.00

AYESHA: THE RETURN OF "SHE," H. Rider Haggard. Virtuoso sequel featuring the great mythic creation, Ayesha, in an adventure that is fully as good as the first book, *She*. Original magazine version, with 47 original illustrations by Maurice Greiffenhagen. 189pp. 6½ x 9¼.
　　　23649-8 Pa. $3.50

UNCLE SILAS, J. Sheridan LeFanu. Victorian Gothic mystery novel, considered by many best of period, even better than Collins or Dickens. Wonderful psychological terror. Introduction by Frederick Shroyer. 436pp. 5⅜ x 8½. 21715-9 Pa. $6.00

JURGEN, James Branch Cabell. The great erotic fantasy of the 1920's that delighted thousands, shocked thousands more. Full final text, Lane edition with 13 plates by Frank Pape. 346pp. 5⅜ x 8½. 23507-6 Pa. $4.50

THE CLAVERINGS, Anthony Trollope. Major novel, chronicling aspects of British Victorian society, personalities. Reprint of Cornhill serialization, 16 plates by M. Edwards; first reprint of full text. Introduction by Norman Donaldson. 412pp. 5⅜ x 8½. 23464-9 Pa. $5.00

KEPT IN THE DARK, Anthony Trollope. Unusual short novel about Victorian morality and abnormal psychology by the great English author. Probably the first American publication. Frontispiece by Sir John Millais. 92pp. 6½ x 9¼. 23609-9 Pa. $2.50

RALPH THE HEIR, Anthony Trollope. Forgotten tale of illegitimacy, inheritance. Master novel of Trollope's later years. Victorian country estates, clubs, Parliament, fox hunting, world of fully realized characters. Reprint of 1871 edition. 12 illustrations by F. A. Faser. 434pp. of text. 5⅜ x 8½. 23642-0 Pa. $5.00

YEKL and THE IMPORTED BRIDEGROOM AND OTHER STORIES OF THE NEW YORK GHETTO, Abraham Cahan. Film *Hester Street* based on *Yekl* (1896). Novel, other stories among first about Jewish immigrants of N.Y.'s East Side. Highly praised by W. D. Howells—Cahan "a new star of realism." New introduction by Bernard G. Richards. 240pp. 5⅜ x 8½. 22427-9 Pa. $3.50

THE HIGH PLACE, James Branch Cabell. Great fantasy writer's enchanting comedy of disenchantment set in 18th-century France. Considered by some critics to be even better than his famous *Jurgen*. 10 illustrations and numerous vignettes by noted fantasy artist Frank C. Pape. 320pp. 5⅜ x 8½. 23670-6 Pa. $4.00

ALICE'S ADVENTURES UNDER GROUND, Lewis Carroll. Facsimile of ms. Carroll gave Alice Liddell in 1864. Different in many ways from final Alice. Handlettered, illustrated by Carroll. Introduction by Martin Gardner. 128pp. 5⅜ x 8½. 21482-6 Pa. $2.50

FAVORITE ANDREW LANG FAIRY TALE BOOKS IN MANY COLORS, Andrew Lang. The four Lang favorites in a boxed set—the complete *Red, Green, Yellow* and *Blue* Fairy Books. 164 stories; 439 illustrations by Lancelot Speed, Henry Ford and G. P. Jacomb Hood. Total of about 1500pp. 5⅜ x 8½. 23407-X Boxed set, Pa. $15.95

HOUSEHOLD STORIES BY THE BROTHERS GRIMM. All the great Grimm stories: "Rumpelstiltskin," "Snow White," "Hansel and Gretel," etc., with 114 illustrations by Walter Crane. 269pp. 5⅜ x 8½.
21080-4 Pa. $3.50

SLEEPING BEAUTY, illustrated by Arthur Rackham. Perhaps the fullest, most delightful version ever, told by C. S. Evans. Rackham's best work. 49 illustrations. 110pp. 7⅞ x 10¾. 22756-1 Pa. $2.50

AMERICAN FAIRY TALES, L. Frank Baum. Young cowboy lassoes Father Time; dummy in Mr. Floman's department store window comes to life; and 10 other fairy tales. 41 illustrations by N. P. Hall, Harry Kennedy, Ike Morgan, and Ralph Gardner. 209pp. 5⅜ x 8½. 23643-9 Pa. $3.00

THE WONDERFUL WIZARD OF OZ, L. Frank Baum. Facsimile in full color of America's finest children's classic. Introduction by Martin Gardner. 143 illustrations by W. W. Denslow. 267pp. 5⅜ x 8½.
20691-2 Pa. $3.50

THE TALE OF PETER RABBIT, Beatrix Potter. The inimitable Peter's terrifying adventure in Mr. McGregor's garden, with all 27 wonderful, full-color Potter illustrations. 55pp. 4¼ x 5½. (Available in U.S. only)
22827-4 Pa. $1.25

THE STORY OF KING ARTHUR AND HIS KNIGHTS, Howard Pyle. Finest children's version of life of King Arthur. 48 illustrations by Pyle. 131pp. 6⅛ x 9¼. 21445-1 Pa. $4.95

CARUSO'S CARICATURES, Enrico Caruso. Great tenor's remarkable caricatures of self, fellow musicians, composers, others. Toscanini, Puccini, Farrar, etc. Impish, cutting, insightful. 473 illustrations. Preface by M. Sisca. 217pp. 8⅜ x 11¼. 23528-9 Pa. $6.95

PERSONAL NARRATIVE OF A PILGRIMAGE TO ALMADINAH AND MECCAH, Richard Burton. Great travel classic by remarkably colorful personality. Burton, disguised as a Moroccan, visited sacred shrines of Islam, narrowly escaping death. Wonderful observations of Islamic life, customs, personalities. 47 illustrations. Total of 959pp. 5⅜ x 8½.
21217-3, 21218-1 Pa., Two-vol. set $12.00

INCIDENTS OF TRAVEL IN YUCATAN, John L. Stephens. Classic (1843) exploration of jungles of Yucatan, looking for evidences of Maya civilization. Travel adventures, Mexican and Indian culture, etc. Total of 669pp. 5⅜ x 8½. 20926-1, 20927-X Pa., Two-vol. set $7.90

AMERICAN LITERARY AUTOGRAPHS FROM WASHINGTON IRVING TO HENRY JAMES, Herbert Cahoon, et al. Letters, poems, manuscripts of Hawthorne, Thoreau, Twain, Alcott, Whitman, 67 other prominent American authors. Reproductions, full transcripts and commentary. Plus checklist of all American Literary Autographs in The Pierpont Morgan Library. Printed on exceptionally high-quality paper. 136 illustrations. 212pp. 9⅛ x 12¼. 23548-3 Pa. $12.50

AN AUTOBIOGRAPHY, Margaret Sanger. Exciting personal account of hard-fought battle for woman's right to birth control, against prejudice, church, law. Foremost feminist document. 504pp. 5⅜ x 8½.
20470-7 Pa. $5.50

MY BONDAGE AND MY FREEDOM, Frederick Douglass. Born as a slave, Douglass became outspoken force in antislavery movement. The best of Douglass's autobiographies. Graphic description of slave life. Introduction by P. Foner. 464pp. 5⅜ x 8½.
22457-0 Pa. $5.50

LIVING MY LIFE, Emma Goldman. Candid, no holds barred account by foremost American anarchist: her own life, anarchist movement, famous contemporaries, ideas and their impact. Struggles and confrontations in America, plus deportation to U.S.S.R. Shocking inside account of persecution of anarchists under Lenin. 13 plates. Total of 944pp. 5⅜ x 8½.
22543-7, 22544-5 Pa., Two-vol. set $12.00

LETTERS AND NOTES ON THE MANNERS, CUSTOMS AND CONDITIONS OF THE NORTH AMERICAN INDIANS, George Catlin. Classic account of life among Plains Indians: ceremonies, hunt, warfare, etc. Dover edition reproduces for first time all original paintings. 312 plates. 572pp. of text. 6⅛ x 9¼.
22118-0, 22119-9 Pa.. Two-vol. set $12.00

THE MAYA AND THEIR NEIGHBORS, edited by Clarence L. Hay, others. Synoptic view of Maya civilization in broadest sense, together with Northern, Southern neighbors. Integrates much background, valuable detail not elsewhere. Prepared by greatest scholars: Kroeber, Morley, Thompson, Spinden, Vaillant, many others. Sometimes called Tozzer Memorial Volume. 60 illustrations, linguistic map. 634pp. 5⅜ x 8½.
23510-6 Pa. $10.00

HANDBOOK OF THE INDIANS OF CALIFORNIA, A. L. Kroeber. Foremost American anthropologist offers complete ethnographic study of each group. Monumental classic. 459 illustrations, maps. 995pp. 5⅜ x 8½.
23368-5 Pa. $13.00

SHAKTI AND SHAKTA, Arthur Avalon. First book to give clear, cohesive analysis of Shakta doctrine, Shakta ritual and Kundalini Shakti (yoga). Important work by one of world's foremost students of Shaktic and Tantric thought. 732pp. 5⅜ x 8½. (Available in U.S. only)
23645-5 Pa. $7.95

AN INTRODUCTION TO THE STUDY OF THE MAYA HIEROGLYPHS, Syvanus Griswold Morley. Classic study by one of the truly great figures in hieroglyph research. Still the best introduction for the student for reading Maya hieroglyphs. New introduction by J. Eric S. Thompson. 117 illustrations. 284pp. 5⅜ x 8½.
23108-9 Pa. $4.00

A STUDY OF MAYA ART, Herbert J. Spinden. Landmark classic interprets Maya symbolism, estimates styles, covers ceramics, architecture, murals, stone carvings as artforms. Still a basic book in area. New introduction by J. Eric Thompson. Over 750 illustrations. 341pp. 8⅜ x 11¼.
21235-1 Pa. $6.95

GEOMETRY, RELATIVITY AND THE FOURTH DIMENSION, Rudolf Rucker. Exposition of fourth dimension, means of visualization, concepts of relativity as Flatland characters continue adventures. Popular, easily followed yet accurate, profound. 141 illustrations. 133pp. 5⅜ x 8½.
23400-2 Pa. $2.75

THE ORIGIN OF LIFE, A. I. Oparin. Modern classic in biochemistry, the first rigorous examination of possible evolution of life from nitrocarbon compounds. Non-technical, easily followed. Total of 295pp. 5⅜ x 8½.
60213-3 Pa. $4.00

PLANETS, STARS AND GALAXIES, A. E. Fanning. Comprehensive introductory survey: the sun, solar system, stars, galaxies, universe, cosmology; quasars, radio stars, etc. 24pp. of photographs. 189pp. 5⅜ x 8½. (Available in U.S. only)
21680-2 Pa. $3.75

THE THIRTEEN BOOKS OF EUCLID'S ELEMENTS, translated with introduction and commentary by Sir Thomas L. Heath. Definitive edition. Textual and linguistic notes, mathematical analysis, 2500 years of critical commentary. Do not confuse with abridged school editions. Total of 1414pp. 5⅜ x 8½.
60088-2, 60089-0, 60090-4 Pa., Three-vol. set $18.50